Radar Detection Theory of Sliding Window Processes

Radar Detection Theory of Sliding Window Processes

Graham V. Weinberg
Defence Science and Technology Group
Adelaide, South Australia

CRC Press
Taylor & Francis Group
Boca Raton London New York

CRC Press is an imprint of the
Taylor & Francis Group, an **informa** business

A SCIENCE PUBLISHERS BOOK

CRC Press
Taylor & Francis Group
6000 Broken Sound Parkway NW, Suite 300
Boca Raton, FL 33487-2742

First issued in paperback 2021

Version Date: 20170316

ISBN-13: 978-0-367-78188-0 (pbk)
ISBN-13: 978-1-4987-6818-4 (hbk)

Library of Congress Cataloging-in-Publication Data

Names: Weinberg, Graham V., author.
Title: Radar detection theory of sliding window processes / Graham V. Weinberg,
Defence Science and Technology Group, Adelaide, South Australia.
Description: Boca Raton : Taylor & Francis, CRC Press, [2017] | "A science publishers book."
| Includes bibliographical references and index. |
Identifiers: LCCN 2017010340 (print) | LCCN 2017012366 (ebook) |
ISBN 9781498768184 (hardback) | ISBN 9781498768191 (e-book) |
ISBN 9781498768184 (acid-free paper)
Subjects: LCSH: Moving target indicator radar. | Radar cross sections. |
Radar--Interference. | Surveillance radar. | Radar--Military applications.
Classification: LCC TK6592.M67 (ebook) | LCC TK6592.M67 .W45 2017 (print) |
DDC 621.3848--dc23
LC record available at https://lccn.loc.gov/2017010340

Visit the Taylor & Francis Web site at
http://www.taylorandfrancis.com

and the CRC Press Web site at
http://www.crcpress.com

For Mila

Preface

For the last quarter of a century the text *CFAR: The Principles of Automatic Detection in Clutter* by Gary and Jing Minkler has been the standard guide on the implementation of sliding window detection processes for radar target detection. The purpose of this new book is to provide a modern perspective on such detectors, in the context of contemporary models of clutter. The focus is primarily on X-band maritime surveillance radar, and so the Pareto distribution is used as the basis for detector development.

It has been written to provide a context to motivate the investigation of such detection processes, as well as providing a literature overview of their historical development. The latter explains why there has been a dearth of analysis of such schemes with the validation and acceptance of the K-distribution for maritime surveillance radar clutter returns. However, the fact that the Pareto model is a suitable alternative has allowed the development of sliding window detectors to be re-examined.

A brief overview of the validation of the Pareto model is provided, including examples of model fitting to real X-band radar clutter returns. However the main focus is on the techniques with which sliding window detection processes can be produced so that they achieve the constant false alarm rate (CFAR) property. The reader will observe that the latter is provided in a fairly general setting, and so can be used to produce CFAR detectors for operation in Weibull distributed clutter as an example. Hence the book is of relevance to researchers interested in the development of non-coherent CFAR detection processes for high frequency radar applications.

The focus is predominantly on the mathematical development of such detectors, and so a guide on the elements of probability and statistics is included in the second chapter. However, a large number of figures are provided to demonstrate the performance of the decision rules produced. This provides the practical engineer with an indication of the relative merits of each of the detection processes studied.

It is assumed that the reader has a basic understanding of the fundamentals of radar and signal processing. However, the subject of non-coherent sliding window detector development is somewhat self-contained and essentially one of analysis of problems in mathematical statistics.

As a result of the research reported in this book, the reader will find that there are many research ideas postulated throughout, with some hints on avenues toward a solution. It is hoped that these provide a basis for future investigations.

January 2017

Graham V. Weinberg
Adelaide, South Australia

Acknowledgements

There are many who have directly and indirectly influenced and motivated the writing of this book. In the first instance my line management at DST Group needs to be acknowledged for their support of the research documented herein. This includes Dr Andrew Shaw and Dr Brett Haywood who read in detail *every* piece of work I wrote on this subject matter. I also appreciate the time allowed to conduct research not only important to the interests of DST Group but in many cases I have been permitted to investigate and publish pieces of work more in line with my interests in mathematical probability and statistics.

There have been many collaborators, both staff members at DST Group as well as former students who have contributed to publications on which this book is based. Their enthusiasm and interest in this work is thus acknowledged, and it has been a privilege to publish in the peer-reviewed literature with a new generation of young researchers.

Dr Andrew Shaw, Dr Brett Haywood, Dr Leigh Powis and Mr Lachlan Bateman (Adelaide University vacation student at DST Group in 2016/2017) read the book in full and provided feedback, which was appreciated considerably.

Although the majority of the research reported here was completed at DST Group, the writing of the book, and its consolidation, has been largely undertaken outside of work hours. Hence the patience and support of my family is appreciated, especially my wife Mila.

Contents

Appendices **325**

List of Figures

List of Tables

Acronyms and Symbols

Acronyms

SWD	:	Sliding window detector
CFAR	:	Constant false alarm rate
CUT	:	Cell under test
CRP	:	Clutter range profile
N	:	Size of CRP
Pfa	:	Probability of false alarm
Pd	:	Probability of detection
CA	:	Cell-averaging
GM	:	Geometric mean
OS	:	Order statistic
SO	:	Smallest of
GO	:	Greatest of
TM	:	Trimmed mean
DOS	:	Dual order statistic
TMin	:	Transformed minimum
SW	:	Switching
Log-t	:	Logarithmic t
Log-t OS	:	Logarithmic t order statistic
GM-DOS	:	Geometric mean dual order statistic
SCR	:	Signal to clutter ratio
ICR	:	Interference to clutter ratio
CCR	:	Clutter to clutter ratio
dB	:	Decibel
BI	:	Binary integration
Inter	:	Interference (with reference to plots in a figure)
HH	:	Horizontal transmit and receive polarisation
HV	:	Horizontal transmit and vertical receive polarisation
VV	:	Vertical transmit and receive polarisation
RCS	:	Radar cross section

MLE	:	Maximum likelihood estimation
SSR	:	Serial shift register
SLD	:	Square law detector
SIRP	:	Spherically invariant random process
DST	:	Defence Science and Technology
CSIR	:	Council for Scientific and Industrial Research

Symbol Description

P	Probability	Weib	Weibull distribution
E	Expectation	M_Ξ	Moment generating function
Var	Variance		of Ξ
Cov	Covariance	$\sum_{j=1}^{N} Z_j$	$Z_1 + Z_2 + \cdots + Z_N$
$\stackrel{d}{=}$	Equality in distribution		
$\stackrel{d}{\to}$	Convergence in distribution	$\prod_{j=1}^{N} Z_j$	$Z_1 \times Z_2 \times \cdots \times Z_N$
α	Pareto shape parameter	$Z_{(k)}$	kth order statistic
β	Pareto scale parameter	$X\|Y$	Random variable X
τ	Threshold multiplier		conditioned on Y
Γ	Gamma function	Z_0	CUT
$\mathcal{G}_I$	Upper incomplete gamma	H_0	Null hypothesis
	function	H_1	Alternative hypothesis
$\mathcal{B}$	Beta function	$\underset{H_0}{\overset{H_1}{\gtrless}}$	Hypothesis test
$\mathcal{B}_I$	Incomplete beta function		
$F_X(t)$	Distribution function of X	R	Real numbers
$f_X(t)$	Density function of X	R^+	Nonnegative real numbers
$\mathcal{P}$	Pareto distributed variable	N	Natural numbers
$\mathcal{L}$	Lomax distributed variable	P_k^N	Permutation coefficient
$\mathcal{W}$	Weibull distributed variable	$\binom{N}{k}$	Combinatorial coefficient
$\mathcal{K}$	K-distributed variable	D_{KL}	Kullback Leibler divergence
Pareto	Pareto distribution	$\|\cdot\|_\infty$	Supremum norm
Exp	Exponential distribution		
Bin	Binomial distribution		

PRELIMINARIES I

Chapter 1

Introduction

1.1 Purpose

The purpose of this book is to provide the radar signal processing researcher with a comprehensive account of recent developments in sliding window non-coherent constant false alarm rate (CFAR) radar detection in a Pareto-type clutter environment. The development and analysis of such detection processes has been facilitated by the elegant nature of the Pareto distributional model. As a result of the model's simplicity it has been possible to develop a rich mathematical framework for the study of CFAR detection schemes. Hence the focus in this book is on the mathematical methods that can be utilised to produce CFAR detection processes for target identification in Pareto distributed clutter. It is hoped that a practical engineer will find this book to be a useful guide on how to design and test in ideal situations such detection processes. In addition to this, as will become apparent to the reader, many of the theoretical developments apply not only to the Pareto distributional model but also to other radar clutter distributions of interest, such as the Weibull.

Although this book is concerned fundamentally with the mathematical development and analysis of sliding window CFAR detectors, its stimulus is the practical problem of small target detection from an airborne X-band high resolution maritime surveillance radar. Such radars have the advantage of permitting the acquisition of detailed information about illuminated targets (Wehner, 1987). However, at high resolution the sea surface admits intermittent strong backscattering, known as sea spikes, which can result in an increased number of false alarms. These sea-spikes result from ocean features such as breaking waves and white caps (West, 2002). Maritime surveillance, conducted at a low grazing angle, results in minimisation of the effects of sea spikes, with a resultant reduction in the radar surveillance region. At medium to high grazing angles there is a significant increase in the available surveillance region, but with an increase in occurrences of sea spikes (Rosenberg, 2013).

Hence it is important to have an accurate model of the sea backscattering to facilitate the design of detectors. The Pareto distribution has been proposed as a new model for X-band sea clutter returns and so is used as a basis for the developments outlined in this book. An extensive validation of the Pareto model will not be provided, since this has already been reported in the open literature. However, some examples of model fitting to the Australian Defence Science and Technology (DST) Group's Ingara radar data will be included. In order to understand the performance of the proposed detection schemes it is necessary to test them in a controlled environment. Consequently simulated Pareto clutter, whose distributional parameters are matched to those acquired from the aforementioned real data, will be used exclusively. Such an approach allows for the investigation of performance when the detector is subjected to interference, as well an examination of the way in which the decision rule regulates the false alarm probability.

Some examples of the performance of the detection schemes, when operating on real data directly with a synthetic target model, have been reported in several recent publications, which will be discussed in the appropriate chapters. Hence the primary role of this book is to provide the reader with an account of the development of non-coherent decision rules developed for operation in Pareto distributed clutter. The type of decision rules to be examined are known as sliding window detectors, which have had a long history of interest with researchers as an alternative to Neyman-Pearson based decision rules. Detectors based upon these were popular in lower resolution radar detector development due to the fact that the clutter could be well modelled by exponential distributions, for which the sliding window approach permits the construction of CFAR detectors. With increased resolution, maritime radar clutter experienced a transition to a more spiky nature, which ultimately resulted in the well-known K-distribution being used as the basis model. As a result of this, there was a shift away from sliding window decision rules, due to the fact that in K-distributed clutter, such detectors lost the CFAR property. However the Pareto distributional model allowed for the CFAR property to be re-acquired with sliding window decision rules. In particular, it is possible to produce such detectors, under the assumption of a Pareto clutter model, which are CFAR with respect to both the shape and scale parameters, without the need for application of approximations. Thus, from a non-coherent detection perspective, these provide a solution to the CFAR management problem. To illustrate the issues one faces with the construction of optimal detectors based upon the Neyman-Pearson Lemma, the reader can consult Appendix A, which provides a simple illustration of detector design issues in a Pareto clutter context.

1.2 Sliding Window Detectors

Sliding window detectors (SWDs) have been in use since the 1960s and provide a somewhat simpler alternative to the Neyman-Pearson based decision rules. These arose out of analysis of plan position indicator (PPI) displays used to show the radar view of the scanned region, as well as plots of intensity measurements of radar returns as a function of range, Doppler or both. The popularity of such *ad hoc* detection processes was due to the fact that, in lower resolution X-band maritime surveillance radar clutter returns, the Gaussian model was appropriate. When viewed from an amplitude squared or intensity perspective, the clutter is exponential in distribution. Hence it became somewhat simple to propose sliding window detection schemes whose probability of false alarm (Pfa) could be set independently of the clutter power, and so allowing for CFAR control. These detectors could be implemented regardless of the underlying target model, thus producing a useful alternative to optimal detectors which require some assumptions to be made regarding the target model, or at least an approximation for it.

Figure 1.1 illustrates a sliding window process, designed to run across a range-Doppler or range-time map (Rohling, 1983). A series of returns is passed into a serial shift register (SSR), where subsampling can be applied in order to provide approximate decorrelation of the returns. The result of this is then passed to a square law detector (SLD) which converts the complex returns to amplitude or intensity measurements. The resulting sequence is then passed to the first stage of the detection process, which separates a cell under test (CUT) from a series of measurements which will be used to produce the estimate of the clutter level. As can be observed, the two sets of clutter measurements, denoted $\{C_1, C_2, \ldots, C_m\}$ and $\{C_{m+1}, C_{m+2}, \ldots, C_{m+r}\}$, have been separated from the test cell with two sets of guard cells, namely $\{G_1, G_2, \ldots, G_k\}$ and $\{G_{k+1}, \ldots, G_{k+n}\}$. Here it is understood that the indices m, r, k and n are positive integers. From the two sets of clutter measurements two functions are applied, denoted f_1 and f_2 respectively, to these to synthesize the clutter measurement. Following this a function g is applied to merge f_1 and f_2 to a single measurement of the clutter level. This is then normalised by τ, which is used to allow adaptive control of the design Pfa in ideal situations. Finally, this is then compared with the CUT and a detection decision can be made. This binary result is recorded, and the sliding window is then continued over the range-Doppler or range-time map, and the results can be applied to a moving target indicator to allow tracking (Minkler and Minkler, 1990).

In the case where the clutter measurements are homogeneous and independent exponentially distributed returns, a detection process such as that illustrated in Figure 1.1, can achieve the CFAR property for a very large class of admissible clutter measurement functions f_1, f_2 and $g(f_1, f_2)$ (Gandhi and Kassam, 1988). This class of functions correspond to those with a pseudo-linearity or scale invariance property, which will be discussed subsequently. The selection of alter-

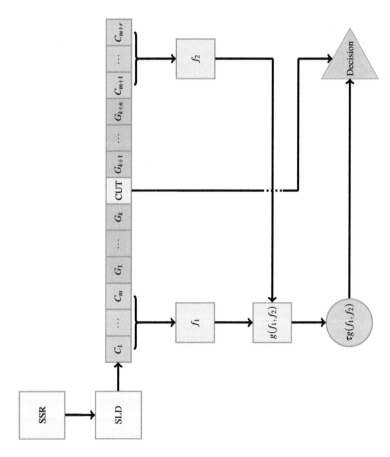

Figure 1.1: Classical sliding window detector schematic. The complex returns are passed to a serial shift register (SSR), followed by square law detection (SLD) and then the series is processed to select a test cell (CUT), which is compared with a processed measurement of the clutter level. The latter is normalised so that the Pfa remains constant in ideal situations. A detection decision is then possible, and the process can be continued, with the results applied to a tracking algorithm.

natives to the averaging case have been proposed due to the fact that as the sliding window is run across the range-Doppler map, the data used for clutter measurement will be subjected to interference and changes in the clutter homogeneity. Interference can be due to secondary targets present in the clutter cells, which can be managed through the selection of an order statistic (OS) for f_1 and f_2, for example. In the context of maritime surveillance radar, changes in the clutter homogeneity can arise from variations in the sea surface. Hence, although a simpler alternative to the Neyman-Pearson based detectors, it is, from a practical engineering design perspective, a very challenging problem to design an effective sliding window detection scheme that achieves CFAR and manages these performance issues. This is a much more difficult exercise when such processes are constructed for higher resolution maritime surveillance radar. In particular, it is a difficult problem to produce such detection processes, with the CFAR property, for operation in Pareto distributed clutter via direct adaptation of the original CFAR decision rules for exponentially distributed clutter. The developments reported in this book will outline the way in which this can be achieved. There are two main approaches which can be used for this purpose. The first is to identify decision rules, which have appeared in the open literature, for which the CFAR property still holds for a Pareto clutter model. The second approach is to apply a transformation methodology, which converts the original exponential-clutter based CFAR rules to operate in the Pareto setting, while preserving the original probability of false alarm and threshold multiplier relationship. The latter method results in Pareto scale parameter dependence in the transformed decision rules. It will be shown that this issue can be rectified by application of certain statistics, which result in an invariant decision rule, and thus acquisition of the full CFAR property.

The implementation in Figure 1.1 can be altered so that the sliding window detector processes the data in a block format. As an example, Figure 1.2 provides a schematic of a detector which processes the data in two dimensions. Hence the clutter cells are denoted $C_{(i,j)}$ and guard cells are $G_{(i,j)}$, where i and j are positive integers. In this case, the CUT is surrounded by a series of guard cells, while the outer measurements are used to obtain the estimate of the clutter level. This is then passed through a normalisation process, with the CUT and the latter compared as before. The implementation can be adjusted in a number of different ways. Recently (Kronauge and Rohling, 2013) examined the development of 2-dimensional sliding window CFAR detection processes.

1.3 Range-Time Intensity Example

In order to provide a context for the application of SWDs introduced in the previous section, an example of radar data with a real target is provided in this section. The South African Council for Scientific and Industrial Research (CSIR) has un-

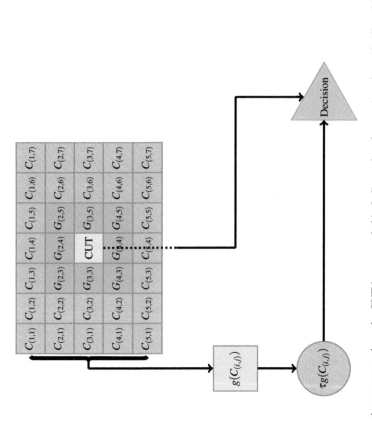

Figure 1.2: A sliding window detection process where the CUT is surrounded in 2-dimensions by a series of guard cells, while the outer cells are used to extract the measurement of the clutter level. It is assumed that the serial shift register and square law detection processes have been applied beforehand. The extracted clutter statistics are normalised as in the detector illustrated in Figure 1.1.

dertaken performance assessments of a custom built as well as an experimental radar, both of which have been deployed during a series of trials over the period 2006–2007. These trials gathered maritime surveillance radar returns, with the main focus being the detection of small boats. The first of these trials involved the Fynmeet Dynamic Radar Cross Section Measurement Facility, which was a joint development between CSIR, the Armaments Corporation of South Africa and the South African Air Force. This facility is located at Measurement Station 3 at the Overberg Test Range (OTB), which is in the Overberg Region on the south coast of South Africa, near Arniston, Western Cape. The radar used in this trial operated in a frequency range from 6.5 GHz to 17.5 GHz.

The second trial deployed an experimental radar on Signal Hill in Cape Town, South Africa. It is reported that the radar used was an experimental, monopulse, pulsed Doppler X-band radar. Further details of these two trials, as well as radar specifications and targets deployed, can be found in (Herselman and Baker, 2007), (Herselman et al., 2008) and (Herselman and de Wind, 2008).

In each of the trials a number of boats were deployed as targets. In the OTB 2006 trial one target used was a CSIR-built speedboat known as a Wave Rider, while in the Signal Hill trial a pencil duck was used. These are essentially small boats with minimal refectivity. Two examples of range-time intensity plots are provided, based upon examples acquired during these trials. It is reported that in both cases, a sea state 1 was observed, with the boat approximately radial inbound, whereas the general direction of the waves was approximately radial outbound. Figure 1.3 corresponds to the OTB 2006 trial, while Figure 1.4 is for the 2007 Signal Hill trial. These two plots show the intensity return at a given range (in metres) and at a given observation time (in seconds). In view of Figure 1.3 the presence of a target can be clearly observed. It is somewhat more difficult to see the target in the case of Figure 1.4, which is due to the fact that the radar is experimental and lower-resolution, as well as variation in the target and difference in the trial location. It is interesting to observe that in Figure 1.4 the presence of the large ship, located at a range of roughly 4 km, and stationary. The sliding window decision process, outlined in the previous section, could be run across each range line, with the results inputted to a tracking algorithm. In order to minimise the effects of target spillover into the clutter cells, a clutter map could be produced, where a bank of clutter measurements are used for this purpose. This bank of measurements can be updated adaptively.

The two targets in Figures 1.3 and 1.4 are distinguishable from the background clutter. This situation is not the case in terms of an airborne maritime surveillance radar, especially if the target is smaller in radar cross section than the moving targets in these figures, and the aircraft is scanning the sea surface from a high altitude. From a military perspective, a surveillance aircraft may be searching for a submarine periscope. Hence it is important for the aircraft to remain at a reasonable altitude, in order to avoid making itself vulnerable to

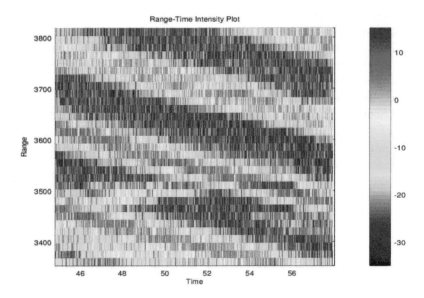

Figure 1.3: Example of a range-time intensity plot for South African OTB 2006 Measurement Trial. In this experiment a small speedboat target is under observation.

missile attacks. Consequently it is necessary for the signal processing detection architecture to be optimised for maximal performance.

1.4   Historical Development

The following account outlines the development of sliding window decision rules as reported in the open literature. It is by no means an exhaustive account, but is meant to provide a survey of the genesis and evolution of this approach.

As a prelude to this discussion it is important to recall the pioneering work of early investigators of radar system performance analysis. A comprehensive discussion of this can be found in the classic radar detection text (Di Franco and Rubin, 1968), on which the following discussion has been based. Prior to the Second World War, radar was an experimental technique and its signal processing mechanisms were rather basic. However, it was realised that it could provide an early warning system for imminent air strikes, which stimulated its development during the Second World War in the United Kingdom. Developments in the theory underlying radar detection resulted in improved system performance, which continued with the development of sophisticated electronics, which improved the physical nature and capabilities of radar systems. The concept of an ideal receiver, and the ambiguity function which provides a measurement of radar

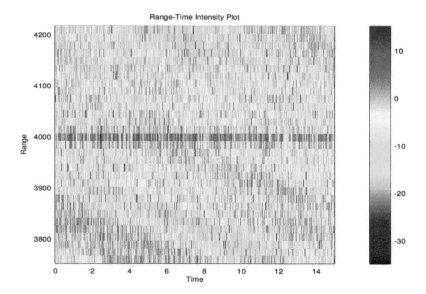

Figure 1.4: An example of a range-time intensity plot for the South African Signal Hill 2007 Measurement Trial, also with a small speedboat target present. In addition to this, there is a stationary vessel at a range of 4 km from the radar.

resolution capabilities, was introduced by (Woodward, 1953), who initiated the application of information theory to radar. Statistical analysis of the mathematical structure of noise encountered by radar systems was first investigated by (Rice, 1944) and (Rice, 1945). The concept of a matched filter is attributed to North in 1943 and documented in (North, 1963), which facilitates the understanding of optimal filter theory for signals embedded in Gaussian-type interference. Marcum was the first to examine the problem of statistical signal processing techniques for target detection in Gaussian type interference (Marcum, 1960). As a result of this the famous Marcum Q-Function, and its generalisation, have provided researchers with the tools for understanding detection in Gaussian interference (Nuttall, 1975). Marcum's original work was further developed by Swerling (Swerling, 1960) and (Swerling, 1965) who extended Marcum's development and introduced models for fluctuating targets. These studies were in the context of Gaussian-type interference, which was a plausible model for radar clutter attributable to the low-resolution capabilities of early radar systems. Due to the additive nature of Gaussian clutter, and the need to control the false alarm rate in such detection processes, properties of sequences of Gaussian intensity measurements were identified as a potential solution to adaptive false alarm control. In the context of exponentially distributed clutter, corresponding to Gaussian intensity measurements, the first reported development of sliding window CFAR detectors is (Finn, 1966) and (Finn and Johnson, 1968), who investigated the fact that a cell-averaging process could be used to produce the measurement of clutter, while maintaining the desired Pfa in ideal scenarios. This was further examined in (Steenson, 1968), while (Schwartz, 1969) investigated the application of invariance theory through the Hunt-Stein Theorem to produce CFAR detectors. The case of interference was examined in (Nitzberg, 1972), who investigated the effects of the environment on detector design. A special implementation of the cell-averaging CFAR was reported in (Gregers Hansen and Ward, 1972), who examined the estimation of the underlying clutter level with a series of samples passed through a logarithmic amplifier, resulting in a greater dynamic range but with slightly inferior detection performance. A dispersive detector was proposed in (Ward, 1972), which uses a matched dispersive delay lines and provides a practical implementation for operational surveillance radars.

In 1973 Goldstein (Goldstein, 1973) reported a property of a certain class of random variables, which could be exploited to produce CFAR detectors in a number of clutter environments, extending from the case of exponentially distributed clutter to log normal and Weibull. The resultant decision rule, known as the log-t detector, provided an interesting method to produce the clutter returns to yield the decision rule, which is based upon the well-known Student-t statistic. Adaptive detection algorithms, for management of multiple target scenarios, was examined in (Rickard and Dillard, 1977), while (Nitzberg, 1979) proposed that relaxing the CFAR requirement could be acceptable in decision rule design, since the CFAR property can never be attained in real scenarios. The relationship between decision rules and map-matching algorithms was investigated in (Novak, 1980).

In order to better accommodate interference in the clutter cells, a number of studies examined the application of greatest-of and smallest-of decision rules, which divided the clutter measurements into two sets, with two resultant sums taken. Then one can take either the larger or smaller of these, and apply the same approach as in a cell-averaging CFAR. This was investigated in (Gregers Hansen and Sawyers, 1980) and (Weiss, 1982). The application of order statistics, to replace the cell-averaging process, was first examined in (Rohling, 1983), which provided a robust solution to managing interference.

A further investigation into the management of interference can be found in (Gandhi and Kassam, 1988), who re-examined the greatest-of and smallest-of CFARs as well as the order statistic approach of (Rohling, 1983). A trimmed mean measure of the clutter level was proposed, which consisted of a sum of the ordered clutter measurements. This was then optimised for application to exponentially distributed clutter.

Order statistic detectors were then examined for the Weibull case in (Levanon and Shor, 1990) and (Shor and Levanon, 1991). Direct application of the sliding window decision rules, developed for the exponentially distributed clutter case to Weibull, resulted in loss of the CFAR property with respect to the shape parameter. To compensate for this (Ravid and Levanon, 1992) examined a maximum likelihood approach. However such an approach usually results in variation in the design Pfa in practice, resulting in complete loss of the CFAR property.

Development of sliding window detection processes for the case of K-distributed clutter has proven even more problematic, due to the complexity of the underlying clutter model and analytic intractability of the resultant decision rules. An example of approximate strategies for this context can be found in (Watts, 1996). However, in the literature, there has been a dearth of exploration of sliding window detection processes with the acceptance of the K-distribution as a more accurate model for X-band maritime surveillance radar clutter returns. However, this has since changed with the emergence of the Pareto model as an alternative to the K-distribution model for clutter. The development of sliding window detection processes, for the case of Pareto distributed clutter, have been documented in a series of recent publications from DST Group. The first approach was reported in (Weinberg, 2013d), known as a transformation approach. This is based upon the principle of memoryless nonlinear transformations, and shows how the sliding window detectors, developed for the exponentially distributed clutter case, can be adapted to the Pareto scenario while preserving the original Pfa and threshold multiplier relationship. The cost in this approach was dependency on the Pareto scale parameter in the decision rule. This was then extended in (Weinberg, 2014a) to a wider range of detection schemes. The direct application of the sliding window detectors, designed for operation in exponentially distributed clutter, to the Pareto clutter setting was studied in (Weinberg, 2015b). A generalisation of the transformation approach to detector design was reported in (Weinberg, 2014b), while a switching approach was considered in (Weinberg, 2014c). A variation of a minimum-based detection process was investigated in (Weinberg, 2015a). More specialised developments and approaches have been reported in (Weinberg, 2016b), (Weinberg, 2017b) and (Weinberg, 2017c). The state of the art, in terms of sliding window detection processes for the Pareto case, is reported in (Weinberg, 2017b), which shows how the CFAR property can be attained for the Pareto case with respect to both clutter parameters. The key to this is to replace the Pareto scale parameter, which appears in the transformed detectors, by a complete sufficient statistic for it. As will be demonstrated in the book, this results in an detector which is invariant under a group of transformations. Future directions are outlined in (Weinberg, 2017c) which shows how to account for correlations in the clutter measurements.

1.5 Mathematical Formulation

The mathematical fomulation of sliding window detectors is now provided. For the purposes of reference, the book (Levanon, 1988) contains a very good account of the development of such detectors in exponentially distributed clutter, while (Minkler and Minkler, 1990) has been the standard textbook on this subject for over a quarter of a century. Additionally, (Levanon, 1988) contains an excellent example demonstrating the sensitivity of the Pfa on clutter power variations. The publication (Gandhi and Kassam, 1988) also contains a very well written account of the design issues that radar engineers experience in the construction of such detection processes. The classic book (Di Franco and Rubin, 1968) provides an excellent account of radar detection in general.

With reference to Figures 1.1 and 1.2, it is common to assume the existence of a series of N clutter statistics which are denoted $Z_1, Z_2, \ldots, Z_N$. These are often referred to as the clutter training statistics or clutter range profile (CRP). The latter terminology will be used throughout. Relative to radar returns it is understood that these are intensity measurements corresponding to amplitude squared measurements from the resultant complex in-phase and quadrature returns. A general assumption adopted throughout the development of sliding window decision rules is that the clutter returns are independently and identically distributed. The reason for this is predominantly one of mathematical tractability, although this can be arranged in practice through subsampling of the returns to minimise correlation. This can be achieved with the serial shift register in Figure 1.1. A cell under test statistic, which is denoted CUT as in Figure 1.1, is represented statistically by Z_0, which is assumed to be separated sufficiently from the CRP with the application of a number of guard cells, as described previously. The idea, as illustrated in Figure 1.1, is to use the CRP to obtain an estimate of the clutter level, and compare it with the CUT, with the objective to determine whether the latter contains a target. This is also done in a way such that the Pfa remains constant in ideal scenarios. This procedure can be formulated in the language of statistical hypothesis testing; the reader can consult (Beaumont, 1980) and (Lehmann, 1986) for further details on hypothesis testing principles. Suppose we denote by H_0 the hypothesis that the CUT does not contain a target. This is referred to as the null hypothesis. Let H_1 be the hypothesis that the CUT does contain a target, which is known as the alternative hypothesis. Here it is understood this target is combined with clutter in the complex domain, and its intensity measurement appears in the CUT. Based upon this, one needs to compare Z_0 with a normalised measurement of the clutter level, $g(Z_1, Z_2, \ldots, Z_N)$, with the normalisation performed so that the resultant Pfa remains fixed in ideal scenarios. The test is to reject H_0 if $Z_0 > \tau g(Z_1, Z_2, \ldots, Z_N)$ for some constant $\tau > 0$. The latter is called a threshold multiplier throughout. Such a test can be expressed using the notation

$$Z_0 \underset{H_0}{\overset{H_1}{\gtrless}} \tau g(Z_1, Z_2, \ldots, Z_N). \tag{1.1}$$

The probability of false alarm is the probability that the hypothesis H_0 is rejected when it is actually true. In statistical hypothesis testing, this is known as a Type I error, or the size of the statistical test. In terms of radar detection, too many false alarms can result in a tracking algorithm missing a true target. Hence this is a critical problem, and the Pfa needs to be minimised. In practice, one usually sets this to an acceptable level, such as 10^{-4} or smaller. In mathematical terms the Pfa is given by

$$\text{Pfa} = \mathbf{P}(Z_0 > \tau g(Z_1, Z_2, \dots, Z_N) | H_0), \tag{1.2}$$

where **P** denotes probability. The probability of detection (Pd) associated with the sliding window detector is the probability that the null hypothesis is rejected when the alternative is actually true. This is given by

$$\text{Pd} = \mathbf{P}(Z_0 > \tau g(Z_1, Z_2, \dots, Z_N) | H_1). \tag{1.3}$$

Type II errors in statistical hypothesis testing result from not rejecting the null hypothesis when the alternative is actually true. The power of a statistical test is the probability of rejecting H_0 when H_1 is true. Hence it is the probability of making a correct detection of a radar target. If an assumption is made regarding the underlying distribution of the statistics in the CRP, then (1.2) can in principle be determined and then τ can be deduced by either analytical or numerical methods. By contrast it is generally difficult to determine (1.3) without *a priori* knowledge of the target model and its intensity when combined with clutter in the complex domain. However, once τ is obtained, the decision rule (1.1) can be applied to produce decisions on the presence or absence of targets. In the next section the case of exponentially distributed clutter is considered.

1.6 Detectors in Exponentially Distributed Clutter

In order to provide motivation for the design of sliding window CFAR detectors the following discussion has been based upon that in (Levanon, 1988). A linear threshold detector compares a test statistic with a threshold to determine the presence of a target. Such detectors are often referred to being ideal, and a discussion of these in the Pareto case can be found in Appendix E. Under the assumption of exponentially distributed clutter, with shape parameter $\lambda > 0$, this decision rule declares the presence of a target if the test statistic Z_0 exceeds $\tau = -\frac{1}{\lambda} \log(\text{Pfa})$, where τ is the detection threshold and Pfa is the probability of false alarm. If one supposes that the threshold is fixed, then an approximation applied to λ will result in a variation in the Pfa. Hence if one denotes Pfa_D as the design Pfa, with λ_D the corresponding clutter parameter, and Pfa_R the resultant Pfa with an approximation λ_R for the original λ applied, then it is not difficult to show that

$$\text{Pfa}_R = \text{Pfa}_D^{\frac{\lambda_R}{\lambda_D}}. \tag{1.4}$$

If one sets the clutter power level change in decibels (dB) to be $\mu := 10 \log_{10} \left(\frac{\lambda_D}{\lambda_R} \right)$ then (1.4) becomes

$$\text{Pfa}_R = \text{Pfa}_D^{10^{-\frac{\mu}{10}}}. \tag{1.5}$$

μ	-10	-2	-1
Pfa_R	10^{-40}	4.5754×10^{-7}	9.2108×10^{-6}
μ	1	2	10
Pfa_R	6.6479×10^{-4}	2.9934×10^{-3}	3.9811×10^{-1}

Table 1.1: Resultant Pfa due to approximations in the distributional shape parameter λ. Here the variation from λ is μ, which is measured in dB. The design Pfa in this example is 10^{-4}.

For a design Pfa of 10^{-4} Table 1.1 provides some examples of the resultant Pfa as μ is varied. As can be observed, there is significant variation from the design Pfa. Such a variation can result in a unacceptable increase in the number of false alarms in a practical radar system. Hence it became of critical importance in radar signal processing to investigate the design of alternative detection schemes which could alleviate this concern, since the same issue is encountered with any detector which requires a clutter parameter approximation.

In the case where the clutter is assumed to be exponentially distributed, it is possible to determine τ from (1.2) for a large number of decision rules. In particular, the well-known cell-averaging detector results from the selection of g as a sum of the elements in the clutter range profile, and has decision rule

$$Z_0 \underset{H_0}{\overset{H_1}{\gtrless}} \tau \sum_{j=1}^{N} Z_j. \tag{1.6}$$

It can be shown that the relationship between the Pfa and τ, given by (1.2), is

$$Pfa = (1+\tau)^{-N}, \tag{1.7}$$

which is independent of the clutter mean parameter. Consequently, the decision rule (1.6) has the CFAR property and is often referred to as the cell averaging (CA) CFAR. The detector (1.6) has a number of remarkable properties. When the target model, in the complex domain, is assumed to be Gaussian and independent of the CRP, the Pd given by (1.3) can be determined in closed form. This can be found in Appendix B, while target models are discussed in Appendix C with a focus on the Swerling fluctuation cases. Throughout the literature it was speculated that the detector (1.6) was optimal within the class of such decision rules, operating in exponentially distributed clutter, and when a Gaussian target model is assumed. Here optimality is used in the sense that the detector has a maximal probability of detection for a given Pfa. The optimality of (1.6) was established in (Gandhi and Kassam, 1994), who showed that this decision rule is a uniformly most powerful test, based upon an analysis of least favourable distributions in composite hypothesis testing.

A second example of a decision rule, designed to operate in exponentially distributed clutter, is Rohling's order statistic detector (Rohling, 1983), which is specified via

$$Z_0 \underset{H_0}{\overset{H_1}{\gtrless}} vZ_{(k)}, \qquad (1.8)$$

where $Z_{(k)}$ is the kth order statistic of the CRP, for a fixed $k \in \{1, 2, \ldots, N\}$. In the case of exponentially distributed returns, it is shown in (Rohling, 1983) that its Pfa is related to the threshold multiplier v through the expression

$$\mathrm{Pfa} = \frac{N!}{(N-k)!} \frac{\Gamma(N-k+v+1)}{\Gamma(N+v+1)}, \qquad (1.9)$$

where $\Gamma(\cdot)$ is the Gamma function. Since (1.9) is independent of the clutter parameter, the decision rule also has the CFAR property, and is referred to as an order statistic CFAR (OS-CFAR). The false alarm probability (1.9) is derived in Appendix B, together with an associated detection probability for the case of a Gaussian target model.

These two decision rules will be used throughout the following, and will be adapted to operate in Pareto distributed clutter. In the first instance the corresponding thresholds will be determined, for application in Pareto clutter, followed by a transformation approach to produce new variants of these detectors, designed for Pareto distributed clutter, while retaining the relationships (1.7) and (1.9). The analysis in the book will develop these detectors so that they attain the full CFAR property with respect to the underlying Pareto clutter model.

1.7 Some Fundamental Concepts

The mathematical developments in this book will demonstrate how non-coherent sliding window CFAR detectors can be produced under the assumption of Pareto distributed clutter. Modelling of maritime radar clutter returns must of necessity include considerations of receiver thermal noise. To account for this as a separate component in a model formulation results inevitably in detector design complexity. Hence the most logical approach is to account for thermal noise via an effective shape parameter, as introduced by Watts (Watts, 1987). The formulation in the latter has been developed for the Pareto case in a number of recent works. The first is (Rosenberg and Bocquet, 2015), who adapt the approach of (Watts, 1987) directly. A second approach, reported in (Alexopoulos and Weinberg, 2015), applies the principles of fractional order calculus to produce a variant of the Pareto model, which then reduces to a standard Pareto distribution whose shape parameter is a function of the optimised fractional order derivative. The latter has been used in a Pareto CFAR context in the analysis of dual order statistic CFAR in (Weinberg and Alexopoulos, 2016). The cost, in the accounting for receiver thermal noise in radar clutter, tends to be a reduction in the resultant Pareto shape parameter. Hence the detection process is operating under a spikier clutter model assumption, resulting in a detection performance reduction. Throughout the book thermal noise will not be considered, since the focus is more on the design and mathematical analysis of CFAR detection processes and their associated issues.

Many analyses of CFAR detection processes examine the CFAR loss incurred from the application of such a detection process. There are several ways in which this can be measured. One approach is to compare the difference between an ideal linear threshold detector with that of the CFAR process under consideration. Ideal linear threshold detectors were introduced in (Watts, 1987), and are based upon the assumption that if the clutter environment is completely known, then there is no need to estimate the level of clutter. Hence a threshold detector can be used, and this provides an upper bound on performance. A more conventional approach to measurement of the CFAR loss is to determine the signal to clutter loss incurred, with respect to standard conventional optimal detectors, by application of the suboptimal CFAR scheme (Minkler and Minkler, 1990). This approach will not be considered in the following.

A *sine qua non* throughout the development of sliding window detection processes is the existence of independent and identically distributed clutter measurements in the CRP. In a practical implementation of any of the detection processes designed under this assumption there is to be expected somewhat uncertain performance. In order to mitigate such issues it is important to perform a subsampling process to yield measurements that have their pairwise correlations minimised. With reference to Figure 1.1 this is the purpose of the serial shift register.

1.8 Structure of the Book

The purpose of this book is to provide a comprehensive guide on the development of sliding window CFAR detection processes, mainly for target detection in Pareto distributed clutter. It outlines a series of ways in which the CFAR property can be attained, as well as overviewing many of the detection processes examined at DST Group over the last few years. It is envisaged that the interested researcher will find new open problems presented here, which have not been fully explored. Additionally, the practical engineer can use the techniques presented in this book to assist in the development of radar detection algorithms. It has been written to complement the comprehensive study of CFAR in (Minkler and Minkler, 1990). Recently, an invaluable text on the topic of modern radar detection has been produced (Greco and De Maio, 2016). The current book is complementary to this, in the sense that it provides a specialised study of non-coherent sliding window decision rules, which is not examined in the latter.

The book begins with Chapter 2 outlining the relevant principles from probability and statistics used throughout the book. Chapter 3 provides an account of the Pareto model for X-band maritime surveillance radar clutter. Included is validation of it for a particular DST Group data set, from which clutter parameters are extracted and used throughout the book. Additionally, Chapter 3 documents many of the useful properties of the Pareto model, which facilitates the development of radar detection schemes. Chapter 4 begins the examination of detection processes by investigating the application of the original sliding window detectors, designed to operate in exponentially distributed clutter, to the Pareto clutter setting directly. A transformation approach for sliding window CFAR detector design is then examined in Chapter 5, which shows how the original detectors, designed for exponentially distributed clutter, can be modified for operation in Pareto distributed clutter, while preserving the way in which the threshold multiplier is set. In the Pareto case it is shown that the transformed decision rules require *a priori* knowledge of the Pareto scale parameter. However, the chapter also shows a rather remarkable result that this can be rectified via application of a complete sufficient statistic. This results in the capability of being able to produce transformed detectors with the full CFAR property, with a minor modification in the original Pfa and threshold multipier relationship.

Chapter 6 examines a minimum-based detector, which possesses a number of interesting properties. In particular, it is shown that it is possible to produce a useful minimum-based detector which regulates the Pfa very well and manages interfering targets. Chapter 7 examines a detection process known as a dual order statistic CFAR, which uses two order statistics as a measurement of the underlying clutter level, while achieving the full CFAR property.

Chapter 8 is concerned with Goldstein's original log-t detector, and demonstrates that it is also CFAR in Pareto distributed clutter. In addition to this, a modification of it, based upon order statistics, is examined and shown to improve on the original log-t detector, while also preserving the CFAR property.

Chapter 9 begins a study of specialised developments, and is concerned with the adaptation of switching-based detectors to the Pareto clutter case. It is shown how the full CFAR property can be attained with such a detector operating in Pareto distributed clutter.

Chapter 10 examines recent developments in binary integration, which can be used to enhance detection performance. Included is a detailed mathematical investigation of optimal design of binary integration processes in general.

Chapter 11 provides a study of detector development for application to the case of correlated clutter returns. It is shown how an OS-based detector needs to have its threshold set to account for correlated clutter. A specialised study is also included, based upon Mardia's multivariate Pareto model.

The second last chapter in the book is concerned with explaining why some of the detectors in Chapter 5 achieve the full CFAR property. Hence Chapter 12 provides an account of the concept of invariance and invariant statistics. It is shown that the decision rules developed in the latter part of Chapter 5 are invariant with respect to a group of transformations. Chapter 12 also establishes some design principles to produce CFAR detectors in Pareto distributed clutter. This is illustrated by the development of some new detection processes.

Finally, Chapter 13 examines the exponential distributional approximation of the Pareto model, in order to derive some rules of thumb on the validity of the approximation. This is important because if the clutter is roughly exponentially distributed then one is justified in applying the large number of CFAR decision rules developed under such a clutter model assumption.

There are also five Appendices included for the interested reader. Appendix A provides an overview of the Neyman-Pearson Lemma, and outlines the issues in its application to non-coherent detector design. Appendix B derives the probability of false alarm for the CA- and OS-CFARs in exponentially distributed clutter, and the corresponding probability of detection in the case of a Swerling I target model. Appendix C provides a brief account of radar cross section and target models. Appendix D gives an overview of classical non-coherent integrators. These are not CFAR decision rules, nor are sliding window type detectors, but are nonetheless useful non-coherent detectors. Finally, Appendix E discusses ideal detectors and examines some of their properties. In particular, it is shown that one of the transformed Pareto detectors limits to an ideal decision rule as the size of the CRP increases without bound.

Chapter 2

Probability and Distribution Theory for Radar Detection

2.1 Outline

It is essential to include a comprehensive overview of the principles of probability and distribution theory that are used in the analysis of sliding window detection processes. This chapter focuses on this, but predominantly from the perspective of continuous random variable theory. However, it is more intuitive to introduce some of the fundamental concepts with reference to discrete random variables. Some probability distributions useful in signal processing are introduced to illustrate the application of the theory. The probability distributions used in the modelling of X-band maritime radar clutter will be discussed in the next chapter in depth.

A number of references are also included for the reader's benefit. Probability theory, developed from a measure-theoretic perspective, can be found in (Billingsley, 1986) and (Shiryaev, 1996), while a more practical perspective to the subject can be found in (Durrett, 1996). For a reference where the subject of probability is developed from a signal processing perspective, one can consult (Shynk, 2012). Properties of random variables and distributions can be found in (Forbes et al., 2010), (Johnson et al., 1994) and (Johnson et al., 1995). The principles of stochastic processes can be sourced from (Ross, 1996), while (Ross, 2013) is an excellent guide on the theory and methodology of random variable simulation.

2.2 Fundamentals of Probability

The term "random variable" is somewhat of a misnomer in probability and statistics, because there is actually nothing random about a random variable. A random variable is a mapping from a sample space to certain types of subsets of this underlying event space. A measure of randomness is then assigned to this function, which provides probabilities associated with admissible events. Although somewhat technical, it is important to set a clear foundation for the analysis to follow. Additionally, it is more intuitive to consider discrete random variables to illustrate the fundamental concepts before embarking on the continuous distribution development. A *probability space* $(\Omega, \mathcal{F}, P)$ consists of a *sample space* Ω, a set of subsets of Ω denoted $\mathcal{F}$ and a *probability measure* P. The sample space describes where the phenomenon of interest exists. As an example, suppose we toss a fair coin three times. Then if we let H_j denote the occurrence of a head on the uppermost face on the jth single toss, and T_j the same but for a tail, then the set of all possible outcomes is $\Omega = \{H_1 \cap H_2 \cap H_3, H_1 \cap H_2 \cap T_3, T_1 \cap H_2 \cap H_3, H_1 \cap T_2 \cap H_3, H_1 \cap T_2 \cap T_3, T_1 \cap H_2 \cap T_3, T_1 \cap T_2 \cap H_3, T_1 \cap T_2 \cap T_3\}$. The structure of this can be illustrated with a tree diagram, showing all the possible outcomes as in Figure 2.1.

In probability theory the set of admissible subsets $\mathcal{F}$ is usually described as the σ−field, which is assumed to be a nonempty set with the properties that it is closed under complementation and unions of countable subsets. To clarify these conditions, one requires that if $S \in \mathcal{F}$ then its complement $S^c \in \mathcal{F}$, and if $\{S_j, j \in \mathbf{N} = \{1, 2, 3, \ldots\}\}$ is a sequence with $S_j \in \mathcal{F}$, then $\cup_{j=1}^{\infty} S_j \in \mathcal{F}$. Returning to our coin tossing example, admissible events can be related to the number of heads obtained. For instance, $\mathcal{F}_1 := \{H_1 \cap H_2 \cap T_3, T_1 \cap H_2 \cap H_3, H_1 \cap T_2 \cap H_3\}$ corresponds to the event of obtaining exactly two heads, while $\mathcal{F}_2 := \{T_1 \cap T_2 \cap T_3\}$ is the event of obtaining no heads.

The probability measure P assigns probabilities to events in $\mathcal{F}$, and is hence a mapping from $\mathcal{F}$ to the unit interval [0, 1]. It is assumed to satisfy the three properties:

1. $P(\Omega) = 1$;

2. $P(A) \in [0, 1]$;

3. $P(A \cup B) = P(A) + P(B)$ if $A \cap B = \Phi$,

where $\Phi := \{\}$ is the empty set. The third condition is referred to as *additivity under mutually exclusive events*. Such events cannot occur simultaneously. As an example, in the toss of a fair coin, the event that the upperhost side is a head, and the event that the same side is a tail, are mutually exclusive. Note that since $A \cap A^c = \Omega$ it follows from the third property above that $P(A) + P(A^c) = 1$ and so

$$P(A^c) = 1 - P(A). \tag{2.1}$$

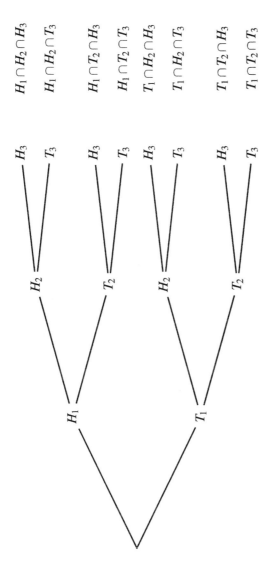

Figure 2.1: A three stage tree diagram illustrating the outcomes when a coin is tossed three times.

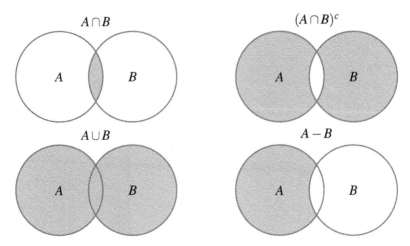

Figure 2.2: Examples of Venn diagrams associated with two sets A and B.

Many of the properties of sets can be determined with the application of Venn diagrams, where the sets are represented by circles and shading applied to illustrate set operations. As an example, Figure 2.2 shows Venn diagrams for four set operations, where it is assumed that the state space consists entirely of the union of A and B. The top left shows $A \cap B$, the top right is for the latter's complement, the bottom left corresponds to the union of sets A and B while the bottom right diagram illustrates $A - B$, which is the complement of B in set A. To illustrate the utility of Venn diagrams, observe that it can be concluded that $(A \cup B) \cap (A \cap B)^c = (A \cup B) - (A \cap B)$ in view of the diagrams in Figure 2.2.

With reference to our coin tossing example, we may assume that the probability of a head is as likely to result as that for a tail. Then $\mathbf{P}(H_j) = \mathbf{P}(T_j) = 1/2$, for example. This can be extended to provide probabilities associated with the two events $\mathcal{F}_1$ and $\mathcal{F}_2$ defined above. To do this one requires the concept of independent events. Two events $\mathcal{A}_1$ and $\mathcal{A}_2$ are said to be statistically independent if

$$\mathbf{P}(\mathcal{A}_1 \cap \mathcal{A}_2) = \mathbf{P}(\mathcal{A}_1)\mathbf{P}(\mathcal{A}_2). \tag{2.2}$$

Hence independence results in multiplication of the respective probabilities. Thus, returning to our coin tossing experiment, since we are assuming independent tosses of the coin, the probability of the event $H_1 \cap H_2 \cap T_3$ is $1/8$, as is for the events $T_1 \cap H_2 \cap H_3$ and $H_1 \cap T_2 \cap H_3$. Since each of these events are mutually exclusive, the probability associated with $\mathcal{F}_1$ defined above is $3/8$.

Suppose the event $A \subset B$. Then we can write $B = A \cap (B - A)$, and hence $B - A = B - (A \cup B)$. Then since A and $B - A$ are disjoint, $\mathbf{P}(B) = \mathbf{P}(A) + \mathbf{P}(B - A)$, from which it follows that

$$\mathbf{P}(B - A) = \mathbf{P}(B) - \mathbf{P}(A) \text{ when } A \subset B. \tag{2.3}$$

Observe that we can write $A \cup B = (A - B) \cup (B - A) \cup (A \cap B)$, where each of the sets are mutually exclusive, forming a disjoint partition of $A \cup B$. This partition becomes clear in view of the diagrams in Figure 2.2. Since $A - B = A - (A \cap B)$ and $B - A = B - (A \cup B)$, it follows by an application of (2.3) that

$$P(A \cup B) = P(A) + P(B) - P(A \cap B). \tag{2.4}$$

This result can be extended to an arbitrary number of events, and is known as the inclusion-exclusion formula. In particular, for a series of N events $\{A_j, j \in \{1, 2, \ldots, N\}\}$, one can show that

$$P\left(\bigcup_{j=1}^{N} A_j\right) = \sum_{k=1}^{N} \left((-1)^{k-1} \sum_{I \subset \{1,2,\ldots,N\}: |I|=k} P\left(\bigcap_{i \in I} A_i\right)\right). \tag{2.5}$$

A *random variable* X is a real valued function defined on the space Ω; this is often written $X : \Omega \longrightarrow \mathbb{R}$. More rigorously, X is a random variable if for every Borel subset B of the real line, the set $\{\omega \in \Omega : X(\omega) \in B\} \in \mathcal{F}$. Returning to our coin tossing example, suppose X counts the number of heads obtained in three independent tosses of a fair coin. Then X is such a mapping. The event $\{\omega \in \Omega : X(\omega) = 2\}$ corresponds to $\mathcal{F}_1$ defined above, while $\{\omega \in \Omega : X(\omega) = 0\}$ is identically the event $\mathcal{F}_2$. The reader familiar with probability will recognise such a random variable is known as discrete and Binomial. A discrete random variable has a countable state space, and a Binomial random variable has its event probabilities as terms in a binomial expansion.

To illustrate this, suppose that X is a random variable counting the number of successes in a series of trials, where each trial is independent. A success in a single trial is denoted by unity, while a failure in the same trial is denoted by zero. If one assumes that p is the probability of success in a single trial, then the probability that X takes a specific value k in the set $\{0, 1, 2, \ldots, N\}$ is the probability of k successes and $N - k$ failures, in all possible orders. Since it is assumed that the trials are independent, the probability of this is

$$P(X = k) = \binom{N}{k} (1 - p)^{N-k} p^k, \tag{2.6}$$

where $\binom{N}{k}$ is the combinatorial coefficient. A random variable with a probability mass function of the form (2.6) is referred to as having a binomial distribution. As a second example of a discrete distribution the well-known Poisson distribution is a model for rare events. If Y has a Poisson distribution with parameter $\lambda > 0$ then its probability mass function is given by

$$P(Y = k) = \frac{e^{-\lambda} \lambda^k}{k!}, \tag{2.7}$$

where $k \in \mathbb{N}$. Suppose that k is fixed, and that $\lambda = Np$, where the latter two parameters are from the probability mass function (2.6). Then by expanding the combinatorial coefficient one can show that (2.6) can be written for each fixed k as

$$P(X = k) = \frac{\left(1 - \frac{1}{N}\right) \cdots \left(1 - \frac{k-1}{N}\right)}{k!} \left(1 - \frac{\lambda}{N}\right)^N \left(\frac{\lambda}{1 - \frac{\lambda}{N}}\right)^k. \qquad (2.8)$$

If it is assumed that $N \to \infty$ and $p \to 0$ so that $\lambda \to \lambda_\infty < \infty$, then based upon (2.8) it follows that $P(X = k) \to P(Y = k)$, where Y has a Poisson distribution with parameter λ_∞. Hence the Poisson distribution results from the limit of a binomial distribution consisting of a large number of trials of rare events (Barbour et al., 1992). Stein approximation (Barbour and Chen, 2005) can be used to show that if X is a random variable with (2.6) as its probability mass function, and Y has a Poisson distribution with parameter $\lambda = Np$, then the difference in distributions can be bounded by

$$|P(X = k) - P(Y = k)| \leq \left(1 - e^{-Np}\right) p, \qquad (2.9)$$

for any $k \in \{0, 1, \ldots, N\}$. Hence the bound (2.9) shows that the limiting behaviour is actually controlled by the magnitude of p.

Conditional probability is an important concept in statistics, and the tools of conditioning are indispensable in statistical signal processing. The simplest way in which to introduce this is in the context of sets. Suppose A and B are two events. Then the *conditional probability* of A given B is defined to be

$$P(A|B) = \frac{P(A \cap B)}{P(B)}, \qquad (2.10)$$

provided $P(B) \neq 0$. The way in which this can be understood is that the sample space has been reduced to B, and the probability that event A occurs is the weighted probability that A occurs within B. For two events A and B note that we can write $A = (A \cup B) \cap (A \cup B^c)$, which is a disjoint partition of A via whether it occurs with B or its complement. Hence with an application of (2.10),

$$P(A) = P(B)P(A|B) + P(B^c)P(A|B^c). \qquad (2.11)$$

This is known as the *Law of Total Probability*, and can be extended easily for a series of events. Also observe that

$$P(B|A) = \frac{P(B \cap A)}{P(A)} = \frac{P(B)P(A|B)}{P(A)}, \qquad (2.12)$$

and with an application of (2.11) to (2.12), we derive *Bayes' Theorem:*

$$P(B|A) = \frac{P(B)P(A|B)}{P(B)P(A|B) + P(B^c)P(A|B^c)}. \qquad (2.13)$$

Suppose X is a random variable, then its *distribution function* is defined to be

$$F_X(A) = \mathbf{P}(\{\omega \in \Omega : X(\omega) \in A\}). \tag{2.14}$$

It is often convenient to select $A = (-\infty, t]$ for some t so that (2.14) becomes

$$F_X(t) := \mathbf{P}(X \le t), \tag{2.15}$$

where the notation used in (2.15) has been relaxed for brevity, as is commonly adopted in probability and statistics. Expression (2.15) is often referred to as the *cumulative distribution function*.

Supposing that X is a random variable, there are several properties of distribution functions worth noting:

1. $F_X(0) = 0$;

2. $\lim_{t \to -\infty} F_X(t) = 0$;

3. $\lim_{t \to +\infty} F_X(t) = 1$;

4. $F_X(t)$ is an increasing function.

The definitions (2.14) and (2.15) apply to a general random variable X, and in many cases of interest, there exists a function f_X which is known as the *probability density* associated with random variable X. Such a function will exist if the distribution function (2.15) is absolutely continuous with respect to Lebesgue measure on the real line (Rudin, 1987). This is known as a Radon-Nikodym derivative in mathematics, and means that we can write the distribution function (2.15) as

$$F_X(t) = \int_{-\infty}^{t} f_X(s) ds. \tag{2.16}$$

Observe that by an application of the Fundamental Theorem of Calculus to (2.16)

$$\frac{d}{dt} F_X(t) = f_X(t), \tag{2.17}$$

showing that the rate of change of the cumulative distribution function is determined via its associated density function.

Generally speaking, a function $g(t) : S \to \mathbf{R}$, where $S \subset \mathbf{R}^+$, is a density if

1. $g(t) \ge 0$ for all $t \in S$;

2. $\int_S g(t) dt = 1$.

Two random variables X and Y, defined on the same probability space, are said to have the same distribution if their distribution functions match everywhere: $F_X(t) = F_Y(t)$ for all t. This is often written $X \stackrel{d}{=} Y$. Equality in distribution is not equivalent to equality of random variables in the usual sense. To illustrate this, consider a random variable R defined on the unit interval $[0, 1]$ with density $f_R(t) = 1$ for $0 \leq t \leq 1$. Then its distribution function is given by $F_R(t) = t$, for $0 \leq t \leq 1$. Such a random variable is known as a *uniform or rectangular random variable*. Now consider the random variable $S = 1 - R$. One can derive its distribution by observing that

$$F_S(t) = \mathbf{P}(S \leq t) = \mathbf{P}(1 - R \leq t) = 1 - \mathbf{P}(R \leq 1 - t) = t, \qquad (2.18)$$

since $0 \leq t \leq 1$. Thus $F_S(t) = F_R(t)$ for all t, and so $S \stackrel{d}{=} R$ or equivalently $1 - R \stackrel{d}{=} R$. However, it is clear that $1 - R \neq R$ unless the random variable R takes the value of $1/2$.

One of the most recognised distributions in probability and statistics is the *normal or Gaussian distribution*. A random variable X has a normal distribution if its density is given by

$$f_X(t) = \frac{1}{\sqrt{2\pi}\sigma} e^{-\frac{1}{2\sigma^2}(t-\mu)^2}, \qquad (2.19)$$

where μ is the distribution's location parameter, which is also its mean, and σ^2 is its shape parameter, which is also its variance. This distribution has support the real line. It does not have a closed form for its distribution function, and so probabilities associated with this distribution are calculated by integrating (2.19) numerically over an appropriate domain. We write $X \stackrel{d}{=} \mathcal{N}(\mu, \sigma^2)$ to specify that X has such a distribution.

The *exponential distribution* is an important model in radar signal processing, as well as in the theory of probability. A random variable X has such a distribution if its density is given by

$$f_X(t) = \lambda e^{-\lambda t}, \qquad (2.20)$$

and distribution function

$$F_X(t) = 1 - e^{-\lambda t} \qquad (2.21)$$

where its nonnegative shape parameter is λ and its support is the nonnegative real line. Exponential random variables possess an interesting *memoryless property*. Suppose s and t are nonnegative real numbers. Then using the definition of conditional probability,

$$\mathbf{P}(X > t + s | X > s) = \frac{\mathbf{P}(X > t + s, X > s)}{\mathbf{P}(X > s)} = \frac{\mathbf{P}(X > t + s)}{\mathbf{P}(X > s)} = e^{-\lambda t} \qquad (2.22)$$

where (2.21) has been used. Clearly (2.22) is $P(X > t)$ and so the conditional probability (2.22) does not depend on s. Hence the exponential distribution has a memoryless property, meaning that if it is known that such a random variable exceeds a fixed value, then the distribution beyond this point has the same identical statistical structure.

Consider a simple transformation of this random variable: define $T = \sqrt{X}$. Then the distribution function of T is

$$F_T(t) = P(X \le t^2) = 1 - e^{-\lambda t^2} \tag{2.23}$$

and differentiating (2.23) results in the density

$$f_T(t) = 2t\lambda e^{-\lambda t^2}. \tag{2.24}$$

A random variable T with such a density is known as *Rayleigh distributed*. Such a random variable also arises as the distribution of the norm on a complex return consisting of zero mean Gaussian components with a common variance. Gaussian processes will not be analysed in detail because they are not directly required for the analysis to follow.

2.3 Transformations

Transformations of random variables is a very useful procedure, and so a brief overview of this is included. Assume that X is a random variable with density $f_X(t)$ and distribution function $F_X(t)$, and that a and b are two constants with $a > 0$. Then since

$$P(aX + b \le t) = F_X\left(\frac{t-b}{a}\right) \tag{2.25}$$

it follows that the density of the linearly transformed random variable $aX + b$ is

$$f_{aX+b}(t) = \frac{1}{a}f_X\left(\frac{t-b}{a}\right). \tag{2.26}$$

Next suppose we would like the density of X^2. In radar signal processing, if X is an amplitude distribution, then its square is called an *intensity model*. Assuming that X is nonnegative it is relatively straightforward to show that the density is given by

$$f_{X^2}(t) = \frac{1}{2}t^{-1/2}f_X(\sqrt{t}). \tag{2.27}$$

Another distribution that often makes an appearance in signal processing is the χ^2-distribution, which is the result of summing a number of independent normal distributions with zero mean and unity variance. Consider $X \overset{d}{=} \mathcal{N}(0,1)$, then by an application of (2.27),

$$f_{X^2}(t) = \frac{1}{2}t^{-\frac{1}{2}}f_X(\sqrt{t}) + \frac{1}{2}t^{-\frac{1}{2}}f_X(-\sqrt{t}) = \frac{1}{\sqrt{2\pi}}t^{-\frac{1}{2}}e^{-\frac{t}{2}}, \tag{2.28}$$

where (2.19) has been applied. This is known as the density of a χ^2-distribution with one degree of freedom. More generally, the sum of k squared standard normal random variables has a χ^2-distribution with k degrees of freedom if its density is given by

$$f(t) = \frac{1}{2^{\frac{k}{2}}\Gamma(k/2)}t^{\frac{k}{2}-1}e^{-\frac{t}{2}}, \tag{2.29}$$

where Γ is the gamma function.

Its distribution function does not have a closed form expression, but can be written in terms of the incomplete gamma function.

Suppose that g is a function and we are interested in the random variable $g(X)$. Suppose g is monotonically increasing, and hence is invertible. Then

$$\mathbf{P}(g(X) \leq t) = \mathbf{P}(X \leq g^{-1}(t)) = F_X(g^{-1}(t)). \tag{2.30}$$

Thus the density of $g(X)$ is

$$f_{g(X)}(t) = f_X(g^{-1}(t))\frac{d}{dt}g^{-1}(t) = \frac{1}{g'(g^{-1}(t))}f_X(g^{-1}(t)), \tag{2.31}$$

where differentiation of the inverse function has been applied.

In the case where g is monotonically decreasing, the argument proceeds in a similar vein as in (2.30) and (2.31), except a negative sign appears in the last expression in (2.31). Consequently, one can show that if g is a monotonic function then the density of $g(X)$ is

$$f_{g(X)}(t) = \frac{1}{|g'(g^{-1}(t))|}f_X(g^{-1}(t)). \tag{2.32}$$

To illustrate the application of (2.32) consider the case where X has an exponential distribution with parameter α and suppose we are interested in the transformation governed by $g(t) = \beta(e^t - 1)$, for some constant $\beta > 0$. Then $g^{-1}(t) = \log(1 + t/\beta)$ and $g'(t) = \beta e^t$. Applying these to (2.32) shows that

$$f_{g(X)}(t) = \frac{\alpha\beta^\alpha}{(t+\beta)^{\alpha+1}}. \tag{2.33}$$

The transformed random variable with density (2.33) is known as having a *Pareto distribution*. More specifically, (2.33) is known as a Pareto Type II distribution. Such distributions will feature significantly in the analysis to follow, since this is a suitable model for X-band maritime surveillance radar clutter. By integration of (2.33) the corresponding distribution function can be shown to be

$$F_{g(X)}(t) = 1 - \left(\frac{\beta}{t+\beta}\right)^{\alpha},$$
(2.34)

for $t \geq 0$. A Pareto Type I distribution can be obtained from the Pareto Type II model specified above by introducing the transformation $W = g(X) + \beta$, from which the distribution function of W is

$$F_W(t) = \mathbf{P}(g(X) + \beta \leq t) = F_{g(X)}(t - \beta) = 1 - \left(\frac{\beta}{t}\right)^{\alpha},$$
(2.35)

where $t \geq \beta$. Differentiation of (2.35) results in the density

$$f_W(t) = \frac{\alpha \beta^{\alpha}}{t^{\alpha+1}},$$
(2.36)

also with the requirement that $t > \beta$.

To illustrate this distribution, Figures 2.3 and 2.4 plot the Pareto density (2.33) and distribution function (2.34) respectively for a series of distributional parameters. The parameters selected in the figures correspond to those obtained from the Pareto fit to the Ingara data, which will be discussed in the following chapter. Case (i) in the two plots is for $\alpha = 4.7241$ and $\beta = 0.0446$, which is common values fitted for the case of horizontally polarised clutter returns. Case (ii) is for $\alpha = 6.7875$ and $\beta = 0.0043$, which is typical for cross-polarisation. Finally case (iii) is for $\alpha = 11.3930$ and $\beta = 0.3440$, which is typical for vertical polarisation. The density plots in Figure 2.3 show the Pareto model's very long tail. The plots of (2.35) and (2.36) have a similar structure, except their support begins at β.

2.4 Moments

Moments of a random variable provide a measure of its central location, or average value in some senses, but extended in definition to continuous random variables. The mean of a continuous random variable X with density f_X is defined to be

$$\mathbf{E}(X) = \int_0^{\infty} t f_X(t) dt,$$
(2.37)

while this can be extended to define the nth order moment via

$$v_n = \mathbf{E}(X^n) = \int_0^{\infty} t^n f_X(t) dt.$$
(2.38)

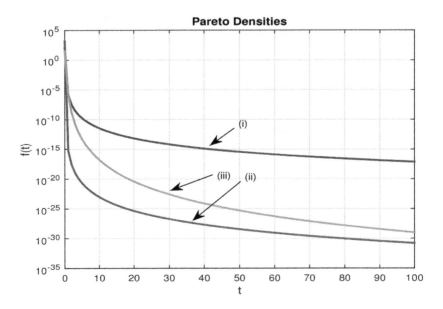

Figure 2.3: A series of Pareto Type II probability density functions. Case (i) is for $\alpha = 4.7241$ and $\beta = 0.0446$, Case (ii) corresponds to $\alpha = 6.7875$ and $\beta = 0.0043$ and Case (iii) is for $\alpha = 11.3930$ and $\beta = 0.3440$. These sets of Pareto parameters coincide with those obtained from fits to DST Group's Ingara radar clutter.

The mean (2.37) is a linear operator since for two random variables X and Y and constants a and b it is not difficult to show that

$$\mathrm{E}(aX + bY) = a\mathrm{E}(X) + b\mathrm{E}(Y). \tag{2.39}$$

The nth central moment about the mean is defined by

$$\mu_n = \mathrm{E}(X - \nu_n)^n = \int_0^\infty (t - \nu_n)^n f_X(t) dt. \tag{2.40}$$

In the case where $n = 2$, (2.40) is well-known to be the distribution's variance, giving a mean square measure of the average deviation from its mean.

Let X be an exponentially distributed random variable with parameter $\lambda > 0$ and density $f_X(t) = \lambda e^{-\lambda t}$. Then its nth order moment is

$$\mathrm{E}(X^n) = \int_0^\infty t^n \lambda e^{-\lambda t} dt = \int_0^\infty \lambda^{-n} w^n e^{-w} dw = \frac{n!}{\lambda^n}, \tag{2.41}$$

where the change or variables $w = \lambda t$ has been applied, together with the definition of the gamma function. The latter is defined by the integral

$$\Gamma(x) = \int_0^\infty t^{x-1} e^{-t} dt, \tag{2.42}$$

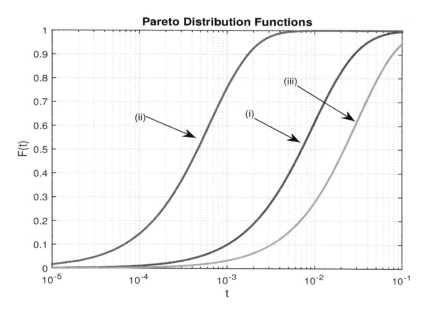

Figure 2.4: Pareto distribution functions corresponding to the densities in Figure 2.3.

which reduces to factorials at discrete t. For the same distribution, the nth central moment about the mean can also be written down. Note that with an application of a binomial expansion

$$E(X - v_n)^n = \sum_{k=0}^{n} \binom{n}{k} E(X^k)(-v_n)^{N-k}, \tag{2.43}$$

allowing the nth central moment to be determined from v_n. Returning to our exponential example, it is shown how its variance can be calculated. Firstly, by applying $n = 2$ to (2.43) it can be demonstrated that

$$\mathrm{Var}(X) := E(X - v_2)^2 = E(X^2) - v_2^2, \tag{2.44}$$

which provides a convenient expression for determining variances of random variables. Referring to (2.41) it is not difficult to see that

$$\mathrm{Var}(X) = \frac{2}{\lambda^2} - \frac{1}{\lambda^2} = \frac{1}{\lambda^2}. \tag{2.45}$$

Moment generating functions (MGFs) are useful tools in probability from the perspective of distributional identification and also for proving limit results. MGFs are similar to Fourier transforms, and as it is often more convenient in signal processing to work in the frequency domain, it is the same in terms of analysing probability distributions. For a continuous random variable, the MGF is defined to be

$$M_X(t) = E(e^{Xt}). \tag{2.46}$$

This is equivalent to a Laplace transformation of the underlying probability density function. By an application of the Taylor expansion for the exponential function in (2.46), it is not difficult to demonstrate that

$$M_X(t) = \sum_{j=0}^{\infty} \mu_j \frac{t^j}{j!}. \tag{2.47}$$

Consequently it follows from (2.46) that

$$E(X^n) = \frac{d^n M_X(t)}{dt^n}\bigg|_{t=0}. \tag{2.48}$$

The value in the MGF is that it can be used to identify distributions as the following Lemma illustrates:

Lemma 2.1
If two random variables X and Y with the same support have MGFs equal everywhere, namely $M_X(t) = M_Y(t)$ for all t, then $X \overset{d}{=} Y$.

MGFs possess some useful properties, as the following Lemma encapsulates:

Lemma 2.2
Suppose random variables X and Y are independent and a is a fixed constant. Then

1. $M_{X+Y}(t) = M_X(t)M_Y(t)$.

2. $M_{aX}(t) = M_X(at)$.

In a similar way to Laplace and Fourier transforms it is possible to recover the underlying probability density function from an MGF by the inversion integral

$$f_X(t) = \frac{1}{2\pi i} \oint_C M_X(z)e^{-tz}dz, \tag{2.49}$$

where the path C a semicircular contour in the complex plain, passing through the points $c - ir$ and $c + ir$, where the limit $r \to \infty$ is taken, for some complex c. However, probabilists tend not to use (2.49) but instead demonstrate convergence or distributional identification in the transformed domain.

For an example, suppose that X has a standard normal distribution. Then its MGF is given by

$$
\begin{aligned}
M_X(t) &= \int_{-\infty}^{\infty} e^{xt} \frac{1}{\sqrt{2\pi}} e^{-\frac{1}{2}x^2} dx \\
&= e^{\frac{t^2}{2}} \frac{1}{\sqrt{2\pi}} \int_{-\infty}^{\infty} e^{-\frac{1}{2}(x-t)^2} dx \\
&= e^{\frac{t^2}{2}}.
\end{aligned}
\tag{2.50}
$$

The usefulness of MGFs in proving limit results is encapsulated in the following:

Lemma 2.3
Assume that $\{F_n, n \in \mathbf{N}\}$ is a sequence of distribution functions and $\{M_n, n \in \mathbf{N}\}$ is a corresponding sequence of MGFs. Also assume distribution function F also has MGF M. Then if $M_n(t) \longrightarrow M(t)$ for all t, then $F_n(t) \longrightarrow F(t)$.

Thus convergence in distribution can be demonstrated via convergence of MGFs. Lemma 2.3 can be used to prove the *Central Limit Theorem*, which states that a sum of independent and identically distributed random variables limits to a Gaussian in distribution. The following is a simplified version which can be proven via analysis of MGFs:

Theorem 2.1
Let $\{X_j, j \in \mathbf{N}\}$ be a sequence of independent and identically distributed random variables with zero mean and unity variance. Define $S_N = \sum_{j=1}^{N} X_j$. Then $\frac{S_N}{\sqrt{N}} \xrightarrow{d} \mathcal{N}(0,1)$, where the latter refers to convergence in terms of distribution functions.

An outline of the proof of Theorem 2.1 is now provided. If we let $M_X(t)$ be the common MGF then by an application of Lemma 2.2 it follows that

$$
M_{\frac{S_N}{\sqrt{N}}}(t) = \left(M_X\left(\frac{t}{\sqrt{N}}\right) \right)^N.
\tag{2.51}
$$

Also based upon (2.47),

$$
M_X\left(\frac{t}{\sqrt{N}}\right) = 1 + \frac{t^2}{2N} + v_3 \frac{t^3}{6N\sqrt{N}} + \cdots
\tag{2.52}
$$

Thus

$$
M_{\frac{S_N}{\sqrt{N}}}(t) = \left(1 + \frac{t^2}{2N} + O\left(\frac{1}{N\sqrt{N}}\right) \right)^N,
\tag{2.53}
$$

where the notation used in (2.53) means that for fixed t, the term $O\left(\frac{1}{N\sqrt{N}}\right)$ decreases at the rate of the reciprocal of $N\sqrt{N}$. Taking logarithms of (2.53) and applying the Taylor series expansion for $\log(1+t)$, it is not difficult to show that

$$\log\left(M_{\frac{S_N}{\sqrt{N}}}(t)\right) = \frac{t^2}{2} + O\left(\frac{1}{\sqrt{N}}\right), \tag{2.54}$$

and taking the limit $N \to \infty$ shows that the MGF limits to (2.50). An appeal to Lemma 2.3 completes the proof.

2.5 Inequalities

Inequalities in probability can provide useful bounds on moments of random variables and on the tail of distributions, and so are of relevance to the analysis to follow. Some of the more useful inequalities are presented here for reference.

Hölder's inequality states that if p and q are two real numbers in the interval $[1,\infty)$ such that $1/p + 1/q = 1$ then

$$E|XY| \le \|X\|_p \|Y\|_q \tag{2.55}$$

for two random variables X and Y, where $\|\cdot\|_p$ is known as the p-norm defined by

$$\|X\|_p = (E|X|^p)^{1/p}. \tag{2.56}$$

In the special case where $p = q = 2$ this reduces to the well-known Cauchy-Schwarz inequality.

Jensen's Inequality states that if ψ is a convex function then

$$\psi(E(X)) \le E(\psi(X)). \tag{2.57}$$

Chebyshev's Inequality supposes that the random variable X has finite mean μ and variance σ^2. For a positive $\varepsilon > 0$ this gives the bound

$$P(|X - \mu| \ge \varepsilon\sigma) \le \frac{1}{\varepsilon^2}, \tag{2.58}$$

which provides a bound on the upper tail region of the random variable $|X - \mu|$. A simple variation of this is known as Markov's Inequality which states that for a random variable X and constant $\gamma > 0$,

$$P(X \ge \gamma) \le \frac{E(X)}{\gamma}. \tag{2.59}$$

2.6 Jointly Distributed Random Variables

The concepts of random variable distributions and densities has been extended to the situation of multiple random variables, and so a brief overview is included. Two random variables (X, Y) are said to have a joint distribution if their distribution function is

$$F_{(X,Y)}(x,y) = P(X \le x, Y \le y). \tag{2.60}$$

If the two univariate random variables are independent then (2.60) reduces to a product of the respective marginal distribution functions. If the random variables admit a joint density, written $f_{(X,Y)}(x,y)$, then the distribution function (2.60) can be written

$$F_{(X,Y)}(x,y) = \int_{-\infty}^{x} \int_{-\infty}^{y} f_{(X,Y)}(x,y) dx dy. \tag{2.61}$$

Assuming both these random variables have support the real line, the marginal distributions can be extracted by integration over the respective random variable. To illustrate, suppose we are interested in the univariate distribution of X. Then

$$f_X(x) = \int_{-\infty}^{\infty} f_{(X,Y)}(x,y) dy \tag{2.62}$$

Transformations can also be applied to jointly distributed random variables. Suppose we have a pair of random variables (X, Y), which are transformed to $(U, V) = (g_1(X), g_2(Y))$, and that (X, Y) has joint density function $f_{(X,Y)}(x,y)$. If we let J be the Jacobian matrix

$$J = \frac{\partial(x,y)}{\partial(u,v)} = \begin{pmatrix} \frac{\partial x}{\partial u} & \frac{\partial x}{\partial v} \\ \frac{\partial y}{\partial u} & \frac{\partial y}{\partial v} \end{pmatrix} \tag{2.63}$$

then the density of the transformation is given by

$$f_{(U,V)}(u,v) = f_{(X,Y)}(x(u,v), y(u,v))|J| \tag{2.64}$$

where $|J|$ is the modulus of the determinate of (2.63).

As a simple example suppose X and Y are independent uniform random variables with joint density $f_{(X,Y)}(x,y) = 1$ on the unit square $0 < x < 1$ and $0 < y < 1$. Consider the transformation defined through the new random variables $U(x,y) = \sqrt{-2\log(x)} \cos(2\pi y)$ and $V(x,y) = \sqrt{-2\log(x)} \sin(2\pi y)$. Then it is relatively straightforward to show that $X(u,v) = e^{-1/2(u^2+v^2)}$ and $Y(u,v) = \frac{1}{2\pi} \arctan(v/u)$. Then it follows that $\frac{\partial X}{\partial u} = -u e^{-1/2(u^2+v^2)}$, $\frac{\partial X}{\partial v} = -v e^{-1/2(u^2+v^2)}$, $\frac{\partial Y}{\partial u} = -\frac{v}{2\pi(u^2+v^2)}$ and $\frac{\partial Y}{\partial v} = \frac{u}{2\pi(u^2+v^2)}$. Applying these to (2.63) and evaluating the modulus of the determinant results in the transformed density

$$f_{(U,V)}(u,v) = \frac{1}{2\pi} e^{-\frac{1}{2}(u^2+v^2)}, \tag{2.65}$$

which can be identified as a standard bivariate Gaussian random variable. In fact, it is clear from (2.65) that both the marginal distributions of U and V are standard independent Gaussian variates.

2.7 Conditional Distributions

The conditional distribution of a continuous random variable X, given another continuous random variable Y is defined through the density

$$f_{X|\{Y=y\}}(x) = \frac{f_{(X,Y)}(x,y)}{f_Y(y)}, \qquad (2.66)$$

which has been adopted as a suitable extension of the discrete case. The corresponding distribution function of X conditioned on $Y = y$ can be obtained by integration of (2.66). This is given by

$$F_{X|\{Y=y\}}(x) = P(X \leq x | Y = y) = \int_{-\infty}^{x} f_{X|\{Y=y\}}(t)dt. \qquad (2.67)$$

Conditional expectations can be defined based upon these conditional distributions. Hence the conditional mean of X given $Y = y$ is defined to be

$$E[X|\{Y=y\}] = \int_{-\infty}^{\infty} t f_{X|\{Y=y\}}(t)dt, \qquad (2.68)$$

with other moments defined similarly. Note that (2.68) is a function of y, and so one can evaluate the mean of (2.68) relative to Y. Hence

$$
\begin{aligned}
E[E[X|\{Y=y\}]] &= \int_{-\infty}^{\infty} f_Y(y) \int_{-\infty}^{\infty} x f_{X|\{Y=y\}}(x)dxdy \\
&= \int_{-\infty}^{\infty} \int_{-\infty}^{\infty} x f_{(X,Y)}(x,y)dxdy \\
&= \int_{-\infty}^{\infty} x f_X(x)dx = E[X], \qquad (2.69)
\end{aligned}
$$

where the density (2.66) has been applied, as well as the fact that integrating a bivariate distribution over one of its variables will eliminate the corresponding component of the bivariate distribution. The result (2.69) is often written in the form

$$E[E[X|Y]] = E[X]. \qquad (2.70)$$

It is possible to derive an expression for the variance of a random variable in terms of conditional expectation, as an analogue of (2.70). In particular, it can be shown that

$$\text{Var}[Y] = E[\text{Var}[Y|X]] + \text{Var}[E[X|Y]]. \qquad (2.71)$$

An interesting and very useful tool based upon (2.66) follows by noting that $f_{(X,Y)}(x,y) = f_{X|\{Y=y\}}(x)f_Y(y)$ and so by integrating this expression over y we obtain

$$f_X(x) = \int_0^\infty f_{X|\{Y=y\}}(x)f_Y(y)dy. \tag{2.72}$$

2.8 Some Special Functions of Random Variables

Let X and Y be two continuous random variables with densities f_X and f_Y respectively. In many applications in sliding window detector analysis the sum of several random variables is of relevance, and so it is useful to be able to construct the distribution of a sum of several random variables. Here we show how this can be done for two random variables. Beginning with the distribution function of $X+Y$,

$$
\begin{aligned}
F_{X+Y}(t) &= P(X+Y \le t) = \int_0^\infty P(X+Y \le t|\{Y=y\})f_Y(y)dy \\
&= \int_0^\infty P(X \le t-y|\{Y=y\})f_Y(y)dy, \tag{2.73}
\end{aligned}
$$

where (2.72) has been applied. In the case where X and Y are independent, the conditional distribution in (2.73) reduces to an unconditional distribution and we can write

$$F_{X+Y}(t) = \int_0^\infty P(X \le t-y)f_Y(y)dy, \tag{2.74}$$

and by differentiating (2.74) one obtains the density

$$f_{X+Y}(t) = \int_0^\infty f_X(t-y)f_Y(y)dy. \tag{2.75}$$

Suppose X and Y are two independent exponentially distributed random variables with parameter λ. Then applying (2.75), the density of the sum is

$$f_{X+Y}(t) = \int_0^t \lambda e^{-\lambda(t-y)}\lambda e^{-\lambda y}dy = \lambda^2 t e^{-\lambda t}. \tag{2.76}$$

One can then apply (2.76), together with a third independent exponentially distributed random variable Z to show

$$f_{X+Y+Z}(t) = \int_0^t f_{X+Y}(t-y)\lambda e^{-\lambda y}dy = \frac{\lambda^3}{2}t^2 e^{-\lambda t}. \tag{2.77}$$

This process can be continued to show that a sum of N exponentially distributed random variables with common parameter λ has density

$$f(t) = \frac{\lambda^N}{(N-1)!} t^{N-1} e^{-\lambda t} \qquad (2.78)$$

which is known as a gamma distribution with parameters N and λ. More generally, a random variable X has a gamma distribution with nonnegative real parameters α and β if its density is given by

$$f_X(t) = \frac{\beta^\alpha}{\Gamma(\alpha)} t^{\alpha-1} e^{-\beta t}. \qquad (2.79)$$

The corresponding distribution function does not possess a closed form but is instead an integral of (2.79) over an appropriate range.

The gamma distribution has a useful scale transformation property: if c is a constant and X has density (2.79) then cX is also gamma distributed, with the same shape parameter but scale parameter β/c.

The result (2.75) can be extended to the case of differences of random variables. Additionally, the same method can be used to derive similar expressions for products and ratios of random variables. As an example, suppose we are interested in the product XY, where these are nonnegative independent random variables. Then

$$
\begin{aligned}
F_{XY}(t) &= \mathrm{P}(XY \leq t) = \int_0^\infty f_Y(s) \mathrm{P}(XY \leq t | Y = s) ds \\
&= \int_0^\infty F_X\left(\frac{t}{s}\right) f_Y(s) ds. \qquad (2.80)
\end{aligned}
$$

The density is thus

$$f_{XY}(t) = \int_0^\infty \frac{1}{s} f_X\left(\frac{t}{s}\right) f_Y(s) ds. \qquad (2.81)$$

Similarly, if X and Y are two nonnegative random variables, the density of $Z = \frac{X}{Y}$ is

$$f_Z(t) = \int_0^\infty y f_X(ty) f_Y(y) dy. \qquad (2.82)$$

Consider, as an example, the case where X and Y are independent gamma random variables; assume that X has density (2.79) with parameters $\alpha = \pi_1$ and $\beta = 1$, and that Y has the same distribution but with $\alpha = \pi_2$ and $\beta = 1$. Then with an application of (2.82), the density of Z is

$$f_Z(t) = \frac{1}{B(\pi_1, \pi_2)} t^{\pi_1 - 1} (1+t)^{-(\pi_1 + \pi_2)}, \qquad (2.83)$$

where $\mathcal{B}$ is the Beta function which is defined in terms of the gamma function (2.42) by

$$\mathcal{B}(\pi_1, \pi_2) = \frac{\Gamma(\pi_1)\Gamma(\pi_2)}{\Gamma(\pi_1 + \pi_2)}. \tag{2.84}$$

A random variable with such a density is known as *beta prime*. As in the gamma distribution case, the beta prime distribution function can be expressed as an integral of its density over an appropriate domain. It can be shown that it reduces to

$$F(t) = I_{\frac{t}{t+1}}(\pi_1, \pi_2), \tag{2.85}$$

where $I_x(\pi_1, \pi_2)$ is the incomplete beta function.

2.9   Order Statistics

Order statistics are a useful tool in the design of sliding window CFAR detection processes because they can be used to produce a measure of the clutter level while censoring potential spurious targets. Useful references on the properties of order statistics are (Arnold, 2014) and (David and Nagaraja, 2003). Suppose that we have a series of random variables $X_1, X_2, \ldots, X_N$, which we suppose are independent and identically distributed with common distribution function F_X. We can order these random variables in an increasing sequence which we denote $X_{(1)}, X_{(2)}, \ldots, X_{(N)}$, where $X_{(j)}$ is the jth largest return. Thus $X_{(1)}$ is the sample minimum while $X_{(N)}$ is its maximum. The distribution of these two statistics are relatively easy to derive. Note that for the minimum

$$P(X_{(1)} \geq t) = P(X_1 \geq t, X_2 \geq t, \ldots, X_N \geq t) = P(X \geq t)^N, \tag{2.86}$$

since the event that the minimum exceeds t is equivalent to requiring each of the individual statistics to exceed t, and the final result in (2.86) follows by the independence and identically distributed assumption. Consequently it follows that

$$F_{X_{(1)}}(t) = 1 - (1 - F_X(t))^N. \tag{2.87}$$

It is interesting to observe that if the sequence of random variables is exponentially distributed, then the minimum will also be exponentially distributed, except its shape parameter is N times the individual common shape parameters. Differentiating (2.87) one derives the corresponding density

$$f_{X_{(1)}}(t) = N(1 - F_X(t))^{N-1} f_X(t). \tag{2.88}$$

The maximum distribution can be determined in a similar fashion; in particular, observing that if the maximum is bounded by t from above, then each of the individual order statistics must be bounded by t and so

$$F_{X_{(N)}}(t) = F_X(t)^N, \tag{2.89}$$

and its density is given by

$$f_{X_{(N)}}(t) = N F_X(t)^{N-1} f_X(t) \tag{2.90}$$

In the exponentially distributed example, the distribution function of the maximum is clearly

$$F_{X_{(N)}}(t) = (1 - e^{-\lambda t})^N = \sum_{k=0}^{N} \binom{N}{k} (-1)^k e^{-\lambda k t}, \tag{2.91}$$

where a binomial expansion has been applied to produce the last term in (2.91). Hence it is clear that the maximum does not have an exponential distribution.

Using a similar line of reasoning, the distribution of a number of functions of order statistics can be derived. The distribution of the kth order statistic has distribution function

$$F_{X_{(k)}}(t) = \sum_{i=k}^{N} \binom{N}{i} [1 - F_X(t)]^{N-i} F_X(t)^i \tag{2.92}$$

and density

$$f_{X_{(k)}}(t) = k \binom{N}{k} F_X(t)^{k-1} [1 - F_X(t)]^{N-k} f_X(t). \tag{2.93}$$

Continuing with the exponentially distributed case, if the sequence of random variables has such a distribution with parameter λ then the density (2.93) becomes

$$f_{X_{(k)}}(t) = \lambda k \binom{N}{k} (1 - e^{-\lambda t})^{k-1} e^{-\lambda t (N-k+1)}. \tag{2.94}$$

It is interesting to consider a change of variables in (2.94) using (2.32). Choose the function $g(t) = e^{-\lambda t}$, which is monotonically decreasing. Then $g'(t) = -\lambda e^{-\lambda t}$ and $g^{-1}(t) = -1/\lambda \log(t)$ so that $g'(g^{-1}(t)) = -\lambda t$. Thus the transformed density is

$$f_{g(X_{(k)})}(t) = k \binom{N}{k} (1 - t)^{k-1} t^{N-k}. \tag{2.95}$$

The density (2.95) can be identified as that of a beta random variable with parameters k and $N - k + 1$. More generally, a random variable with nonnegative parameters α and β, has a beta distribution if its density is given by

$$f(t) = \frac{1}{B(\alpha, \beta)} t^{\alpha-1} (1 - t)^{\beta-1}, \tag{2.96}$$

where $0 < t < 1$. The beta density (2.95) can also be identified as that of the kth order statistic for a series of uniformly distributed random variables on the unit interval.

It is also interesting to consider a change of variables in the beta distribution, by letting $Z = \frac{X}{1-X}$, where X has density (2.96). It can be shown that

$$f_Z(t) = \frac{1}{(1+t)^2} f_X\left(\frac{t}{t+1}\right), \tag{2.97}$$

where f_X is the density (2.96). After simplification, the density of Z can be shown to be exactly (2.83), showing the beta prime distribution is also related to the beta distribution.

The joint distribution of a pair of order statistics can also be derived. For $i < j$, the joint density of $X_{(i)}$ and $X_{(j)}$ is given by

$$f_{(X_{(i)},X_{(j)})}(x,y) = \frac{N!}{(i-1)!(j-i-1)!(N-j)!} F_X(x)^{i-1} \times$$

$$[F_X(y) - F_X(x)]^{j-i-1} [1 - F_X(y)]^{N-j} \times$$

$$f_X(x)f_X(y), \tag{2.98}$$

where $x \le y$. The distribution function of the order statistic of a series of measurements can also be obtained, for the case where the measurements are independent but not necessarily identically distributed. Assume that we have a series of statistics $X_1, X_2, \ldots, X_N$ as before, except L are from a distribution with distribution function $F_1(t)$ and $N - L$ with distribution function $F_2(t)$, then the distribution function of the kth order statistic is given by

$$F(t) = \sum_{i=k}^{N} \sum_{j=\max(0,i-L)}^{\min(i,N-L)} \binom{N-L}{j}\binom{L}{i-j} \times$$

$$F_2^j(t)[1 - F_2(t)]^{N-L-j} \times$$

$$F_1^{i-j}(t)[1 - F_1(t)]^{L-i+j}. \tag{2.99}$$

2.10 Uniform Distributions and Simulation

The idea of simulating a random variable is to generate observations from its state space such that when analysed in a large collection, it will possess the statistical structure of the underlying distribution. Hence a histogram will slowly converge to the form of the data's density function as the histogram's window size decreases. For a sufficiently large sample size generated from the distribution the sample statistics should also limit to the population model's statistics. A comprehensive guide to simulation is (Ross, 2013).

In the performance analysis of sliding window CFAR detection systems it is necessary to simulate clutter from a particular model, and hence it is useful to overview methods with which this can be achieved. Since the clutter models of interest are continuous, the simulation of returns can be achieved using the inverse distribution function method. Suppose that X is a random variable with distribution function $F_X(t)$. Then since this is invertible, the random variable $Y := F_X^{-1}(R)$ exists, where R is uniformly distributed on the unit interval. Then the distribution function of Y is given by

$$F_Y(t) = \mathbf{P}(F_X^{-1}(R) \leq t) = \mathbf{P}(R \leq F_X(t)) = F_X(t), \qquad (2.100)$$

where monotonicity of F_X and its inverse have been employed, together with the fact that a distribution takes values in the unit interval. Hence $Y \overset{d}{=} X$ and equivalently

$$F_X^{-1}(R) \overset{d}{=} X. \qquad (2.101)$$

Thus to generate a realisation from a random variable X, one can generate a random number in the unit interval and evaluate the inverse distribution function at this realisation. To illustrate this consider the exponential distribution function $F_X(t) = 1 - \exp(-\lambda t)$, for some $\lambda > 0$. The inverse distribution function is $F_X^{-1}(t) = -1/\lambda \log(1-t)$. Thus, according to (2.101), $-1/\lambda \log(1-R) \overset{d}{=} X$, and hence to simulate the exponential distribution, one may generate a series of random numbers $r_1, r_2, \ldots, r_N$ from the unit interval and then evaluate $-1/\lambda \log(r_j)$, noting that R and $1 - R$ have the same uniform distribution. Assuming these random numbers are independent, the resultant series of exponentially distributed variables will also be independent. Generating a correlated series is more of a challenge, which can also be achieved using the same principle.

Note that for a random variable X, $F_X(X) \overset{d}{=} R$, which can be shown quite easily as in the derivation (2.100). Hence suppose Y is a random variable with distribution function F_Y. Then based upon these results, if we define a function

$$\eta(t) := F_Y^{-1}(F_X(t)) \qquad (2.102)$$

then it follows that $\eta(X) \overset{d}{=} Y$. Thus if we can generate a series of returns from X that are correlated, we can apply them to (2.102) to produce a correlated series of returns that will be distributed marginally according to Y. It is relatively simple to generate a correlated series of Gaussian distributed random variables, and so this can be used in conjunction with (2.102).

The difficulty is one would like to generate a series distributed according to Y with a pre-determined correlation structure. This would necessitate an ability to determine the corresponding correlation for the distribution X. This has been investigated in (Tough and Ward, 1999), and a more recent solution has been investigated in (Weinberg and Gunn, 2011).

2.11 Properties of Estimators

It is useful to recall some properties of estimators. The following has been based upon the development of the theory of estimators in (Beaumont, 1980), which is an excellent reference on mathematical statistics.

Assume that we have a series of random variables $\{X_1, X_2, \ldots, X_N\}$, and that this sample is from a population with an unknown parameter θ. A *statistic* for θ is a function of the sample that does not depend on θ, which can be written $H(X_1, X_2, \ldots, X_N)$. If this statistic is used to estimate θ on the basis of a sample of observations, then it is referred to as an *estimator*. Suppose that the sample has joint density function $f_{(X_1, X_2, \ldots, X_N)}(x_1, x_2, \ldots, x_N | \theta)$. Then the statistic H is said to be *sufficient* for θ if the conditional distribution of the sample given H does not depend on θ. This definition can be extended to multiple unknown parameters as follows. Suppose that the series $\{X_1, X_2, \ldots, X_N\}$ of random variables is from a population with parameters $\{\theta_j, j \in J \subset \mathbf{N}\}$. Then the statistics $H_j(X_1, X_2, \ldots, X_N)$, $j \in J \subset \mathbf{N}$ are *jointly sufficient* for the θ_j if the conditional distribution of the sample conditioned on the statistics H_j does not depend on any of the θ_j.

Returning to the univariate case, the statistic is an *unbiased estimator* of θ if its mean is exactly θ. The *Rao-Blackwell Theorem* states that if X and Y are jointly distributed random variables with finite moments, and we let $\mu(X) = E[Y|X]$ be the conditional mean then $\mathbf{E}(\mu(X)) = \mathbf{E}(Y)$ and $\mathrm{Var}(\mu(X)) \leq \mathrm{Var}(Y)$. This states that $\mu(X)$ is an unbiased estimator of $\mathbf{E}(Y)$ which has smaller variance than the estimator Y. The consequence of this is that if we are searching for a minimum variance unbiased estimator of a parameter θ then we must restrict attention to functions of sufficient statistics, otherwise the Rao-Blackwell Theorem can be used to improve on them.

Assume that H is a statistic, and g a function of it. Then if $\mathbf{E}(g(H)) = 0$ implies that $g(H) = 0$ almost surely then H is said to be complete. Two random variables are said to be equal almost surely if they are equal except on sets of measure zero. The importance of completeness is related to the idea of a minimally sufficient statistic. H is minimally sufficient if its components are functions of any other sufficient statistic. It can be shown that a sufficient statistic that is also complete is minimally sufficient. The *Cramér Lower Bound* provides the minimal achievable variance of estimators of a parameter. If an estimator can be found which has this variance, and it is unbiased, then it is known as the *minimum variance unbiased estimator*. Suppose $X_1, X_2, \ldots, X_N$ are statistics with joint density $f_{(X_1, X_2, \ldots, X_N)}(x_1, x_2, \ldots, x_N | \theta)$ and that H is a statistic that is an unbiased estimator of θ. Then

$$\mathrm{Var}(H) \geq \frac{1}{\left(\mathbf{E}\frac{\partial}{\partial \theta} \log f_{(X_1, X_2, \ldots, X_N)}(x_1, x_2, \ldots, x_N | \theta)\right)^2}, \qquad (2.103)$$

provided the region where the density f is nonzero does not depend on θ.

To illustrate these concepts, suppose we have a random sample from a distribution with density $f(t|\theta) = e^{\theta - t}$, for $t \geq \theta$. Then changing variables to $Y = X - \theta$, it is relatively straightforward to show that Y has an exponential distribution with parameter 1. Consider the statistic $H(X_1, X_2, \ldots, X_N) = X_{(1)}$, the minimum of the sample. Then since the minimum of a random sample from an exponential distribution with unity mean is exponentially distributed with parameter N, it follows that $H = \theta + E$, where $E \stackrel{d}{=} \mathrm{Exp}(N)$, and consequently

$$f_H(t|\theta) = Ne^{-N(t-\theta)}, \tag{2.104}$$

for $t \geq \theta$. Thus the joint density of $X_1, X_2, \ldots, X_N$, conditioned on $H = h$, has density

$$f_{(X_1, X_2, \ldots, X_N | H)}(x_1, x_2, \ldots, x_N | h, \theta) = \frac{\prod_{j=1}^{N} e^{\theta - x_j}}{f_H(h)}$$

$$= \frac{1}{N} \exp\left(-\sum_{j=1}^{N} x_j + Nh \right), \tag{2.105}$$

where $h = x_{(1)}$. Since the density (2.105) does not depend on θ, H is sufficient for θ.

Next suppose that ψ is a function and that $\mathrm{E}[\psi(H)] = 0$. Then this implies that

$$\int_{\theta}^{\infty} \psi(t)Ne^{-N(t-\theta)}dt = 0. \tag{2.106}$$

Therefore (2.106) implies that

$$\int_{\infty}^{\theta} \psi(t)e^{-Nt}dt = 0, \tag{2.107}$$

and by differentiating (2.107) with respect to θ we deduce that $\psi(\theta)e^{-\theta N} = 0$ for all θ and thus $\psi(\theta) = 0$ for all θ. Hence H is a complete sufficient statistic for θ, and therefore it is minimally sufficient for θ.

Due to the expression derived for H it follows that $\mathrm{E}(H) = \theta + \frac{1}{N}$, so that H is not unbiased. However, one can instead consider the estimator $\tilde{H} = H - \frac{1}{N}$, which will be unbiased for θ. Its variance is $\mathrm{Var}(\tilde{H}) = \frac{1}{N^2}$. Since the joint density is $f_{(X_1, X_2, \ldots, X_N)}(x_1, x_2, \ldots, x_N | \theta) = e^{N\theta - \sum_{j=1}^{N} x_j}$, it follows that by taking logarithms and differentiating with respect to θ, one obtains the minimum variance bound $\frac{1}{N^2}$, and since the density of each element of the sample is nonzero regardless of θ, it can be inferred that the estimator $\tilde{H}$ attains the minimum variance bound, implying it is the minimum variance unbiased estimator of θ.

There are some useful results available concerning statistics independent of complete sufficient statistics. In particular, (Basu, 1955) and (Williams, 1957) prove the following result:

Lemma 2.4

Suppose $X_1, X_2, \ldots, X_N$ is a random sample from a distribution with population parameter θ. Suppose that H is a complete sufficient statistic for θ. Also suppose that Y is a secondary statistic, which may be a function of H, but which has a distribution that does not depend on θ. Then Y is independent of H.

An elegant and surprisingly simple proof can be found in (Beaumont, 1980). Lemma 2.4 can be used to show pairs of random variables are independent, which in some cases can otherwise prove to be an extremely difficult exercise.

Returning to the example considered above, since H is a complete sufficient statistic for θ, and $X = \theta + Y$, where Y has an exponential distribution with parameter 1, the range is defined to be $X_{(N)} - X_{(1)} = Y_{(N)} - Y_{(1)}$, and the latter does not depend on θ. Hence Lemma 2.4 implies that $X_{(1)}$ and $X_{(N)} - X_{(1)}$ are independent.

So far nothing has been said on the subject of construction of estimators. One of the most popular methods is *maximum likelihood estimation* (MLE) of unknown population parameters. Suppose we have a random sample $x_1, x_2, \ldots, x_N$, from a population with unknown parameter θ. We define the likelihood function

$$L(x_1, x_2, \ldots, x_N | \theta) = f_{(X_1, X_2, \ldots, X_N)}(x_1, x_2, \ldots, x_N | \theta) = \prod_{j=1}^{N} f_{X_j}(x_j | \theta), \quad (2.108)$$

where the latter product results from an independent and identically distributed assumption. Hence the likelihood function is the joint density of the sample. Maximum likelihood estimation selects the value of θ that maximises (2.108). Thus one sets $\frac{\partial \log L}{\partial \theta} = 0$ to determine the MLE for θ.

As an illustration, consider a random sample obtained from a distribution with density $f_X(x | \theta) = \frac{1}{\theta} e^{-\frac{t}{\theta}}$, for $t \geq 0$. Then $L(x_1, x_2, \ldots, x_N) = \frac{1}{\theta^N} e^{-\frac{\sum_{j=1}^{N} x_j}{\theta}}$ and so $\log(L) = -N \log(\theta) - \frac{\sum_{j=1}^{N} x_j}{\theta}$. Thus $\frac{\partial \log(L)}{\partial \theta} = -\frac{N}{\theta} + \frac{\sum_{j=1}^{N} x_j}{\theta^2}$ and setting this to zero shows that the MLE of θ is $\frac{\sum_{j=1}^{N} x_j}{N}$.

2.12 Spherically Invariant Random Processes

When a series of radar returns are modelled in the complex domain, it is necessary to not only consider the univariate marginal distribution of the clutter in the intensity domain, but to also account for correlations in the time series of returns. Much research has been devoted to the validation of such a model for radar clutter modelling, and it has been found that the complex nature of maritime clutter can be represented as a *spherically invariant random process* (SIRP). These processes were first investigated in (Vershik, 1964) who was examining classes of random processes that shared attributes of Gaussian sequences. As discussed in (Wise and Gallagher, 1978), the SIRP class of processes is the most general case of stochastic processes for which optimal minimum mean square error estimators admit linear solutions. Subsequently, (Yao, 1973) reported a representation theorem for finite dimensional distributions of SIRPs, which then set the foundation for its introduction as a model for clutter in radar signal processing. This application and validation to radar clutter has been examined in (Conte and Longo, 1987), as well as in (Rangaswamy, 1983). This representation of clutter as a SIRP facilitated the design of coherent detection processes and consequently resulted in many years of active research in this area. Recent studies of coherent multi-look detection, where the resultant intensity model is Pareto, include (Shang and Song, 2011), (Sangston et al., 2012), (Weinberg, 2013b), (Weinberg, 2013c) and (Weinberg, 2016c).

In the context of non-coherent sliding window radar detection examined in this book, SIRP theory is only necessary for the analysis of the cell under test of the detection processes. Hence only a brief outline of SIRPs is required. As a SIRP, the clutter is modelled as a product of a nonnegative random variable S, termed the texture, and a correlated Gaussian process, known as the speckle. Hence if the clutter vector under consideration is of length N, each element of it is given by $c_j = SG_j$, where the vector $G = (G_1, G_2, \ldots, G_N)$ is a zero vector mean complex Gaussian process with $N \times N$ covariance matrix Σ, which is assumed to be positive definite. For the case of the N-dimensional clutter vector c, the probability density of it can be shown to be

$$f_{\mathbf{C}}(\mathbf{c}) = (2\pi)^{-N/2}|\Sigma|^{-1/2}h_N\left(\mathbf{c}^H\Sigma^{-1}\mathbf{c}\right), \qquad (2.109)$$

where the function h_N is given by

$$h_N(p) = \int_0^\infty s^{-N}e^{-\frac{p}{2s^2}}f_S(s)ds \qquad (2.110)$$

where $|\Sigma|$ is the determinant of Σ and $\mathbf{c}^H$ is the Hermitian transpose of $\mathbf{c}$. Thus (2.109) provides a convenient representation for the clutter model's density function. It is outlined in (Rangaswamy et al., 1993) that the univariate marginal distributions of $\mathbf{C}$ can be extracted from (2.110). Toward this aim, suppose C_j is the modulus of the jth element of $\mathbf{C}$. Then

$$f_{C_j}(t) = \frac{t}{\sigma^2}h_2\left(\frac{t^2}{\sigma^2}\right), \qquad (2.111)$$

where $\sigma^2 = (1/2)\mathrm{E}(C_j^2)$.

As an example, consider the case where the texture is generated by a distribution with density

$$f_S(s) = \frac{2\beta^\alpha}{\Gamma(\alpha)}s^{-2\alpha-1}e^{-\beta s^{-2}}, \qquad (2.112)$$

where α and β are nonnegative distributional parameters. A distribution with density (2.112) is known as inverse gamma. Observe that with the transformation $T = S^{-2}$,

$$F_T(t) = P(S \geq t^{-1/2}) = 1 - F_S(t^{-1/2}). \tag{2.113}$$

Hence by differentiating (2.113), one acquires the density function

$$f_T(t) = 1/2t^{-3/2}f_S(t^{-1/2}) = \frac{\beta^\alpha}{\Gamma(\alpha)}t^{\alpha-1}e^{-\beta t}, \tag{2.114}$$

so that S^{-2} has a gamma distribution, explaining the somewhat incorrect terminology. The distribution function can also be expressed in terms of the incomplete gamma function. In particular,

$$F(t) = \frac{1}{\Gamma(\alpha)} \int_{\beta t^{-2}}^{\infty} \phi^{\alpha-1}e^{-\phi}d\phi. \tag{2.115}$$

The amplitude version of the inverse gamma distribution will be required in the analysis to follow, whose density can be shown to be

$$f(t) = \frac{\beta^\alpha}{\Gamma(\alpha)}t^{-\alpha-1}e^{-\beta t^{-1}}. \tag{2.116}$$

By applying the density (2.112) to (2.110) it can be demonstrated that with a change of variables that

$$h_N(p) = \frac{\beta^\alpha \Gamma(\alpha + N/2)}{\Gamma(\alpha)} \frac{1}{[p/2 + \beta]^{\alpha + N/2}}, \tag{2.117}$$

and so

$$h_2(p) = \frac{\alpha\beta^\alpha}{[p/2 + \beta]^{\alpha+1}}. \tag{2.118}$$

An application of (2.118) to (2.111) results in

$$f_{C_j}(t) = \frac{t}{\sigma^2} \frac{\alpha\beta^\alpha}{\left[\frac{t^2}{2\sigma^2} + \beta\right]^{\alpha+1}}, \tag{2.119}$$

and by the transformation $\Phi = C_j^2/(2\sigma^2)$ the density (2.119) reduces to

$$f_\Phi(t) = \frac{\alpha\beta^\alpha}{[t + \beta]^{\alpha+1}}. \tag{2.120}$$

The resultant distribution with density (2.120) can be recognised as Pareto Type II, based upon (2.33). Here we have demonstrated that the Pareto model arises as the univariate distribution of a compound Gaussian process with inverse gamma texture. This provides validation of the Pareto distribution since it fits into the currently accepted framework for radar clutter models.

where α and β are nonnegative distribution parameters. A continuous density [2.1] is shown as a survivor function. Observe that

$$P(t) = P(S > t) = \cdots$$

Note: By differentiating [2.1] we get a corresponding density function

$$P(t) = P(S = t) = \cdots$$

Chapter 3

Distributions for X-Band Maritime Surveillance Radar Clutter

3.1 Introduction

This chapter presents a survey of the statistical distributions applied to model X-band maritime surveillance radar clutter returns. Included is also a discussion, from a historical perspective, of the evolution of such models, which has coincided with improvements in radar resolution. However, most of the chapter will be focused on the Pareto distributional model. This will not only include an outline of its validation, but also a specific example will be used to demonstrate its validity for DST Group's Ingara radar clutter. Following this a general overview of the various types of Pareto distributions will be included. A synopsis of some of the main properties of Pareto models will be presented, which can facilitate the design of sliding window detectors. The chapter concludes with a brief discussion of parameter estimation methods.

The importance of modelling radar clutter backscattering is that it permits the specification of a model on which the design of radar detection processes can be based. However, the validity of fit to data can never be a precise science, since every data set will exhibit different artefacts and statistics. Consequently a good distributional fit to one set of data may not necessarily imply a similar result for other analogous data sets. This has been shown to be the case in terms of fitting distributions to X-band radar clutter returns. As will be explained lower resolution returns could be modelled fairly accurately, while with improved radar resolution the fits of distributions tended to be problematic, due to the increased spikiness of the resultant data. Nonetheless tight fits to X-band high resolution maritime clutter returns is certainly achievable in the upper and lower tail regions of the empirical data, which is of primary concern to radar engineers (Skolnik, 2008).

3.2   Early Models for Clutter

In the early development of radar systems and their associated detection architecture, observed clutter returns were modelled adequately, in the complex domain, by Gaussian processes. In terms of amplitude this means the returns are modelled as Rayleigh, or in intensity by exponentially distributed random variables. The Gaussian assumption was valid because these early radar systems were low resolution. Based upon this assumption of clutter and noise modelled by Gaussian processes, Marcum (Marcum, 1960) and Swerling (Swerling, 1960) were able to construct an elegant theory of radar detection. With the discovery of the CFAR property of sliding window detection processes operating in exponentially distributed intensity clutter, there was a huge body of research produced towards yielding improved detection performance by varying the fundamental detection process used in the sliding window scheme: see (Gandhi and Kassam, 1988) and the references contained therein.

With the systematic evolution of radar systems, and corresponding improvements in radar resolution, it was found that new models of the backscatter were required because improved resolution resulted in spikier returns in the range resolution cells. These returns tended to have a heavier tail, in terms of its empirical distribution function, than that under a Gaussian model assumption. The problem with an inaccurate model for clutter is that the resultant detection process can yield an increase in the number of false alarms.

The non-Gaussian nature of maritime surveillance clutter was first reported in (Ballard, 1966), and then confirmed by (Trunk and George, 1970). In the latter, analysis of radar clutter from two separate trials were used to demonstrate the non-Gaussian property of sea clutter. In both trials, an X-band airborne radar was deployed by the United States' Naval Research Laboratory (NRL), surveying the sea surface at a grazing angle of $4.7°$. Analysis of the data from the first trial showed that the clutter in amplitude deviated significantly from Rayleigh as the sea state increased. The second trial reported in (Trunk and George, 1970) was conducted 200 miles off Virginia Capes in 26 to 31 knot winds, with the same airborne radar operating at the same grazing angle. In this second trial both horizontal and vertical polarised returns were collected. Not only was the clutter's deviation from Rayleigh in amplitude observed as the sea state increased, but it was also observed that in horizontal polarisation the deviation was more extreme.

As a result of the observed non-Rayleigh amplitude statistics of sea clutter, (Ballard, 1966), proposed that the amplitude could be modelled as a log-normal process, whose standard deviation was related to the distribution of the radar's illuminated patch area (Trunk and George, 1970). Although (Ballard, 1966) introduced the idea of applying a log-normal process, (Trunk and George, 1970) set the foundation for radar detection under such a clutter model assumption.

A random variable Y has a log-normal distribution if its density is given by

$$f_Y(t) = \frac{1}{\sqrt{2\pi}\sigma_L}t^{-1}e^{-\frac{1}{2\sigma_L^2}\log^2\left(\frac{t}{\mu_L}\right)}, \tag{3.1}$$

where $t \geq 0$ and μ_L and σ_L are its distributional parameters. To see how this is related to a Gaussian density, consider the change of variables $X = \log(Y)$ with resultant density

$$f_X(t) = e^t f_Y(e^t) = \frac{1}{\sqrt{2\pi}\sigma_L}e^{-\frac{1}{2\sigma_L^2}(t-\log(\mu_L))^2}, \tag{3.2}$$

which implies that Y has a Gaussian distribution with mean μ_L and variance σ_L^2. Since $Y = e^X$ this explains the way in which the log-normal produces a longer tail, and so a more accurate model of higher resolution clutter returns.

Based upon this clutter model, a large number of detection processes have been proposed and analysed; see for example (Goldstein, 1973), (Bucciarelli et al., 1985), (Farina et al., 1986) and (Al-Hussaini, 1988). However, the log-normal model does not admit an elegant theory of radar detectors as in the Gaussian case, and hence there was interest in the search for a reasonable compromise.

3.3 The Weibull Distribution

The Weibull model was introduced into radar signal processing as an alternative to the Log-Normal and Rayleigh amplitude distributions for not only maritime surveillance radar clutter at X-band and low grazing angles, but also to model land-based clutter returns at X- and L-band. This model is a natural extension of the Rayleigh distribution, introducing a power shape parameter into the clutter model.

A random variable $\mathcal{W}$ has a Weibull distribution if its density is given by

$$f_W(t) = \frac{k}{\lambda}\left(\frac{t}{\lambda}\right)^{k-1}e^{-\left(\frac{t}{\lambda}\right)^k}, \tag{3.3}$$

where $t \geq 0$, and k is the Weibull shape parameter while λ is its scale parameter. Its distribution function is given by

$$F_W(t) = 1 - e^{-\left(\frac{t}{\lambda}\right)^k}, \tag{3.4}$$

where $t \geq 0$. As reported in (Schleher, 1976), the Weibull model has been validated for maritime surveillance clutter returns at X-band at a grazing angle of between 1° and 30°. It is also noted that as the grazing angle increases, the amplitude model limits to Rayleigh (Schleher, 1976).

Validation of the same model for L-Band clutter returns has been reported in (Sekine et al., 1981), where it was shown to fit ground clutter returns from a long-range air route surveillance radar. Consequently research effort was invested in the development of detection schemes under such a clutter model. Examples include the works by (Rifkin, 1994), (Anastassopoulos and Lampropoulos, 1995) and (Pourmottaghi et al., 2012). Although the Weibull model resulted in more tractable detection schemes, it was not a suitable model for high resolution maritime surveillance radars operating at medium to high grazing angles. Hence it was still of relevance to investigate the development of an appropriate sea clutter model for all grazing angles.

3.4 K-Distribution

The K-Distribution was introduced into radar signal processing through considerations of the structure of the sea backscattering. It was proposed in (Jakeman and Pusey, 1976), who derive the distribution through considerations of the scatterers and statistical properties of the radar cross section on the surface area illuminated by a radar beam. In particular, it is derived under the assumption of the illuminated area consisting of a finite number of discrete scatterers, each of which give randomly phased contributions of the fluctuating amplitude in the radar's far field.

A random variable $\mathcal{K}$ is said to have a K-distribution with nonnegative shape and scale parameters v and c if its density is

$$f_{\mathcal{K}}(t) = \frac{2c}{\Gamma(v)} \left(\frac{ct}{2}\right)^v K_{v-1}(ct), \tag{3.5}$$

for $t \geq 0$, where Γ is the Gamma function, and K_v is the modified Bessel function of the second kind of order v (Watts, 1985). The corresponding distribution function is

$$F_{\mathcal{K}}(t) = \mathbf{P}(\mathcal{K} \leq t) = 1 - \frac{(ct)^v K_v(ct)}{2^{v-1}\Gamma(v)}. \tag{3.6}$$

The K-distribution also has a compound representation, which gives the model an interpretation which is consistent with observed characteristics of sea clutter. Real sea clutter has been observed to consist of two components. These are generally referred to as the texture and speckle respectively. The texture models an observed slowly varying component in the clutter, while the speckle models a fast varying component. Thus the clutter model is a modulated Gaussian process. To illustrate this mathematically, the texture T is assumed to follow a distribution with density

$$f_T(t) = \frac{2b}{\Gamma(v)}(bt)^{2v-1}e^{-b^2t^2},\tag{3.7}$$

where $t \geq 0$, which can be identified as the square root of a gamma distribution, while the speckle component is Rayleigh and assumed to have conditional density

$$f_{S|T}(s|t) = \frac{\pi s}{2t^2}e^{-\frac{\pi s^2}{4t^2}},\tag{3.8}$$

for $s \geq 0$ and a fixed $t \geq 0$. A random variable $\mathcal{K}$ has a K-distribution if it can be expressed in the product model form

$$\mathcal{K} = TS\tag{3.9}$$

where T and S have the densities (3.7) and (3.8) respectively. In order to derive the density of (3.9), one can apply conditional probability to show

$$f_{\mathcal{K}}(x) = \int_0^\infty f_{S|T}(x|t)f_T(t)dt\tag{3.10}$$

and it can be shown that (3.10) reduces to (3.5) with an application of (3.7) and (3.8), where $c = 2b\sqrt{\pi/4}$.

Validation of the K-distribution as a model for X-band maritime surveillance radar has been documented in a number of sources. In the case of low grazing angle sea clutter, (Hair et al., 1991) discusses the K-distribution fit to sea clutter obtained from several test sites on the South Coast of England. The radar was located at heights from 15 m to 65 m above sea level and measurements at ranges of up to 2 km were recorded. Due to the selection of heights of the radar and surveillance range, the resultant grazing angle varied from 0.42° to 2.9°, and the sea states ranged from 1 to 3. Three frequency bands were used, namely 3.5 GHz (S-band), 10 GHz (X-band) and 16 GHz (J-band), and measurements over all polarisations were obtained. The conclusion of (Hair et al., 1991) was that the K-distribution model was a very good fit to the data in all scenarios considered.

A second validation is documented in (Nohara and Haykin, 1991), who reported on the results of sea clutter analysis based upon data obtained by the Canadian McMaster University's IPIX radar. This radar's original application was for iceberg detection, and its name was an acronym for *Ice Multiparameter Imaging X-Band Radar*. During upgrades in the early to mid 1990s the focus of the radar's application changed and so it became known as the *Intelligent Pixel Processing Radar*. This radar is an X-band instrumentation quality coherent dual polarised radar. A data gathering exercise was conducted at Cape Bonavista, Newfoundland in Canada during 1989. The radar was fixed with a parabolic dish antenna 22 m above sea level. To characterise sea clutter, measurements were taken pertaining to three pulse widths at three different bearings, over all polarisations.

The range was 6 km in 2.5–3 m seas, with wind ranging from 24 to 26 knots in a Northerly direction. The swell direction was approximately 110° to 120°, and data was obtained representing sea states 0 to 6. The conclusions reached were that the K-distribution provided a good fit to the sea clutter in all cases.

An empirical validation of the compound representation of sea clutter is reported in (Ward, 1981), who analysed sea clutter obtained from an airborne non-coherent pulsed X-band radar, operating with vertical polarisation. The aircraft flew at 220 knots at an altitude of 1600 ft and 6000 ft, observing sea states ranging from 1 to 5, at ranges of 20, 16, 12, 8 and 4 nautical miles. It was found that the K-distribution fitted this data well as in the previous examples.

Consequently, the K-distribution was found to be an excellent model for sea clutter at X-band, regardless of the range and sea state. Additionally, it was a good model over all polarisations. This stimulated the development of radar detection schemes under a K-distribution clutter model assumption. Examples of developments can be found in (Watts, 1985), (Watts, 1996), (Bucciarelli et al., 1996) and (Gini, 1999).

3.5   The Pareto Class of Distributions

The Pareto distribution has appeared within the last decade in radar signal processing as an alternative to the K-distribution. This model was first examined in (Balleri et al., 2007), in the context of modelling sea clutter returns at X-band. The perspective taken in (Balleri et al., 2007) was to examine the fit of the inverse gamma texture to real data. Consequently the authors focused on efficient texture parameter estimation, and although they analysed the fit of a Pareto amplitude model, did not identify it as such. The authors validated their model using data gathered in a trial using the IPIX radar. This data was gathered in 1998 at a test site located on Lake Ontario, in Grimsby Canada, with the radar located at a height of 20 m. The radar operated in vertical transmit and receive (VV), horizontal transmit and receive (HH) and vertical transmit and horizontal receive (VH) polarisations, at a frequency of 8.9–9.4 GHz, with a pulse repetition frequency of 1000 Hz and pulse length of 0.06 μ s, resulting in a range resolution of 9 m. It was reported that the Pareto fit improved on that of the Weibull, K and log-normal in all cases examined, especially in the HH polarised case. The type of Pareto model fitted to the data was one with density given by (2.33), so that it was a Pareto Type II model.

A second validation of the Pareto model for sea clutter is given in (Farshchian and Posner, 2010), who describe and analyse sea clutter returns obtained during an NRL-led trial in 1994, located in Kauai, Hawaii. The radar operated at a frequency of 9.5–10 GHz with a pulse repetition frequency of 2000 Hz. The radar operated in both HH and VV polarisations; however, the radar used was not dual polarised and so these were collected separately. The radar was at a height of 23 m above sea level, so that the grazing angle was 0.22° and the radar range was 5.74 km for VV and 6.11 km for HH-polarisation. The data analysed in (Farshchian and Posner, 2010) focused on the up wind direction, which is generally the most spiky. The wind speed was roughly 9 m/s and the largest wave height was roughly 3 m, so that the sea state was approximately 4. The results of the trial was conclusive evidence that at a low grazing angle, the Pareto model outperformed the Weibull, log-normal and K-distributions. Additionally, the model was compared to mixtures of Weibull and K, and shown to perform better than Weibull mixtures, while having comparable performance to a K-mixture model. Given the latter is a three to four parameter model, the performance of the two parameter Pareto model was determined to be excellent. The form of Pareto distribution examined in this data fitting exercise was a generalised Pareto model, to be discussed subsequently. The important point is that members of the Pareto family of distributions have provided a good fit to this second data set.

A third validation for the Pareto model has been provided by DST Group, based upon their Ingara radar. Ingara is an experimental fully polarimetric airborne multi-mode X-band imaging radar developed by DST Group (Stacy et al., 1996), which was deployed in a Raytheon Beech 1900C aircraft during a number of trials (Stacy et al., 2005). Figure 3.1 shows this aircraft with the surveillance antenna mounted under its fuselage. A specific trial of interest was conducted in 2004, in the Southern Ocean near Port Lincoln in South Australia. The radar operated with a frequency of 10.1 GHz, with a pulse length of 20 μs, and LFM transmitted bandwidth of 200 MHz. This permitted a range resolution of 0.75 m. Ingara operated in a circular spotlight mode, surveying the same patch of ocean at different azimuth angles, and at different incidence angles. Sea states varied from 2 to 5, while wind speeds varied from 6.1 to 13.2 m/s. The data gathered in this trial was designated by a number of runs, where 1024 range compressed samples of roughly 920 pulses were collected, at 5° azimuth angle increments. Extensive analysis of the data collected, and further details on the trial, have been reported in (Stacy et al., 2006), (Crisp et al., 2007), and (Crisp et al., 2009). One particular run, denoted run 34 683, which has been investigated extensively in (Dong, 2006). This data set was collected on August 16, 2004, at around 10:52 am local time, and consists of 1024 range compressed samples of 821 pulses, obtained at an incidence angle of 51.3°, and at an altitude of approximately 2314 m (Dong, 2006). Hence the radar operated from a medium to high grazing angle. The wind speed, at the mid-data collection time on the trial day, was reported to be 7.1 m/s, in a direction of 47°, with a wave height of 2.4 m in a direction of 211°. The upwind direction was 227°, downwind at 47°, while the cross wind directions were 137° and 317° respectively. The fit of the K-distribution, as well as log-normal, Weibull and a number of mixture distributions were assessed in (Dong, 2006). The model giving the best fit, especially in HH-polarisation, was a mixture of K-distributions, known as the KK-distribution. This was re-examined in (Weinberg,

Figure 3.1: Raytheon Beech 1900C aircraft showing the Ingara radar antenna mounted under its fuselage.

2011a), who considered the Pareto fit to the Ingara data. It was shown that the Pareto model fitted the Ingara data very well, confirming the results of (Balleri et al., 2007) and (Farshchian and Posner, 2010) for the medium to high grazing angle case. To illustrate this, the Pareto model (2.120) was fitted to run 34 683, as well as the K- and Weibull distributions. The data set selected was at an azimuth angle of 225°, which is approximately upwind and so represents the most spiky case. The log-normal and KK-distributions were not considered because the former is well known to be inferior to the K-distribution while the latter is a minimium of three parameter clutter model which is considered excessive.

Table 3.1 documents the parameter estimates obtained by MLE for each fitted distribution, and over the three polarisations. Figures 3.2 and 3.3 show the lower and upper tail region fits for the HH-polarised case; Figures 3.4 and 3.5 are for the cross polarisation case, while Figures 3.6 and 3.7 are for vertical transmit and receive polarisation. In terms of the Ingara data, cross polarisation corresponds to horizontal transmit and receive (HV). These figures show the superiority of the Pareto fit to the data. In most cases it is difficult to perceive the error in the Pareto fit to the data set.

There has also been further examination of the Ingara data at DST Group, which has been reported in the literature. Particularly notable contributions include (Rosenberg and Bocquet, 2013) and (Rosenberg and Bocquet, 2015) who present a multi-look formulation of the Pareto model, which accounts for receiver thermal noise via an effective shape parameter. Accounting for thermal noise, also via an effective shape parameter, was considered in (Alexopoulos and

Polarisation	Model	Shape Parameter	Scale Parameter
	Pareto	$\alpha = 4.7241$	$\beta = 0.0446$
HH	K	$v = 4.58$	$c = 40$
	Weibull	$k = 0.8716$	$\lambda = 0.0111$
	Pareto	$\alpha = 6.7875$	$\beta = 0.0043$
HV	K	$v = 5.8$	$c = 180$
	Weibull	$k = 0.9082$	$\lambda = 0.0007$
	Pareto	$\alpha = 11.3930$	$\beta = 0.3440$
VV	K	$v = 9.629$	$c = 32$
	Weibull	$k = 0.9439$	$\lambda = 0.0322$

Table 3.1: Clutter model estimates based upon Ingara data set run 34 683, showing maximum likelihood estimates for each of the clutter models.

Weinberg, 2015), who derived a fractional order Pareto distribution. This is based upon fractional calculus considerations, and it was shown that the effective shape parameter is related to the fractional order derivative.

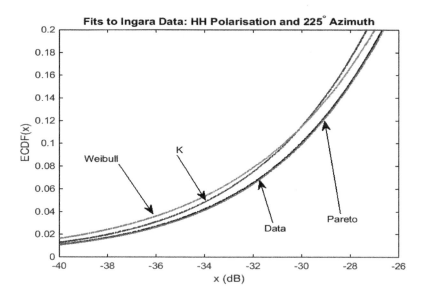

Figure 3.2: Distribution fits to a subset of the Ingara data, showing the fit in the lower tail region (horizontal polarisation). The Pareto Type II model provides the most accurate fit to the data.

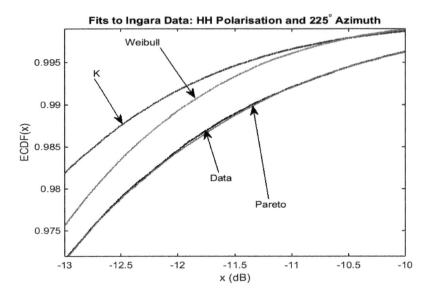

Figure 3.3: Fit to the same data as for Figure 3.2, except in the upper tail region. The Pareto model again provides the best fit to the data.

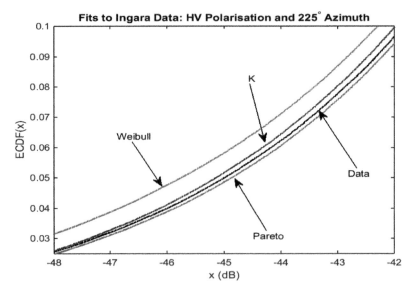

Figure 3.4: Fits in the lower tail for the Ingara data set, with HV polarisation. In this case the Pareto and K-distribution are both providing suitable fits to the data.

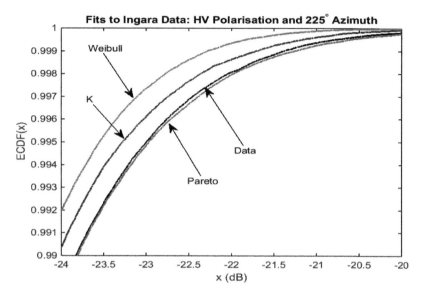

Figure 3.5: Fits to the Ingara data set in the upper tail region, for the case of cross-polarisation. Here it is clear that the Pareto Type II model has the best fit.

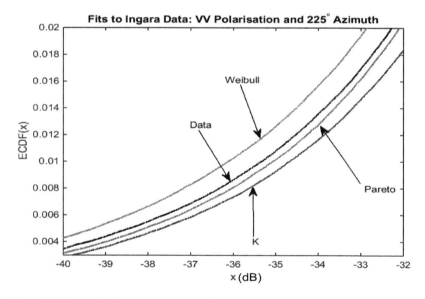

Figure 3.6: Fits to the Ingara clutter set, in the lower tail region, and in the vertically polarised case. One observes that the best fit is provided by the Pareto Type II model, but the latter has a discrepancy in its fit.

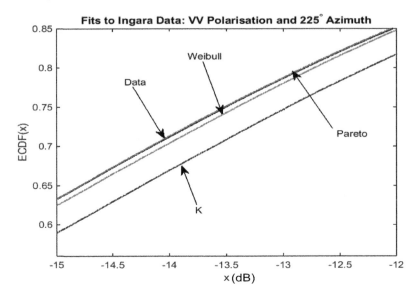

Figure 3.7: Fits in the upper tail region, in the vertically polarised case. Here one observes that the Pareto Type II model provides a very good fit to the data. The Weibull distribution is better than the K-distributional fit due to the polarisation.

3.6 Pareto Type Distributions

This section introduces the class of Pareto distributions. A comprehensive overview of such distributions can be found in (Arnold, 1983), which has been partially updated in (Arnold, 2014). Suppose that $\mathcal{P}$ is a Pareto distributed random variable. We now examine several common statistical structures for such a random variable. The most basic form of a Pareto distribution, known as a Pareto Type I Model, has density function given by

$$f_{\mathcal{P}}(t) = \frac{\alpha \beta^{\alpha}}{t^{\alpha+1}},$$
(3.11)

where $t \geq \beta > 0$ and $\alpha > 0$. This has been derived previously. The shape parameter governs the distribution's shape, while the scale parameter determines where its support begins. Hence such a random variable has support $[\beta, \infty)$. By integration of (3.11), the corresponding distribution function is given by

$$F_{\mathcal{P}}(t) = 1 - \left(\frac{\beta}{t}\right)^{\alpha},$$
(3.12)

also for $t \geq \beta$. This is the most fundamental form of the Pareto distribution. If the random variable $\mathcal{P}$ has distribution function (3.12) this will be written $\mathcal{P} \overset{d}{=} \text{Pareto}(\alpha, \beta)$.

The second type of Pareto distribution introduces a third parameter into the model. It has density

$$f_P(t) = \frac{\alpha \beta^{\alpha}}{(\beta + t - \gamma)^{\alpha+1}}, \tag{3.13}$$

where $t \geq \gamma$. The latter parameter is referred to as the distribution's location parameter. The corresponding distribution function is

$$F_P(t) = 1 - \left(\frac{\beta}{\beta + t - \gamma} \right)^{\alpha} \tag{3.14}$$

where $t \geq \gamma \geq 0$. This model is known as Pareto Type II, and reduces to the Pareto Type I model when $\gamma = \beta$. Observe that with the choice of $\gamma = 0$ the density (3.13) and distribution function (3.14) have support $[0, \infty)$, and this case is known as a Lomax distribution (Lomax, 1954). Recall that this distribution appeared earlier in (2.33) and (2.120).

Another variant of the Pareto distribution introduces a fourth parameter into the model and has density

$$f_P(t) = \frac{\alpha}{\kappa \beta} \left(\frac{t - \gamma}{\beta} \right)^{\frac{1}{\kappa}-1} \left[1 + \left(\frac{t - \gamma}{\beta} \right)^{\frac{1}{\kappa}} \right]^{-\alpha-1}, \tag{3.15}$$

and distribution function

$$F_P(t) = 1 - \left[1 + \left(\frac{t - \gamma}{\beta} \right)^{\frac{1}{\kappa}} \right]^{-\alpha}, \tag{3.16}$$

where the support of both these functions is $t \geq \gamma$, and all distributional parameters are assumed nonnegative. Additionally, it is necessary to impose the constraint that $\kappa > \alpha$ to ensure the distribution function (3.16) converges to unity as $t \to \infty$. Such a model is known as Pareto Type IV; the special case where $\alpha = 1$ is referred to as Pareto Type III.

As a final variant of the Pareto model, the Generalised Pareto distribution is introduced. Here we consider only the version which has a semi-infinite support. The version with a finite support is not really useful from a radar signal processing perspective. Such a distribution has density

$$f_P(t) = \frac{1}{\beta} \left(1 + \alpha \left(\frac{t - \gamma}{\beta} \right) \right)^{-\left(\frac{1}{\alpha}+1 \right)}, \tag{3.17}$$

where $\alpha > 0$ is the shape parameter, $\beta > 0$ is its scale parameter and $\gamma \in R$ is the distribution's location parameter. The density (3.17) has support $t \geq \gamma$. The corresponding distribution function is

$$F_P(t) = 1 - \left(1 + \frac{\alpha}{\beta} (t - \gamma) \right)^{-\frac{1}{\alpha}}. \tag{3.18}$$

Clearly the distribution function (3.18) is a special case of the Pareto Type IV model (3.16). In the discussion to follow, attention will be restricted to the Pareto Type II distribution, with the choices of $\gamma = 0$ and $\gamma = \beta$, so that in both cases the model has two parameters only. The motivation for this is that this model encompasses the key Pareto distributions to be analysed, and represent three parameter models. Increasing the number of degrees of freedom may yield tighter fits to real data, but this may add computational complexity and potential loss of the CFAR property with respect to new parameters. Generally speaking a two parameter clutter model is considered an acceptable limit, while the Pareto Type IV Model, with four degrees of freedom, is considered excessive. Additionally, note that the Pareto Type II model is a special case of the Generalised Pareto distribution.

Henceforth we will refer to a random variable $\mathcal{P}$ having a Pareto distribution with parameters α, β and γ if its density and distribution function are given by (3.13) and (3.14) respectively. This will be denoted by $\mathcal{P} \stackrel{d}{=} \text{Pareto}(\alpha, \beta, \gamma)$.

3.7 Properties of the Pareto Distribution

This section examines some properties of the Pareto distribution under consideration, which will facilitate the development of radar detection schemes to follow. To begin, it is not difficult to show that the inverse of (3.14) is given by

$$F_{\mathcal{P}}^{-1}(t) = \gamma + \beta \left((1-t)^{-\frac{1}{\alpha}} - 1 \right). \tag{3.19}$$

Based upon this, we have the following result:

Lemma 3.1
A random variable $\mathcal{P}$ with the Pareto distribution can be expressed as a function of an Exponentially distributed random variable $X \stackrel{d}{=} \text{Exp}(\alpha)$ through the expression

$$\gamma + \beta \left(e^X - 1 \right) \stackrel{d}{=} \mathcal{P}(\alpha, \beta, \gamma). \tag{3.20}$$

The proof follows by first noting that the random variable on the left hand side of (3.20) has support $[\gamma, \infty)$ and then showing its distribution function is exactly (3.14). The result (3.20) will be referred to as the Pareto-exponential duality property. In the simplier case where a random variable Z has a Pareto Type I distribution, the duality property simplifies to

$$\beta e^X \stackrel{d}{=} Z, \tag{3.21}$$

where X has the same exponential distribution.

Since the analysis to follow will restrict attention to Pareto Type I and II distributions, we derive expressions for their moments. An early study on the derivation of moments of Pareto distributions is discussed in (Malik, 1966). Observe that for a Pareto Type I distribution, assuming $X \stackrel{d}{=} \text{Pareto}(\alpha, \beta)$,

$$\text{E}(X^n) = \int_{\beta}^{\infty} \alpha \beta^n t^{n-\alpha-1} dt = \frac{\alpha \beta^n}{\alpha - n}, \tag{3.22}$$

provided $\alpha > n$. Hence the mean is given by

$$\text{E}(X) = \frac{\alpha \beta}{\alpha - 1}, \tag{3.23}$$

while its variance is

$$\text{Var}(X) = \text{E}(X^2) - (\text{E}(X))^2 = \frac{\alpha \beta^2}{(\alpha - 2)(\alpha - 1)^2}. \tag{3.24}$$

For the case of a Lomax distribution, with shape and scale parameters α and β respectively, since $X = Y - \beta$ where $Y \stackrel{d}{=} \text{Pareto}(\alpha, \beta)$ it follows that the Lomax mean is

$$\text{E}(X) = \frac{\beta}{\alpha - 1}, \tag{3.25}$$

while its variance is also given by (3.24). The nth order moment of the Lomax distribution can also be derived from (3.22) by observing that since

$$X^n = (Y - \beta)^n = \sum_{k=0}^{n} \binom{n}{k} Y^{n-k} (-\beta)^k, \tag{3.26}$$

and applying (3.22) it follows that

$$\text{E}(X^n) = \alpha \beta^n \sum_{k=0}^{n} \binom{n}{k} (-1)^k \frac{1}{\alpha - n + k} \tag{3.27}$$

where it is sufficient to assume $\alpha > n$.

Another useful property of Pareto Type II distribution is the following:

Lemma 3.2
Suppose that $\lambda \in \mathbf{R}$ is a fixed constant. If $X \stackrel{d}{=} \mathcal{P}(\alpha, \beta, \gamma)$ then $\lambda X \stackrel{d}{=} \mathcal{P}(\alpha, \lambda \beta, \lambda \gamma)$.
Hence the class of Pareto distributions is closed under multiplication by a constant.

The proof of Lemma 3.2 follows directly from Lemma 3.1. Pareto Type I distributions are also closed under power transformations of random variables:

Lemma 3.3
Suppose that X is a random variable with a Pareto distribution with shape parameter

α, and scale and location parameters $\beta = \gamma$. Then for any $k \in \mathbf{R}^+$, the random variable X^k also has a Pareto distribution, with shape parameter $\frac{\alpha}{k}$ and scale and location parameter β^k.

The proof of Lemma 3.3 follows from the fact that in the case where $\beta = \gamma$, the decomposition (3.20) reduces to βe^X and raising this to the power of k shows the resultant distribution is also Pareto Type I with shape parameter α/k and scale parameter β^k. The next result discusses the distribution of a minimium of a series of independent and identically distributed Pareto random variables:

Lemma 3.4

Let $\{X_j, j \in \{1,2,\ldots,N\}\}$ be a series of Pareto distributed random variables with shape parameter α and scale and location parameter $\beta = \gamma$. Then the minimum of this series of random variables is also Pareto distributed, with shape parameter αN and scale and location parameters β.

To prove this result, through an application of (2.87), the distribution function of the minimum is

$$F_{X_{(1)}}(t) = 1 - \left(\frac{\beta}{t}\right)^{\alpha N} \tag{3.28}$$

from which the desired result follows. A useful result that follows from this is expressed in the following Corollary:

Corollary 3.1

The minimum order statistic defined in Lemma 3.4 is a complete sufficient statistic for β.

To prove Corollary 3.1, the density of $X_1, X_2, \ldots, X_N$ given $X_{(1)} = s$ and β is

$$f_{(X_1,X_2,\ldots,X_N | X_{(1)}=s,\beta)}(x_1,x_2,\ldots,x_N | s,\beta) = \frac{f_{(X_1,X_2,\ldots,X_N)}(x_1,x_2,\ldots,x_N)}{f_{X_{(1)}}(s)}$$

$$= \frac{\prod_{j=1}^{N}\left(\frac{\alpha\beta^\alpha}{x_j^{\alpha+1}}\right)}{\frac{\alpha N \beta^{\alpha N}}{s^{\alpha N+1}}}$$

$$= \frac{\alpha^{N-1} s^{\alpha N+1}}{\prod_{j=1}^{N} x_j^{\alpha+1} N}, \tag{3.29}$$

where $s = x_{(1)}$. Since the density (3.29) does not depend on β, it follows that $X_{(1)}$ is sufficient for β.

For the completeness property proof, suppose for some function ζ that

$E(\zeta(X_{(1)})) = 0$. Then by applying the density for the minimum,

$$\int_{\beta}^{\infty} \zeta(t) \frac{\alpha N \beta^{\alpha N}}{t^{\alpha N+1}} = 0. \tag{3.30}$$

Hence

$$\int_{\infty}^{\beta} \zeta(t) t^{-\alpha N-1} dt = 0. \tag{3.31}$$

Differentiating (3.31) with respect to β implies that $\zeta(\beta)\beta^{-\alpha N-1} = 0$ and so $\zeta(\beta) = 0$ for all β. Hence $X_{(1)}$ is also complete.

The order statistic for a series of Pareto distributed random variables can be expressed in terms of beta distributions as the following Lemma shows:

Lemma 3.5
For the same sequence of random variables defined in Lemma 3.4, if we let $X_{(k)}$ be the kth order statistic for the sample, where $1 \leq k \leq N$, then the random variable W defined by

$$W = \left(\frac{\beta}{X_{(k)}}\right)^{\alpha} \stackrel{d}{=} \text{Beta}(N-k+1,k). \tag{3.32}$$

To prove this result, note that if X_j has a Pareto distribution with shape and scale parameters α and β respectively then $X_j = \beta e^{\alpha^{-1} Z_j}$, where $Z_j \stackrel{d}{=} \text{Exp}(1)$. Based upon this it is not difficult to show the order statistics of samples from these distributions is related by $X_{(k)} = \beta e^{\alpha^{-1} Z_{(k)}}$. Consequently W defined in (3.32) is equivalent to $e^{-Z_{(k)}}$. It follows that the density of W is

$$f_W(t) = \frac{1}{t} f_{Z_{(k)}}(-\log(t)), \tag{3.33}$$

and an application of (2.94) reduces the above to the aforementioned beta distribution's density, as required.

Sums of independent and identically distributed Pareto random variables are of considerable interest in sliding window detector design, as the most fundamental detectors are based upon these. The following Lemma illustrates the fact that there is no simple closed form expression for Pareto sums:

Lemma 3.6
Suppose that $\{X_j, j \in \{1,2,\ldots,N\}\}$ is a series of Pareto distributed random variables with shape parameter α and scale and location parameter $\beta = \gamma$. Define

$$S_N = \sum_{j=1}^{N} X_j. \tag{3.34}$$

Then the distribution function of S_N is given by the multidimensional integral

$$P(Z \le t) = 1 - \left(\frac{\beta}{t}\right)^{\alpha} \int_0^1 \int_0^1 \cdots \int_0^1 \left(1 - \beta t^{-1} \sum_{j=2}^{N} \omega_j^{-\frac{1}{\alpha}}\right)^{-\alpha} d\omega_2 d\omega_3 \dots d\omega_N,$$

(3.35)

for $t \ge \beta N$.

To demonstrate this, observe that we can generate a Pareto random variable from $\beta R^{-\frac{1}{\alpha}}$, where R is uniformly distributed on the unit interval. Then a sum of N such random variables is equivalent to $\sum_{j=1}^{N} R_j^{-\frac{1}{\alpha}}$, and so by conditioning on $N-1$ of these random variables one obtains

$$\begin{aligned}
P(Z \le t) &= \int_0^1 \int_0^1 \cdots \int_0^1 P\left(R_1^{-\frac{1}{\alpha}} + \sum_{j=2}^{N} w_j^{-\frac{1}{\alpha}} \le \frac{t}{\beta}\right) dw_2 dw_3 \dots dw_N \\
&= \int_0^1 \cdots \int_0^1 P\left(R_1 \ge \left(\frac{t}{\beta} - \sum_{j=2}^{N} w_j^{-\frac{1}{\alpha}}\right)^{-\alpha}\right) dw_2 \dots dw_N
\end{aligned}$$

(3.36)

and the final result follows by applying the distribution function of R_1 to (3.36).

In view of (3.35) it is interesting to note that the distribution function is asymptotically equal to that of the univariate Pareto distribution in the case where $t \longrightarrow \infty$ or when $\alpha \longrightarrow \infty$.

Products of independent and identically distributed Pareto random variables appear in the design of sliding window detectors via the transformation approach to be introduced in a subsequent chapter. Such products were first studied in (Malik, 1970a), and the following Lemma shows that its distribution is related to a χ^2 model:

Lemma 3.7
Let $\{X_j, j \in \{1, 2, \dots, N\}\}$ be the same sequence of random variables defined in Lemma 3.4, and define the product

$$P_N = \prod_{j=1}^{N} X_j.$$

(3.37)

Then

$$2\alpha \log\left(\frac{P_N}{\beta^N}\right) \overset{d}{=} \chi_{2N}^2.$$

(3.38)

An outline of the proof of this result now follows. Observe that the product can be written $P_N = \beta^N e^Z$, where Z has a gamma distribution with shape parameter N and scale parameter α. The final result follows by noting that such a random variable is equivalent to $\frac{1}{2\alpha}$ times a χ^2 distribution with $2N$ degrees of freedom, and applying this to (3.38).

The following result is a useful property of the normalised sum of Pareto distributed random variables. It is attributable to (Srivastava, 1965) who used it as a characterisation of Pareto distributions:

Lemma 3.8

Suppose again that $\{X_j, j \in \{1,2,\ldots,N\}\}$ is the sequence of Pareto random variables in Lemma 3.4. Then the statistic

$$Z_N = \sum_{j=1}^{N} \log\left(\frac{X_j}{X_{(1)}}\right) \tag{3.39}$$

is independent of the minimum $X_{(1)}$.

The proof follows from the fact that the minimum is a complete sufficient statistic for β and the sum normalised by the minimum does not depend on β, and applying Lemma 2.4.

The Pareto distribution can also be characterised in a similar way through order statistics, as the following Lemma illustrates:

Lemma 3.9

Suppose we have the same series of random variables $\{X_j, j \in \{1,2,\ldots,N\}\}$, and that $X_{(r)}$ and $X_{(s)}$ are two order statistics such that $1 \leq r < s \leq N$. Then a necessary and sufficient condition for the underlying distribution to be Pareto Type I is that $X_{(r)}$ and $\frac{X_{(s)}}{X_{(r)}}$ are independent.

This result is due to (Ahsanullah, 1973), to which the reader is referred for the proof.

A final characterisation of the Pareto distribution is now investigated, which has been discussed in (Revankar et al., 1974). As is pointed out in the latter, the Pareto Law is one describing income distribution, and states that "the logarithm of the percentage of units with an income exceeding some value is a negatively sloped linear function of the logarithm of that value" (Revankar et al., 1974). However, the original Pareto law did not account for what is known as "under reporting" of income and so it became of interest to understand the Pareto model when it was assumed that values of it exceed a fixed level (Klein, 1962). Hence (Revankar et al., 1974) investigated characterisations of the Pareto model in terms of conditional expectations. The following is the main result:

Lemma 3.10

A nonnegative random variable X with finite mean has a Pareto distribution if and only if $E(X|X > x) = h + gx$, *for some constants h and g, where* $g > 1$ *and for all x.*

To prove this result, we begin with necessity. Note that

$$F_{X|\{X>x\}}(t) = \frac{F_X(t) - F_X(x)}{1 - F_X(x)}, \tag{3.40}$$

provided $t > x$ so that the density is

$$f_{X|\{X>x\}}(t) = \frac{f_X(t)}{1 - F_X(x)}, \tag{3.41}$$

also provided $t > x$. Hence the conditional mean is

$$
\begin{aligned}
E(X|\{X > x\}) &= \int_x^\infty \frac{t f_X(t)}{1 - F_X(x)} dt \\
&= \left(\frac{x}{\beta}\right)^\alpha \int_x^\infty \alpha \beta^\alpha t^{-\alpha} dt \\
&= \frac{\alpha}{\alpha - 1} x, \tag{3.42}
\end{aligned}
$$

which shows that in the case of a Pareto Type I distribution the desired form of the conditional is attained. If we instead consider the case of a Lomax distribution, so that $Y \overset{d}{=} L(\alpha, \beta)$, then since $Y = X - \beta$ where $X \overset{d}{=} \text{Pareto}(\alpha, \beta)$ then

$$E(Y|Y > y) = E(X|\{X > \beta + y\}) = \left(\frac{\alpha}{\alpha - 1}\right) y + \frac{\alpha\beta}{\alpha - 1}, \tag{3.43}$$

where (3.42) has been applied. Hence the Lomax model has the desired conditional expectation.

For the proof of sufficiency, suppose that

$$\int_x^\infty \frac{t f_X(t)}{1 - F_X(x)} dt = h + gx, \tag{3.44}$$

for some constants h and $g > 1$ and for all x, where the first expression in (3.42) has been applied, and the nonnegative random variable X has density $f_X(t)$ and distribution function $F_X(t)$. Then we can write

$$-\int_\infty^x t f_X(t) dt = (1 - F_X(t))(h + gx). \tag{3.45}$$

Let $H(x) = 1 - F_X(x)$ so that $H'(x) = -f_X(x)$ and thus by differentiating (3.45) and arranging, one obtains the differential equation

$$\frac{H'(x)}{H(x)} = -\frac{g}{h + (g - 1)x}. \tag{3.46}$$

Finally, by integrating (3.46) one obtains the solution of the form

$$H(x) = (h + (g-1)x)^{-\frac{g}{g-1}} e^{\lambda},$$ (3.47)

where λ is a constant. Further manipulation of (3.47) shows that the resultant distribution function is

$$F_X(t) = 1 - \left(\frac{\lambda_1}{\lambda_2 + t} \right)^{\lambda_3}$$ (3.48)

where $\lambda_i > 0$, showing that the resultant distribution is Pareto Type I or II.

3.8 Parameter Estimation

The problem of parameter estimation in the Pareto distribution is now examined, focusing on the Pareto Type I model. Useful early studies on this subject include (Quandt, 1966) and (Malik, 1970b). There has been considerable interest in recent years on this subject matter. Notable contributions include (Weinberg, 2013e), (Bocquet, 2015a), (Bocquet, 2015b), (Mezache et al., 2016) and (Hu et al., 2016). Since parameter estimation is not a central focus in this book only a brief account of Pareto parameter estimation is included, which will be based upon maximum likelihood estimation. The interested reader is encouraged to consult the aforementioned references for the state of the art in Pareto parameter estimation.

Suppose that we have a random sample $x_1, x_2, \ldots, x_N$ from such a distribution. Then the likelihood function is

$$L(\alpha, \beta) = \prod_{j=1}^{N} \frac{\alpha \beta^\alpha}{x_j^{\alpha+1}} = \alpha^N \beta^{\alpha N} \left(\prod_{j=1}^{N} x_j \right)^{-\alpha-1}.$$ (3.49)

The logarithmic likelihood of (3.49) is

$$\log(L(\alpha, \beta)) = N \log(\alpha) + \alpha N \log(\beta) - (\alpha+1) \sum_{j=1}^{N} \log(x_j).$$ (3.50)

Minimising (3.50) with respect to β merits the choice of $\hat{\beta} = x_{(1)}$, while differentiating (3.50) with respect to α and setting it to zero shows that

$$\hat{\alpha} = \frac{N}{\sum_{j=1}^{N} \log \left(\frac{x_j}{x_{(1)}} \right)},$$ (3.51)

where the MLE estimate of β has been applied. Since the minimum is Pareto distributed, its mean is $\left(\frac{\alpha N}{\alpha N - 1}\right)\beta$, it follows that $\left(\frac{\alpha N - 1}{\alpha N}\right)X_{(1)}$ is the minimum variance unbiased estimator of β, in view of Lemma 3.4. However, such an estimator requires *a priori* knowledge of the Pareto shape parameter, and so it is often more convenient to use the minimum instead, since for large N it is asymptotically unbiased.

Although not required for the analysis to follow, it is interesting to examine the distribution of these estimators. The distribution of the minimum has already been discussed, so it is of interest to investigate the distribution of the estimator of the Pareto shape parameter. This was first done in (Malik, 1970b), who demonstrated the following:

Lemma 3.11
The maximum likelihood estimator of the Pareto shape parameter has an inverse gamma distribution, with shape parameter $N - 1$ and scale parameter αN and density corresponding to the amplitude form (2.116). In addition to this it is independent of the MLE of the Pareto scale parameter.

The independence component of the proof is a consequence of Lemma 3.8, while the derivation of the distribution of the MLE was first produced in (Malik, 1970b). The latter applies a moment generating function approach to derive the distribution, but it turns out that this is redundant. Hence a modified proof is included below for interest.

Following the proof in (Malik, 1970b), observe that if $X_1, X_2 \ldots, X_N$ is a series of Pareto distributed random variables then

$$\sum_{j=1}^{N} \log\left(\frac{X_j}{X_{(1)}}\right) = \sum_{j=1}^{N} \log\left(\frac{X_{(j)}}{X_{(1)}}\right) = \sum_{j=2}^{N} \log\left(\frac{X_{(j)}}{X_{(1)}}\right) \tag{3.52}$$

since summing the entire series of random variables is equivalent to summing the order statistic counterparts, and since the sum is normalised by the minimum, the first logarithmic term vanishes. Next note that for a series of random variables Z_j

$$\sum_{j=2}^{N} \log\left(\frac{Z_j}{Z_1}\right) = \log(Z_2) + \log(Z_3) + \cdots + \log(Z_N) - (N - 1)\log(Z_1) \tag{3.53}$$

so that it is clear that the ordering of $Z_2, \ldots Z_N$ does not affect the sum in (3.53). Hence if we assume that $Z_1 < Z_2, \ldots Z_1 < Z_N$ or equivalently $Z_{(1)} = Z_1$ and apply this to (3.52), one concludes in terms of the Pareto random variables that

$$\sum_{j=2}^{N} \log\left(\frac{X_{(j)}}{X_{(1)}}\right) = \sum_{j=2}^{N} \log\left(\frac{X_j}{X_1}\right) \tag{3.54}$$

where the latter is conditioned on the event $E := \{X_1 = x_1, X_{(1)} = x_1\}$. Due to independence, the joint distribution of $X_2, X_3, \ldots, X_N$ conditioned on E is

$$f_{(X_2, X_3, \ldots, X_N) | E}(x_2, x_3, \ldots, x_N) = \frac{f_{X_2}(x_2) \ldots f_{X_N}(x_N)}{(1 - F_{X_1}(x_1))^{N-1}}$$

$$= \frac{\alpha x_1^\alpha}{x_2^{\alpha+1}} \ldots \frac{\alpha x_1^\alpha}{x_N^{\alpha+1}}, \tag{3.55}$$

where the univariate density is denoted f_{X_j} and the univariate distribution function is F_{X_1}. Based upon (3.55) it is clear that the distribution of each X_j conditioned on E has a Pareto distribution with shape parameter α and scale parameter x_1. Thus the distribution of $\log(X_j/X_1)$ conditioned on E is exponential with shape parameter α and is independent of x_1. Hence the sum

$$\sum_{j=2}^{N} \log\left(\frac{X_j}{X_1}\right) \tag{3.56}$$

conditioned on E has a gamma distribution with shape parameter $N-1$ and scale parameter α. Since the minimum $X_{(1)}$ is a complete sufficient statistic for β the sum (3.56) is independent of $X_{(1)}$. Therefore it follows that the unconditional distribution has the same gamma distribution, meaning (3.56) must also have this gamma distribution regardless of E. For brevity, if we let W be the sum (3.56) then the distribution function of the MLE is

$$\mathbf{P}(\hat\alpha \leq t) = \mathbf{P}(W \geq NT^{-1}) = 1 - F_W(Nt^{-1}), \tag{3.57}$$

and applying the gamma density to the derivative of (3.57) one can show the density of $\hat\alpha$ is

$$f_{\hat\alpha}(t) = \frac{\alpha^{N-1} N^{N-1}}{\Gamma(N-1)} t^{-N} e^{-\alpha N t^{-1}}, \tag{3.58}$$

which can be recognised as that of the appropriate inverse gamma density, completing the proof. This is a somewhat more natural way in which to prove (3.58) without the need for moment generating functions, as in the proof in (Malik, 1970b).

3.9 Pareto Model Adopted for Detector Development

Before proceeding with the development of sliding window CFAR detection schemes some comments and discussion on the assumed Pareto clutter model is necessary. Observe that in view of (3.20) if Z has a Pareto distribution with density (2.33) then one can write $Z = \beta e^X - \beta$, where X has an exponential distribution with shape parameter α. If β is relatively small then one can apply the approximation $Z \approx \beta e^X$. This has been found to be a reasonable assumption in the case of DST Group's Ingara data, since the scale parameter β is always between 0 and 1. This is especially true in the cases of horizontal and cross polarisation, as demonstrated by the estimates provided in Table 3.1. Pareto distributional fitting to the Ingara data has shown that as the polarisation is switched from horizontal to cross and then vertical transmit and receive, the estimated shape parameter increases from 3 to around 20, with the scale parameter varying from around 0.001 to 0.2 (Weinberg, 2011a). For large α it is reasonable to assume exponentially distributed clutter, as will be discussed in Chapter 13. In such cases there exists a large number of sliding window decision processes, as documented in Chapter 1.

These considerations imply that it will be acceptable to assume a Pareto Type I model for the following development and analysis, since Ingara parameter fits will be used throughout the following. As such, and for convenience, the Pareto clutter parameters documented in Table 3.1 will be used in conjunction with this model, to provide a basis on which to test the radar detection schemes developed throughout the book.

It is also important to comment on the compound Gaussian model used to generate the signal plus clutter in the complex domain, for the CUT as well as for the case of interfering targets in the CRP. Since the Ingara clutter parameter estimates will be used throughout it follows that the scale parameter is smaller than unity. Hence one can apply the compound Gaussian model in Section 2.12 as an approximation to generate the complex equivalent of the Pareto Type I model.

FUNDAMENTAL DETECTION PROCESSES

Chapter 4

Adaptation of Exponential Detectors to Pareto Type I Distributed Clutter

4.1   Introduction

This chapter begins the investigation of sliding window detection processes for clutter modelled by a Pareto Type I distribution. As remarked previously a common approach, coinciding with the validation of new clutter models, is to adapt the original sliding window detectors, developed for the case of exponentially distributed clutter, directly to the new clutter environment of interest. This approach then requires determination of the corresponding threshold multiplier for the adapted decision rules. Such an approach has been employed successfully in the case of Weibull distributed clutter, as demonstrated in (Anastassopoulos and Lampropoulos, 1995) and (Levanon and Shor, 1990) for example. The cost in such an approach is that the CFAR property is usually lost with respect to one of the distributional parameters. In the case of the Weibull clutter model, the threshold multiplier is usually dependent on the shape parameter. Hence (Anastassopoulos and Lampropoulos, 1995) apply a maximum likelihood estimate in an attempt to compensate for this shortcoming. A similar issue will be shown to occur in the case of the Pareto Type I clutter model. Hence throughout this chapter it will be necessary to assume *a priori* knowledge of the Pareto shape parameter. One of the major conclusions to be reached in this chapter is that the underlying approach is not an ideal way in which to derive sliding window detectors. The results in this chapter have been based upon the analysis reported in (Weinberg, 2015b).

4.2   General Considerations

A general discussion and mathematical analysis is undertaken firstly before specification of a series of adapted decision rules. This is, in part, to illustrate the issues with the direct adaptation approach underlying the central theme of this chapter. For simplicity, it is useful to consider the standard minimum-based decision rule, which is specified by

$$Z_0 \underset{H_0}{\overset{H_1}{\gtrless}} \tau Z_{(1)}, \tag{4.1}$$

where the CUT statistic is Z_0, the clutter range profile is $Z_1, Z_2, \ldots, Z_N$ and $Z_{(1)}$ is the minimum of the latter. Assuming detection in Pareto distributed clutter the Pfa of (4.1) is

$$
\begin{aligned}
\text{Pfa} \;&=\; \mathbf{P}(Z_0 > \tau Z_{(1)} | H_0) = \mathbf{P}(Y > \log(\tau) + W) \\[2mm]
&=\; \int_0^\infty \alpha N e^{-\alpha N t} \mathbf{P}(Y > \log(\tau) + t) dt \\[2mm]
&=\; \int_0^\infty \alpha N e^{-\alpha N t} e^{-\alpha \log(\tau) - \alpha t} dt = \frac{N}{N+1} \tau^{-\alpha},
\end{aligned}
\tag{4.2}
$$

where $Y \overset{d}{=} \text{Exp}(\alpha)$, $W \overset{d}{=} \text{Exp}(\alpha N)$ and the duality property (3.21) has been utilised. Hence it is clear that the detector will not attain the CFAR property with respect to the Pareto shape parameter. However, it is certainly CFAR with respect to β. Unless *a priori* knowledge of α is assumed, an estimate must be applied in order to determine a threshold, for a given Pfa, from (4.2). However, the consequence of applying an approximation for α will be an inevitable change in the resultant Pfa.

   In order to acquire an insight into this, observe that by inversion of (4.2) the threshold τ is

$$\tau = \left(\frac{N}{N+1} \text{Pfa}^{-1} \right)^{\frac{1}{\alpha}} \tag{4.3}$$

so that if an approximation for α is applied to (4.3), the resultant threshold is

$$\widehat{\tau} = \left(\frac{N}{N+1} \text{Pfa}^{-1} \right)^{\frac{1}{\widehat{\alpha}}}, \tag{4.4}$$

where $\widehat{\alpha}$ is the approximation for α. To understand the impact this will have on the resultant Pfa, one can apply (4.4) to (4.2) to conclude that the resultant Pfa is

$$\widehat{\text{Pfa}} = \left(\frac{N}{N+1} \text{Pfa}^{-1} \right)^{\left(1-\frac{\alpha}{\widehat{\alpha}}\right)} \text{Pfa}, \tag{4.5}$$

which has been written to show the multiplicative factor which results in a deviation from the desired Pfa. As an example, consider the case where $N = 32$ and Pfa $= 10^{-6}$. Then if we assumed that $\widehat{\alpha} = 1.01\alpha$ then (4.5) yields $\widehat{\text{Pfa}} = 1.1462\text{Pfa}$, while if $\widehat{\alpha} = 0.9\alpha$ then $\widehat{\text{Pfa}} = 0.21618\text{Pfa}$. Hence an overestimation of α results in an increase in the resultant Pfa, while an underestimation results in a reduction in the Pfa. With worse approximations, the resultant Pfa can be affected more severely. To illustrate this, if α is underestimated by $1/2$ then the resultant Pfa reduces to the order of 10^{-12}, while if it overestimated by double then the resultant Pfa increases the order of 10^{-3}. Hence it is clear that in order to achieve CFAR in a practical situation, the decision rule (4.1) will not be useful unless α is known *a priori* or can be approximated with a small error.

The adaptation of CFAR detectors from the Exponentially distributed clutter case to the Pareto scenario can be understood with the following mathematical analysis, adapted from (Weinberg, 2015b). Consider the general detection scheme

$$Z_0 \underset{H_0}{\overset{H_1}{\gtrless}} \tau g(Z_1, Z_2, \ldots, Z_N) \tag{4.6}$$

operating in Pareto Type I distributed clutter, for some scale invariant clutter measurement function g and threshold multiplier $\tau > 0$. Suppose that the sequence of random variables X_j ($j \in \{0, 1, 2, \ldots, N\}$) have independent exponential distributions with parameter α, and density f_{X_j}. Then the Pfa of (4.6) is

$$
\begin{aligned}
\text{Pfa} \quad &= \quad P(Z_0 > \tau g(Z_1, \ldots, Z_N)) \\[2ex]
&= \quad P\left(\beta e^{X_0} > \tau g\left(\beta e^{X_1}, \beta e^{X_2}, \ldots, \beta e^{X_N}\right)\right) \\[2ex]
&= \quad P\left(X_0 > \log\left(\frac{\tau}{\beta} g\left(\beta e^{X_1}, \beta e^{X_2}, \ldots, \beta e^{X_N}\right)\right)\right) \\[2ex]
&= \quad \int_0^\infty \cdots \int_0^\infty f_{X_1}(x_1) \ldots f_{X_N}(x_N) \times \\[1ex]
&\qquad P\left(X_0 > \log\left(\frac{\tau}{\beta} g\left(\beta e^{x_1}, \beta e^{x_2}, \ldots, \beta e^{x_N}\right)\right)\right) dx_1 \ldots dx_N \\[2ex]
&= \quad \alpha^N \int_0^\infty \cdots \int_0^\infty e^{-\alpha \sum_{j=1}^N x_j} e^{-\alpha \log\left(\left(\frac{\tau}{\beta} g(\beta e^{x_1}, \beta e^{x_2}, \ldots, \beta e^{x_N})\right)\right)} dx_1 \ldots dx_N \\[2ex]
&= \quad \alpha^N \tau^{-\alpha} \int_0^\infty \cdots \int_0^\infty e^{-\alpha \sum_{j=1}^N x_j} \left(g\left(e^{x_1}, e^{x_2}, \ldots, e^{x_N}\right)\right)^{-\alpha} dx_1 \ldots dx_N,
\end{aligned}
$$

$$(4.7)$$

where statistical conditioning has been applied, as well as the duality property (3.21) and the scale invariance of g. Based upon (4.7) it is clear that it will not be possible to eliminate the factor $\tau^{-\alpha}$ through a specific choice of g, without introducing clutter parameters into the formulation of the detector (4.6). However, it is clear from (4.7) that the decision rule will achieve the CFAR property with respect to β. Hence the direct adaptation of the exponential CFAR processes to the Pareto case will cost in terms of losing the CFAR property with respect to the shape parameter. Nonetheless, it is worth considering the development of detectors based upon this approach. Hence the remainder of this chapter will examine three specific cases of (4.6), beginning with a generalisation of the minimum (4.1).

4.3 The Order Statistic Detector

The logical extension of the minimum detector (4.1) is to consider the case where the clutter is measured by an arbitrary order statistic. Thus the appropriate decision rule is given by

$$
Z_0 \underset{H_0}{\overset{H_1}{\gtrless}} \tau Z_{(j)} \tag{4.8}
$$

for some fixed index $1 \leq j \leq N$. An OS is very useful as a clutter measure as it can avoid spikier clutter measurements, as well as minimisation of the effects of multiple interfering targets that may appear in the CRP cells (Shor and Levanon, 1991). In addition, it can be used to manage successfully the effects of clutter transitions (Gandhi and Kassam, 1988). In order to apply (4.8) it is necessary to determine τ, which requires determination of the Pfa. Towards this aim, it is shown in (Weinberg, 2013e) and in Chapter 3 that the OS for a series of independent and identically distributed Pareto random variables can be written

$$Z_{(j)} = \beta T_j^{-1/\alpha}, \tag{4.9}$$

where $T_j \overset{d}{=} \text{Beta}(N - j + 1, j)$, a beta distribution with parameters $N - j + 1$ and j. To construct explicitly the false alarm probability and threshold multiplier relationship, observe that by applying (4.9) and conditional probability

$$
\begin{aligned}
\text{Pfa} \quad &= \quad \mathbf{P}(\beta e^{X_0} > \tau \beta T_j^{-1/\alpha}) = \mathbf{P}(X_0 > \log(\tau T_j^{-1/\alpha})) \\
&= \quad \int_0^1 f_{T_j}(t) e^{-\alpha \log(\tau t^{-1/\alpha})} dt = \tau^{-\alpha} \mathbf{E}(T_j) = \frac{N - j + 1}{N + 1} \tau^{-\alpha}. \tag{4.10}
\end{aligned}
$$

By inverting (4.10), the threshold τ for this case is

$$\tau = \left(\frac{N - j + 1}{N + 1} \text{Pfa}^{-1} \right)^{1/\alpha}. \tag{4.11}$$

The cost in terms of applying an approximation for α in (4.11) has been documented in (Weinberg, 2015b). Essentially the worse the approximation of α the further the deviation from the design Pfa. To investigate this, suppose that α is approximated by $\hat{\alpha}$. Hence, the suboptimal threshold multiplier is given by

$$\hat{\tau} = \left(\frac{N - k + 1}{N + 1} \text{Pfa}^{-1} \right)^{1/\hat{\alpha}}. \tag{4.12}$$

The effect on the resultant Pfa can be investigated as in the case of the minimum. An application of (4.12) to (4.10) shows that the resultant Pfa is

$$\widehat{\text{Pfa}} = \left(\frac{N - k + 1}{N + 1} \text{Pfa}^{-1} \right)^{1 - \alpha/\hat{\alpha}} \text{Pfa}. \tag{4.13}$$

Similar conclusions, in terms of the effect of an approximation of α on the resultant Pfa, can be reached as in the analysis of the minimum.

4.4 The Cell-Averaging Detector

Next the direct adaptation of a cell-averaging detector is examined; recall that such a detector is specified via the decision rule

$$Z_0 \underset{H_0}{\overset{H_1}{\gtrless}} \tau \sum_{j=1}^{N} Z_j, \tag{4.14}$$

where as before Z_0 is the CUT and the Z_j are elements of the CRP. As for the OS-detector the key is to determine τ for application in (4.14). With an application of the Pareto duality property one can write, for each j, $Z_j = \beta e^{X_j}$, where each $X_j \overset{d}{=} \text{Exp}(\alpha)$, and performing a change of variables with $z_j = e^{y_j}$, the Pfa of (4.14) can be shown to reduce to

$$\text{Pfa} = \alpha^N \tau^{-\alpha} \int_1^\infty \cdots \int_1^\infty z_1^{-\alpha-1} \dots z_N^{-\alpha-1} (z_1 + \cdots + z_N)^{-\alpha} dz_1 \dots dz_N. \tag{4.15}$$

Next, by a change of variables $w_j = z_j^{-\alpha}$, the integrals in (4.15) become

$$\text{Pfa} = \tau^{-\alpha} \int_0^1 \cdots \int_0^1 \left(w_1^{-1/\alpha} + \dots w_N^{-1/\alpha} \right)^{-\alpha} dw_1 \dots dw_N. \tag{4.16}$$

It is convenient to define an auxiliary function

$$\zeta(\alpha) = \left(\int_0^1 \cdots \int_0^1 \left(w_1^{-1/\alpha} + \dots w_N^{-1/\alpha} \right)^{-\alpha} dw_1 \dots dw_N \right)^{1/\alpha}. \tag{4.17}$$

Then by inversion of (4.16) the threshold τ for application in (4.14) is given by

$$\tau = \zeta(\alpha) \text{Pfa}^{-1/\alpha}. \tag{4.18}$$

In order to evaluate the multidimensional integral defined by $\zeta(\alpha)$, it is relatively simple to use Monte Carlo methods. An analysis of the behaviour of (4.17) can be found in (Weinberg, 2015b). The key findings are that this function tends to decrease with increasing N, while for fixed N, it is a sharply increasing function of α, which then tends to flatten out very quickly as α increases. Examples of this function can be found in Figure 4.1. However it is clear that there is an increased computational cost in the adaptive application of the CA-detector (4.14), attributable to the need to evaluate (4.17).

In addition to this complexity issue (4.18) requires *a priori* knowledge of α as before, and similar issues arise with an approximation of α applied to (4.18). As in the previous case, one can show that the resultant Pfa, arising from the approximation

$$\widehat{\tau} = \zeta(\widehat{\alpha}) \text{Pfa}^{-1/\widehat{\alpha}}. \tag{4.19}$$

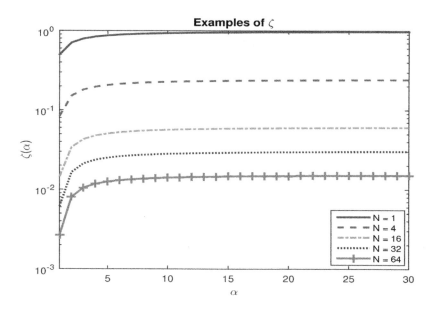

Figure 4.1: Examples of the function ζ, as a function of the Pareto shape parameter α, for a series of CRP lengths N.

is given by

$$\widehat{\mathrm{Pfa}} = \left(\left(\frac{\zeta(\alpha)}{\zeta(\hat{\alpha})} \right)^{\alpha} \mathrm{Pfa}^{(\alpha/\hat{\alpha})-1} \right) \mathrm{Pfa}, \tag{4.20}$$

where the above is written in a form to demonstrate the deviation from the design Pfa. As discussed in (Weinberg, 2015b) there can be a significant deviation in (4.20) from the design Pfa due to the behaviour of the function (4.17).

4.5 The Geometric Mean Detector

As a third example of an adapted decision rule, one can consider a geometric mean (GM) decision process, which is specified through

$$Z_0 \underset{H_0}{\overset{H_1}{\gtrless}} \tau \prod_{j=1}^{N} Z_j^{\frac{1}{N}}, \tag{4.21}$$

where the CUT and elements of the CRP are as specified previously. By an application of a logarithmic transformation, together with the Pareto-exponential duality property, the probability of false alarm of (4.21) can be shown to reduce to

$$\mathrm{Pfa} = \mathbf{P} \left(X_0 > \log \left(\tau e^{\sum_{j=1}^{N} X_j / N} \right) \right), \tag{4.22}$$

where X_0 and each $X_j \overset{d}{=} \text{Exp}(\alpha)$. Since $X_j/N \overset{d}{=} \text{Exp}(\alpha N)$, it follows that $Q = \sum_{j=1}^{N} X_j/N \overset{d}{=} \text{Gamma}(N, \alpha N)$, where the latter distribution is gamma. Assuming that f_Q is its density it can be shown that (4.22) becomes

$$
\begin{aligned}
\text{Pfa} \quad &= \quad \int_0^\infty f_Q(t) P(X_0 > \log(\tau e^t)) dt = \tau^{-\alpha} \int_0^\infty f_Q(t) e^{-\alpha t} dt \\
&= \quad \tau^{-\alpha} \frac{(\alpha N)^N}{\Gamma(N)} \int_0^\infty t^{N-1} e^{-\alpha(N+1)t} dt = \tau^{-\alpha} \left(\frac{N}{N+1} \right)^N. \quad (4.23)
\end{aligned}
$$

Consequently, by inversion of (4.23), the threshold multiplier τ is given by

$$
\tau = \left(\frac{N}{N+1} \right)^{N/\alpha} \text{Pfa}^{-1/\alpha}, \quad (4.24)
$$

which again requires α to be known. It is demonstrated in (Weinberg, 2015b) that an inaccurate estimate of α, which when applied to (4.24), will cause a serious deviation from the design Pfa, as for the other two adapted decision rules analysed previously.

Again, by introducing an estimate of α, the resultant probability of false alarm can be shown to be given by

$$
\widehat{\text{Pfa}} = \left(\left(\frac{N}{N+1} \right)^N \text{Pfa}^{-1} \right)^{1-(\alpha/\hat{\alpha})} \text{Pfa}, \quad (4.25)
$$

from which similar conclusions can be deduced on the effects of approximations of α on the design Pfa.

The next three sections examine the performance of these three detectors, when run on simulated Pareto Type I distributed clutter, and where it is assumed that full knowledge of the Pareto shape parameter is available. The numerical parameter test set, which will be used throughout the book, will also be clarified.

4.6   Performance in Homogeneous Clutter

Detector performance begins with the homogeneous clutter case, and with clutter simulated from a Pareto Type I distribution whose shape and scale parameters are matched to the Ingara data set run 34 683. A synthetic target model is also used throughout, which is distributed according to a Swerling I fluctuation model. Such target models are discussed in Appendix C; the reader can also refer to (D.6) in Appendix D for further clarification. For uniformity throughout the book's numerical analysis the design Pfa is set to 10^{-4}, which allows for 10^6 Monte Carlo runs to be sufficient for estimation of the resultant probabilities of false alarm. In addition to this the length of the CRP will be set predominantly to $N = 32$. The clutter parameters are documented in Table 3.1 in Chapter 3, with the figures in the following indicating which parameters are used via reference to the underlying polarisation of the data set, for brevity. In order to provide an indication of detector performance the probability of detection is plotted as a function of the target signal to clutter ratio (SCR), which has been estimated using 10^6 Monte Carlo runs.

It is necessary to clarify the way in which the SCR is calculated throughout the book. This is measured as the signal power relative to that of the clutter. Since Swerling target models will be used throughout, consider the random variable A whose density is given by (D.6) in Appendix D, which is the model for the target amplitude. Then by defining the random variable $Z = \left(\frac{A}{A_{rms}}\right)^2$, where A_{rms} is specified in Appendix D, it is not difficult to show that Z has a gamma distribution whose shape and scale parameters are both m. Consequently the mean of Z is unity and so the mean of A^2 is exactly A_{rms}^2. Since the signal power is measured as the mean square of its amplitude, it follows that the signal power, for the Swerling models described by (D.6) is exactly the square of A_{rms}.

Recall that since the applications are specialised to the Ingara data clutter fits the compound Gaussian model with inverse Gamma texture is used to generate the clutter in the complex domain for the CUT and interfering targets. Since the Pareto Type II model is in intensity its power is given by its mean, and so it is relatively simple to show that the clutter power is given by $\frac{\beta}{\alpha-1}$. Thus, based upon these considerations, it follows that the standard SCR is given by

$$SCR_S = \frac{(\alpha - 1)A_{rms}^2}{\beta}, \qquad (4.26)$$

while if SCR_D is the SCR in dB, then it is related to (4.26) via

$$SCR_S = 10^{0.1SCR_D}. \qquad (4.27)$$

As a benchmark of performance, an ideal linear threshold detector's performance is also included, which provides an upper bound for the decision rules studied. Such a detector assumes that both Pareto clutter parameters are known *a priori* implying that there is no need to estimate the overall level of clutter by a function g as in (4.6). Under such an assumption a fixed threshold detector should give the best performance. Appendix E can be consulted for further details on ideal decision rules.

Figure 4.2 provides an example of detector performance in the case of HH polarisation. The figure shows the Pareto ideal detector, detector (4.8), where OS1 refers to $j = N - 3 = 29$, OS2 is for $j = N - 2 = 30$ and OS3 corresponds to $j = N - 1 = 31$. Decision rule (4.14) is referred to as CA, while (4.21) is GM. Figure 4.2 shows that there is no significant difference, in terms of performance, between these decision rules. Figure 4.3 is a magnification of Figure 4.2, which provides some insight into the difference in performance. It is clear from this

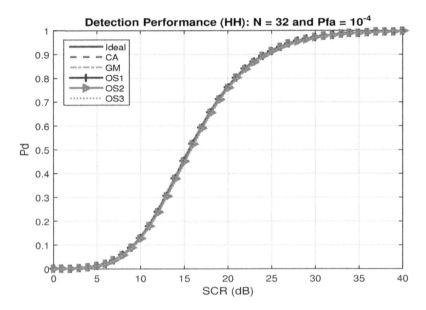

Figure 4.2: Performance of the three decision rules in homogeneous clutter, coinciding with horizontal polarisation. The plot shows the ideal decision rule as well as three examples of the OS-based detector.

figure that the GM is slightly better than the CA, while the OS-based detectors sequentially degrade in performance as their index is increased.

The same situation is repeated in Figures 4.4 and 4.5, except in the case of HV polarisation. The results are similar to the HH polarisation case as shown.

The final example of performance in homogeneous clutter is illustrated in Figure 4.6, together with Figure 4.7, where the latter is a magnification of the former. It is clear from these results that in vertical polarisation the performance is somewhat similar.

In summary it is apparent that in the case of homogeneous clutter the decision rules perform almost ideally, regardless of the polarisation. However, this situation is changed when the CRP is subjected to interference.

4.7 Effect of Interfering Targets

The performance of these detectors becomes much more interesting when independent interfering targets are inserted into the clutter range profile. These can occur, in a sliding window detection process, for a number of reasons. The first is that the target in the CUT may be range spread, resulting in target power spillover into adjacent clutter range cells. This is likely to occur in the situation

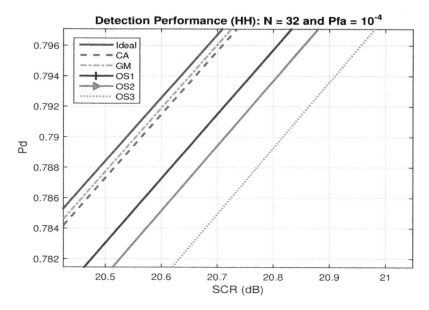

Figure 4.3: Magnification of Figure 4.2 to clarify performance.

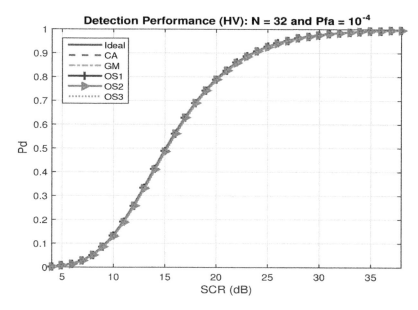

Figure 4.4: Performance in homogeneous clutter, coinciding with cross-polarisation.

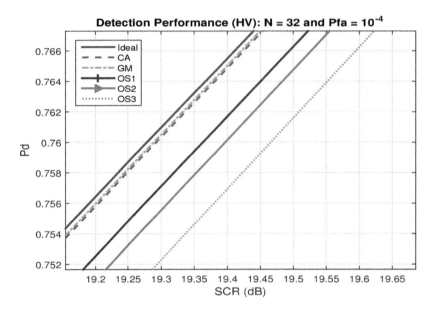

Figure 4.5: Enlargement of the previous figure.

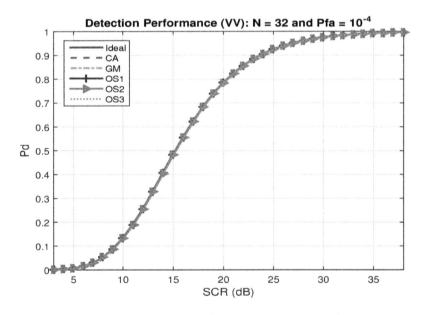

Figure 4.6: Detector performance in homogeneous vertically polarised clutter.

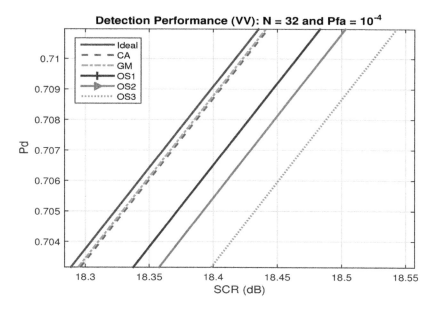

Figure 4.7: A specific magnification of Figure 4.6.

where an insufficient number of guard cells are applied on either side of the CUT. A second reason for the appearance of interfering targets in the CRP may be due to auxiliary targets present in the range-time plot over which the sliding window detector is operating. For a third explanation of such phenomena, it may be possible that a jammer is saturating the CRP with higher powered clutter in an attempt to confound the detection process. Hence it is important, in the design of sliding window decision rules, to design a detection process which is robust to such interference.

In order to simulate such effects to test the robustness of the proposed decision rules, Swerling I target models, with a fixed SCR are inserted into a cell in the CRP and combined with the clutter in the complex domain. Throughout the Pfa and N are fixed as in the homogeneous case. The OS-based detector uses $j = N - 2 = 30$ throughout. Such a choice means that the OS-based decision rule is able to manage up to two strong point interfering targets, since in such cases the OS will not select these as the measurement of the clutter level. The GM and CA based decision rules do not allow for such censoring of strong interference and so are expected to degrade in performance in the presence of strong interference. Figure 4.8 shows performance in HH polarisation, with Figure 4.9 providing an enlargement. In this case two interfering targets are used, with SCR of 20 and 80 dB respectively. Throughout the book the latter SCRs will be referred to as the interference to clutter ratio (ICR) to differentiate it from the SCR of a target in the CUT. The ICR is also obtained in the same way in which the SCR for the target

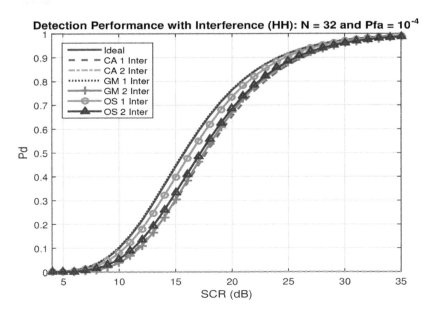

Figure 4.8: Detector performance in the presence of interference, provided by the insertion of one or two independent Swerling I target models into the CRP. The two interfering targets have ICR of 20 and 80 dB respectively. For brevity interference is abbreviated to "inter" throughout.

in the CUT is calculated, based upon (4.26). Detector performance is illustrated when each detector is subjected to only the first interfering target, and then when each is subjected to the two interfering targets simultaneously. Here we observe a change in the performance when compared to the case of no interference. The GM-detector has the best performance in the case of a single interfering target of 20 dB, followed by the OS detector and then the CA. When subjected to the second interfering target, the CA detector saturates as shown, while the GM is slightly worse than the OS. It has been found that a GM-based detector is usually robust to interference, provided there is only one such target and its SCR is less than 20 dB.

In a real radar system an interfering target with ICR of 80 dB would almost certainly spread to adjacent range cells, and would usually occur with multiple range cells also affected with varying degrees of interference levels. The example in Figure 4.8 is provided to indicate the effects of several different interfering point targets on the sliding window detection process.

To examine this further, the second interfering target has had its SCR reduced to 20 dB in Figures 4.10 and 4.11. This example illustrates the fact that the CA-detector has the worst performance with increasing interference, while the GM-detector seems to perform very well despite the two 20 dB interfering targets. The OS-based decision rules have consistent performance as before.

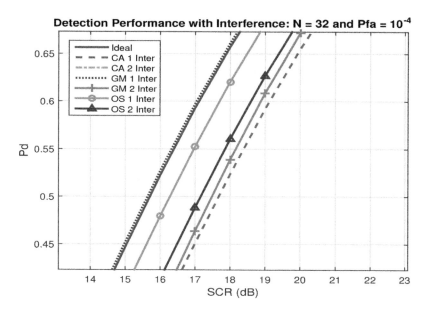

Figure 4.9: Magnification of Figure 4.8 to illustrate detector performance.

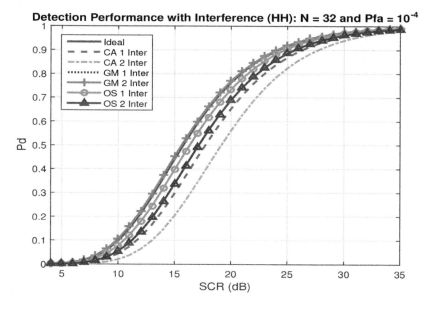

Figure 4.10: Detector performance in the presence of two 20 dB interfering targets, generated by independent Swerling I fluctuations.

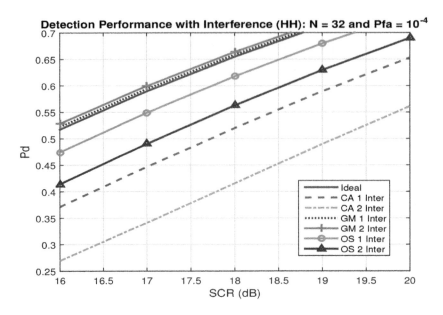

Figure 4.11: Enlargement of Figure 4.10.

Figures 4.12 and 4.13 show what happens when both the two interfering targets have SCR of 40 dB. Clearly the CA detector has the worst performance as before. Figure 4.13 shows that with one 40 dB interfering target the GM and OS detectors have the same performance roughly, while when subjected to the two interfering targets, the GM detector is still slightly better than the OS.

As a final example, it is informative to investigate detection performance when the interference is quite strong. Hence the SCR of the two interfering targets has been increased to 80 dB in Figures 4.14 and 4.15. For this scenario the CA-detector saturates in both cases, while the OS-detector performs much better than the GM counterpart. In fact for the case of two interfering targets, the OS-based detector performs better than either of the GM detector cases.

Based upon these numerical simulations it is clear that the OS-based detector (4.8) provides a robust solution to interfering targets in the CRP, with very little detection performance loss relative to an ideal decision rule. The GM-detector (4.21) is also able to manage interference relatively well, while the CA-based detector (4.14) tends to saturate in the presence of interference. The latter result is not surprising due to the fact that an interfering target will increase the clutter measurement function for a CA-detector. In the case of a GM-based decision rule, since this is equivalent to a logarithmic sum, weak interference is essentially suppressed.

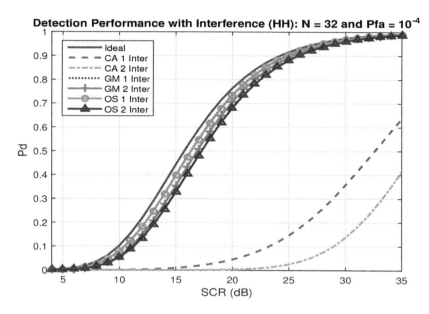

Figure 4.12: Performance with two 40 dB interfering targets.

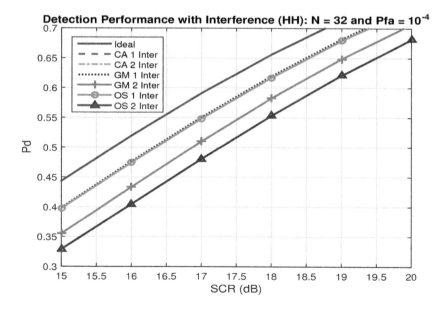

Figure 4.13: Magnification of Figure 4.12.

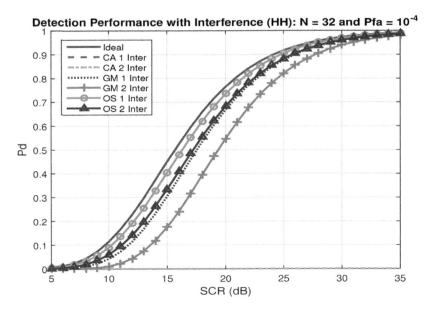

Figure 4.14: Detector performance with two independent 80 dB interfering targets.

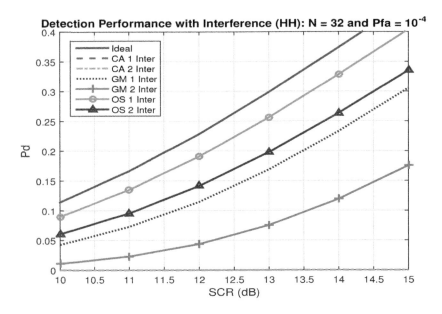

Figure 4.15: Figure 4.14 enlarged.

4.8 Clutter Transitions

Sliding window detectors are designed so that they maintain a constant false alarm rate in practice, by adaptive adjustment of the threshold multiplier τ. These detectors are designed to operate in the ideal situation of homogeneous independent clutter. Hence in situations where there is a variation in the clutter power level, the adaptive threshold may become suboptimal, and result in variations in the resulting Pfa. In order to understand how a proposed detection process manages such an issue, it is run in a scenario where the resulting Pfa is estimated as the number of higher powered clutter cells increases. This is done systematically through the CRP, and when the mid-point of the CPR is saturated with higher powered clutter, the CUT is also considered to be saturated with higher power clutter. This results in a characteristic jump in plots of the resultant Pfa as a function of the number of higher powered clutter cells. In order to estimate the resultant Pfa, Monte Carlo estimation is employed, with 10^6 runs, where the design Pfa is 10^{-4}. It is usually difficult to gain good estimates using such an approach, and increasing the Monte Carlo sample size can result in computational complexity issues. In some cases, throughout the analysis to follow in subsequent chapters, it is possible to avoid Monte Carlo techniques altogether, and derive analytical expressions to compute the resultant Pfa.

In the following $N = 32$ and two OS-based detectors are used, with $j = N - 2 = 30$ and $j = N - 1 = 31$ respectively. Although the plot is for a discrete function it is plotted as a continuous line, which assists in visual interpretation of the results.

Figure 4.16 is for the case of a 0.5 dB power increase and with HH polarisation, while Figure 4.17 is for HV polarisation, and Figure 4.18 corresponds to VV polarisation. Throughout the book the clutter power level increase will be referred to as the clutter to clutter ratio (CCR). There are several ways in which this can be calculated, with a simple convenient method applied throughout the book. Since two Pareto models are being compared, it is assumed that clutter power level increases result from variation in the Pareto shape parameter, while the scale parameter remains fixed. In addition to this, the underlying Pareto models are based upon a Pareto Type I form for convenience. The clutter power level is measured by the second moment of the Pareto model. Based upon this, if X_1 has a Pareto Type I distribution with shape parameter α and scale β, while X_2 is assumed to have a similar Pareto model structure, except with shape parameter α_I then since the mean squares are $\frac{\alpha\beta^2}{\alpha-2}$ and $\frac{\alpha_I\beta^2}{\alpha_I-2}$ respectively, the clutter power ratio is given by

$$\mathrm{CCR} = \frac{\alpha_I}{\alpha_I - 2} \frac{\alpha - 2}{\alpha},\tag{4.28}$$

from which it can be shown that

$$\alpha_I = \frac{2}{1 - \left(\frac{\alpha-2}{\alpha}\right) \times 10^{-0.1\mathrm{CCR}}}.\tag{4.29}$$

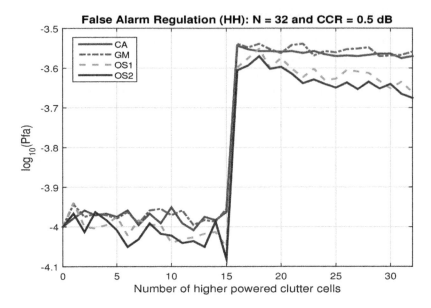

Figure 4.16: False alarm regulation with HH polarisation.

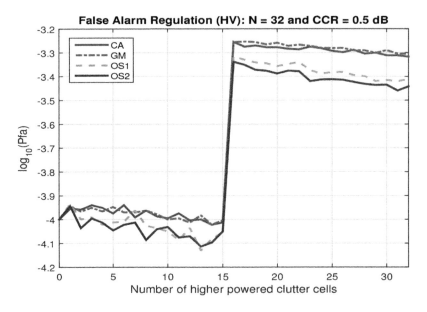

Figure 4.17: False alarm regulation with HV polarisation.

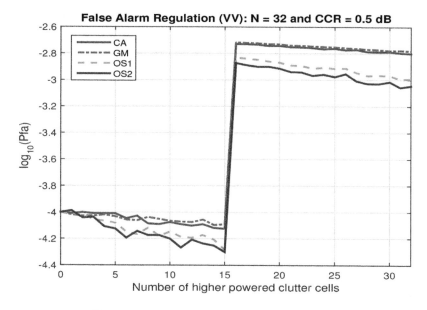

Figure 4.18: False alarm regulation with VV polarisation.

Hence, in terms of clutter power level increases, for an increase of CCR in standard units, one can select the second Pareto model to have shape parameter given by (4.28). An expression similar to (4.27), for a CCR measured in dB, can also be specified.

The OS detector with $j = 30$ is denoted OS1 while that with $j = 31$ is OS2. What is clear from these three figures is that as the polarisation is changed from HH to VV, the resultant Pfa after the CUT is affected tends to increase. Prior to the CUT being affected, the detectors tend to manage the Pfa almost similarly. Once the CUT is affected, the two OS-based decision rules tend to have the least increase in resultant Pfa.

Figures 4.19–4.21 repeat the same scenario with the exception that the clutter power is increased by 5 dB. The same phenomenon is repeated as for the previous case, with the worst Pfa regulation occurring in VV-polarisation. The larger clutter power increase has resulted in an increase in the resultant Pfa in all polarisations. When the clutter power level is increased even more, the situation is worsened.

As a final example, the situation is repeated with a smaller clutter power increase than the previous two cases. Here the case of a 0.2 dB clutter power increase is investigated, with Figures 4.22–4.24 providing the relevant results. The same behaviour as in previous cases is observed, but on a smaller scale than in the previous two scenarios.

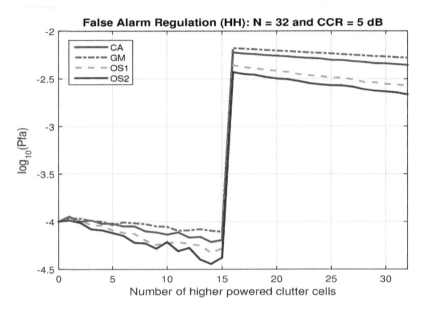

Figure 4.19: Second example of false alarm regulation (HH polarisation).

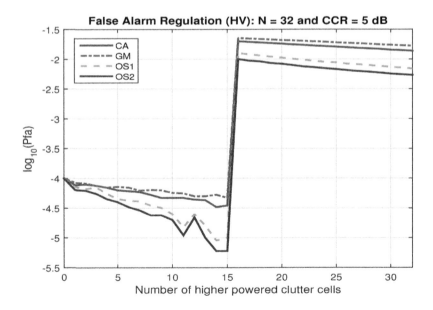

Figure 4.20: Second example of false alarm regulation for the case of cross polarisation.

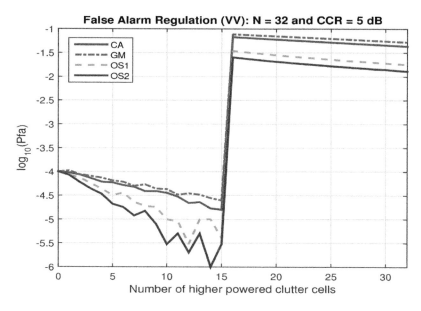

Figure 4.21: Second example of false alarm regulation in vertical polarisation.

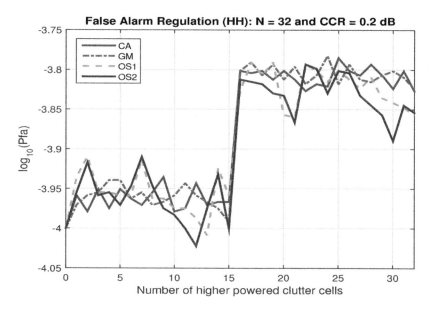

Figure 4.22: Final example of Pfa regulation in horizontal polarisation.

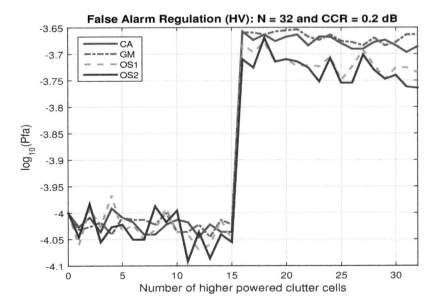

Figure 4.23: Final example of Pfa regulation in cross polarisation.

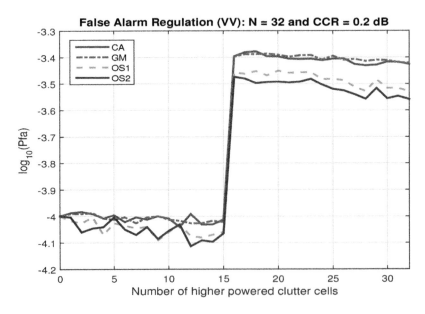

Figure 4.24: Final example of Pfa regulation in vertical polarisation.

4.9 Conclusions

The direct adaptation of CFAR detectors from the exponential distributed clutter case to the Pareto scenario was examined in this chapter. It was shown that the CFAR property could not be attained with respect to the Pareto shape parameter, but certainly could be achieved with respect to the Pareto scale parameter. Three detectors were investigated, based upon a CA, GM and OS-decision rule. In homogeneous clutter the decision rules tended to perform comparably. When subjected to interference, the GM-based detector worked very well in managing interference until the level of interference increased significantly. The OS detector had the greatest resilience to increasing levels of interference. By contrast the CA-based detectors had terrible performance in interference, resulting in detector saturation when the SCR of interference increased substantially.

In terms of regulating the Pfa, each of the decision rules tended to perform similarly until the CUT became affected by higher powered clutter. Once this occurred the OS-based decision rules had the least deviation from the design Pfa. The polarisation also had an impact on the regulation of the Pfa, with a more significant departure from the design Pfa in VV polarised clutter, especially when compared with the false alarm regulation results in HH polarisation.

Chapter 5

A Transformation Approach for Radar Detector Design

5.1 Introduction

This chapter introduces a transformation approach for detector design, where the sliding window CFAR detectors for exponential intensity clutter models are adapted for use in any clutter environment of interest. This approach was first introduced in (Weinberg, 2013d), where some of the standard detection processes designed for exponentially distributed clutter were transformed to the Pareto Type I clutter scenario. This resulted in decision rules which preserved the original exponential-based detector's Pfa and threshold multiplier relationship. The approach provided a more convenient method for the design of detectors since it removed the need to apply numerical estimation procedures to determine detector thresholds. Consequently, many of the exponential-CFAR detectors were transformed to the Pareto domain in (Weinberg, 2014a). However, this approach resulted in the CFAR property only being preserved with respect to the Pareto shape parameter for the Pareto Type I model. This is attributable to the fact that the Pareto scale parameter appeared in the detector's structure as a result of the transformation. It can be assumed that the Pareto scale parameter is known *a priori*, which is not an unreasonable assumption, and was a practice adopted in the work reported in (Weinberg, 2013d) and (Weinberg, 2014a). A few years later it was demonstrated in (Weinberg, 2016b) that the transformation approach would yield a CFAR detector in the transformed domain only if the new clutter model was from a family of one parameter models. This meant, for example, a one parameter Lomax distribution could be used instead of a two parameter Pareto model. Although this yielded improved detection results, the underlying single parameter model was found to be somewhat insufficient for real X-band clutter. Consequently, further analysis of the problem of scale parameter dependence in transformed detectors yielded a new solution. It was found that an application of a complete sufficient statistic for the Pareto scale parameter in the transformed domain yielded a detector possessing the CFAR property with respect to both Pareto distributional parameters (Weinberg, 2017b). This chapter outlines these developments in detail, with illustrations of detector performance in simulated clutter. It will also be shown in Chapter 12 that the acquisition of the CFAR property is due to the fact that the new decision rules are invariant statistics.

5.2   The Transformation Approach

Based upon the analysis of the previous chapter it is clear that a better method for producing decision rules is required, since direct adaptation of the exponential-based CFARs do not yield detectors which have the full CFAR property. A convenient and systematic technique for this purpose can be based upon the memoryless nonlinear transformation. This was first applied in (Weinberg, 2013d) and generalised subsequently in (Weinberg, 2014b), allowing the transformation of any of the exponentially-based CFAR detection processes to any given clutter environment of interest. The key to the transformation approach is exploitation of (2.102), which allow one to begin with the exponential-based CFAR detectors Pfa expression, and then transform it to an expression involving the desired clutter model. This then allows one to specify a decision rule, for operation in the desired clutter environment, which preserves the original Pfa expression.

Suppose that

$$Z_0 \overset{H_1}{\underset{H_0}{\gtrless}} \tau g(Z_1, Z_2, \ldots, Z_N), \tag{5.1}$$

is a generic sliding window detection scheme, designed to operate in independent and identically distributed exponential clutter with a shape parameter $\lambda > 0$. Here Z_0 is again the CUT while the statistics $Z_1, Z_2, \ldots, Z_N$ are members of the clutter range profile. As before $\tau > 0$ is the threshold multiplier and g is a scale invariant function of the CRP statistics. Recall that the probability of false alarm of (5.1) is given by

$$\text{Pfa} = \mathbf{P}(Z_0 > \tau g(Z_1, Z_2, \ldots, Z_N) | H_0). \tag{5.2}$$

As remarked previously, throughout the signal processing literature the selection of g usually entails a choice that is pseudo linear, or scale invariant, in the sense that $g(\lambda(z_1, z_2, \ldots, z_N)) = \lambda g(z_1, z_2, \ldots, z_N)$. With such selections it can be shown that the expression (5.2) does not vary with the exponential distribution's shape parameter, so that the detection process (5.1) has the CFAR property for a large class of admissible functions g (Gandhi and Kassam, 1988).

To see this, note that under H_0 that each $Z_j = \lambda^{-1}W_j$, where W_j has an exponential distribution with unity mean. Then the Pfa (5.2) becomes

$$\text{Pfa} \quad = \quad P\left(\lambda^{-1}W_0 > \tau g\left(\lambda^{-1}W_1, \lambda^{-1}W_2, \ldots, \lambda^{-1}W_N\right)\right)$$

$$= \quad P\left(W_0 > \tau g\left(W_1, W_2, \ldots, W_N\right)\right), \tag{5.3}$$

where the scale invariance property of g has been applied. Since the variables in (5.3) do not depend on λ the Pfa is independent of this parameter and the claimed result is proven.

Since the same scale invariance assumption on g will be adopted here, we can without loss of generality assume that the exponential's distribution's mean is unity.

Assume that Y is a random variable representing the clutter in which one would like to perform detection. Let $\{Y_j, j \in \mathbf{N}\}$ represent independent and identically distributed copies of Y. Supposing that F_Z and F_Y are the distribution functions of Z and Y respectively, the following technical lemma shows how Z and Y can be interrelated statistically:

Lemma 5.1

Suppose that Y and Z are two random variables with distribution functions F_Y and F_Z respectively. Define a function

$$H(t) = F_Y^{-1}(F_Z(t)), \tag{5.4}$$

which is well-defined and differentiable. Then the random variable $H(Z) = F_Y^{-1}(F_Z(Z)) \overset{d}{=} Y$.

This result has already been discussed in Chapter 2, in the context of simulation of dependent sequences of random variables. In addition to this (5.4) is analysed extensively in (Weinberg, 2014b) and has the following properties, where the densities of Y and Z are f_Y and f_Z respectively:

1. $\dfrac{dH}{dt} = \dfrac{f_Z(t)}{f_Y\left(F_Y^{-1}(F_Z(t))\right)} > 0.$

2. $H(t)$ is increasing.

3. $H^{-1}(t) = F_Z^{-1}(F_Y(t)).$

4. $H(t) = F_Y^{-1}(1 - e^{-t})$ for transformation from the exponential case.

5. $H^{-1}(t) = -\log(F_Y^c(t))$, also for transformation from the exponential distribution scenario.

These results can be proved quite simply. Note that Property 2 follows from Property 1. Since H is increasing, one can apply it within the Pfa expression (5.2) to modify it so that it is expressed in terms of the clutter model Y. This then permits the specification of a decision rule, for operation in clutter modelled by Y, such that the Pfa expression (5.2) still remains applicable.

To see this, observe that based upon (5.2) and the properties of H,

$$\text{Pfa} \quad = \quad P(H(Z_0) > H(\tau g(Z_1, Z_2, \ldots, Z_N) | H_0))$$

$$= \quad P(Y_0 > H(\tau g(H^{-1}(Y_1), H^{-1}(Y_2), \ldots, H^{-1}(Y_N)) | H_0)), \quad (5.5)$$

where $Y_0 := H(Z_0)$. In view of (5.5) one can specify the detector

$$Y_0 \overset{H_1}{\underset{H_0}{\gtrless}} H(\tau g(H^{-1}(Y_1), H^{-1}(Y_2), \ldots, H^{-1}(Y_N))) \quad (5.6)$$

where Y_0 is the CUT and $\{Y_1, Y_2, \ldots, Y_N\}$ is the CRP, which is a transformed version of (5.1), designed to operate in clutter modelled by Y while preserving (5.2). This provides one with a convenient and systematic method with which detectors can be produced. The issue of interest is whether the function H introduces clutter parameter dependency into the decision rule (5.6). These results are now specialised to the Pareto Type I model setting. Since the inverse of the Pareto Type I distribution function is $F_Y^{-1}(t) = \beta(1-t)^{-\frac{1}{\alpha}}$ it follows that

$$H(t) = \beta e^{\frac{t}{\alpha}} \quad (5.7)$$

and

$$H^{-1}(t) = \alpha \log\left(\frac{t}{\beta}\right) \quad (5.8)$$

so that by applying these to (5.6), and applying the assumed scale invariance of g, the general form of (5.6) for the Pareto Type I clutter model case is

$$Y_0 \overset{H_1}{\underset{H_0}{\gtrless}} \beta e^{\tau g\left(\log\left(\frac{Y_1}{\beta}\right), \log\left(\frac{Y_2}{\beta}\right), \ldots, \log\left(\frac{Y_N}{\beta}\right)\right)}. \quad (5.9)$$

Note that the decision rule (5.9) is dependent on β but clearly will attain the CFAR property with respect to the Pareto shape parameter. It is now possible to specify a large number of decision rules, based upon the specification of g in (5.9), which is in turn dictated by the underlying CFAR decision rules from the exponentially distributed clutter case.

The most basic choice for g corresponds to the selection of g as a sum of its arguments, resulting in the transformation of the CA-CFAR (1.6). With such a selection it is not difficult to show that (5.9) reduces to

$$Y_0 \overset{H_1}{\underset{H_0}{\gtrless}} \beta^{1-N\tau} \prod_{j=1}^{N} Y_j^{\tau} \quad (5.10)$$

with Pfa given by (1.7). Detector (5.10) is referred to as a geometric mean (GM) detector, and it is CFAR with respect to α but requires *a priori* knowledge of β. A second obvious choice for g is that of an order statistic, where $g(y_1, y_2, \ldots, y_N) = y_{(j)}$ for some $1 \leq j \leq N$. With such a choice, (5.9) becomes

$$Y_0 \underset{H_0}{\overset{H_1}{\gtrless}} \beta^{1-\tau} Y_{(j)}^{\tau}. \tag{5.11}$$

Its threshold multiplier is determined via the expression (1.9), where ν in the latter is replaced with τ. As for the decision rule (5.10), the detector (5.11) requires *a priori* knowledge of β but is CFAR with respect to α.

Next, smallest-of (SO) and greatest-of (GO) transformed detectors are introduced, which have been considered in (Weinberg, 2014a). These entail the assumption that the clutter range profile length is even, which is then divided into two parts. In each part the GM is computed, and then either the smaller of the two or the larger is used as the measurement of clutter level. Hence g is selected to be

$$g(y_1, y_2, \ldots, y_{2M}) = \mathcal{M}\left(\sum_{j=1}^{M} y_j, \sum_{j=M+1}^{2M} y_j\right), \tag{5.12}$$

based upon the exponentially distributed clutter case, where $\mathcal{M}$ is either the minimum for SO or maximum for GO. With this choice for g, the detector (5.9) becomes

$$Y_0 \underset{H_0}{\overset{H_1}{\gtrless}} \beta^{1-M\tau} \left(\begin{array}{c} \min \\ \max \end{array} \left(\prod_{j=1}^{M} Y_j, \prod_{j=M+1}^{N} Y_j\right)\right)^{\tau}, \tag{5.13}$$

where *min* is used for SO-CFAR and *max* is for GO-CFAR, and for simplicity it has been assumed that N is even, and $M = N/2$. The false alarm probabilities are given by

$$\text{Pfa}_{\text{GO}} = 2(1+\tau)^{-M} - \text{Pfa}_{\text{SO}}, \tag{5.14}$$

where

$$\text{Pfa}_{\text{SO}} = 2 \sum_{j=0}^{M-1} \binom{M+j-1}{j} (2+\tau)^{-(M+j)}. \tag{5.15}$$

The derivation of (5.14) and (5.15) is discussed in (Weinberg, 2014a) and can be found in (Gandhi and Kassam, 1988) in the context of detection in exponentially distributed clutter. The GO-detector censors the lower half of the clutter statistics, resulting in a detector that can better manage false alarm power increases during clutter power transitions. Additionally it is robust to interfering targets appearing in the lower subset of the CRP. The SO detector censors the upper half of the clutter range profile, resulting in a detector robust to interference appearing in the upper subset of clutter statistics.

A trimmed mean (TM) detector can also be constructed, which is based upon censoring applied to ordered clutter statistics in the GM detector (5.10). In particular, the corresponding decision rule is

$$Y_0 \underset{H_0}{\overset{H_1}{\gtrless}} \beta^{1-(N-T_1-T_2)\tau} \left(\prod_{j=T_1+1}^{N-T_2} Y_{(j)} \right)^{\tau}, \tag{5.16}$$

where T_1 statistics are trimmed from the lower OS, while T_2 are trimmed from the top. By applying the fact that if $Z_{(k)}$ is the kth OS for a series of exponentially distributed random variables with parameter α, then the kth OS for a series of Pareto random variables is given by $Y_{(k)} = \beta e^{Z_{(k)}}$, and it can be shown that the TM detector has probability of false alarm given by

$$\text{Pfa} = \prod_{j=1}^{N-T_1-T_2} M_{V_j}(\tau), \tag{5.17}$$

where the function

$$M_{V_1}(\tau) = \frac{N!}{T_1!(N-T_1-1)!(N-T_1-T_2)} \times$$
$$\sum_{j=0}^{T_1} \frac{\binom{T_1}{j}(-1)^{T_1-j}}{\left(\frac{N-j}{N-T_1-T_2}\right)+\tau}, \tag{5.18}$$

and for $2 \leq j \leq N - T_1 - T_2$

$$M_{V_j}(\tau) = \frac{a_j}{a_j + \tau} \tag{5.19}$$
$$\text{where } a_j = (N-T_1-j+1)/(N-T_1-T_2-j+1).$$

Hence the detector (5.16) has the CFAR property with respect to α. Expression (5.17) can be found in (Gandhi and Kassam, 1988), from the Gaussian intensity equivalent. The proof is omitted for brevity.

Observe that with the choice of $T_1 = T_2 = 0$, the detector (5.10) is recovered, while for the choice of $T_1 = j - 1$ and $T_2 = N - j$, the jth OS-detector (5.11) is produced, for $1 \leq j \leq N$. From a theoretical perspective, a detector that censors T_1 lower and T_2 upper clutter statistics can be used to filter out an expected number of higher-powered interfering targets. It should also be able to tolerate clutter transitions to a certain point. As in the previous cases, τ is determined from (5.17) with numerical inversion.

5.3 Examples of Detector Performance

An examination of the performance of these transformed detectors is now included, where it will be assumed that one has *a priori* knowledge of the Pareto scale parameter. As in the numerical analysis of the previous chapter, Pareto Type I clutter is simulated with parameters matched to the Ingara data set which has formed the basis for the numerical analysis considered thus far. Thus the parameter estimates are taken from Table 3.1. For consistency, the Pfa will again be set to 10^{-4} with $N = 32$. Here performance analysis is limited to target detection in homogeneous clutter with no interference. As the transformed detector analysis is developed later in this chapter, the cases of interference and clutter transitions will be considered.

As a first example, Figure 5.1 plots a series of the decision rules operating in the case of horizontally polarised clutter. The detectors include the ideal as an upper bound on performance, the GM detector (5.10), the OS detector (5.11) which uses $k = N - 2$, the GO and SO detectors (5.13) as well as the TM detector (5.16). For the latter, three sets of parameters are used: $T_1 = T_2 = 1$, $T_1 = T_2 = 2$ and $T_1 = T_2 = 3$. Figure 5.2 provides an enlargement of Figure 5.1. The GM detector (5.10) has the best performance, followed by the three TM detectors as shown, although the degree of difference is very minute. These are followed by the GO detector, then the OS and finally the SO detector.

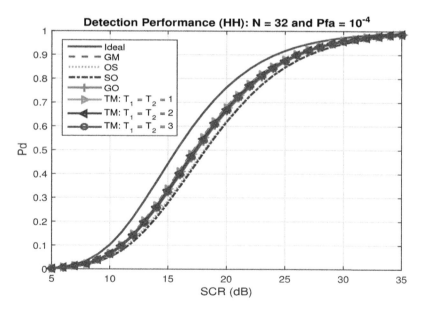

Figure 5.1: Performance of five transformed detectors, in the case of horizontal polarisation, compared with an ideal detector.

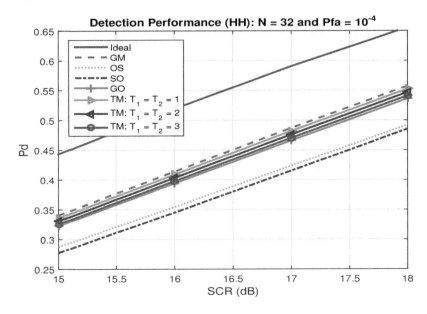

Figure 5.2: An enlargement of Figure 5.1 to clarify performance.

The same results are repeated in the case of cross polarisation (Figures 5.3 and 5.4) and vertical polarisation (Figures 5.5 and 5.6).

These results show that the transformed detectors introduce a detection loss when compared with the decision rules in the previous chapter. This is due to the fact that the detectors in Chapter 4 assume *a priori* knowledge of the Pareto shape parameter, while those introduced in this chapter are CFAR with respect to the Pareto shape parameter. These results demonstrate the critical nature of α in terms of its impact on detection performance. Although the SO, GO and TM detectors can be useful in practice they will not be analysed any further in the following developments. The interested reader will note that subsequent analysis can be applied to produce full CFAR variants of these detectors. Before proceeding with the development of detectors a brief digression into the CFAR property is undertaken, with a view to establishing conditions under which a transformed detector will retain the full CFAR property.

5.4   Preservation of the CFAR Property

It is clear from the decision rules introduced in the last section that the CFAR property is only achieved with respect to the Pareto shape parameter. As has been reported in (Weinberg, 2014b) when the exponential CFARs are transformed for operation in Weibull distributed clutter, the situation is reversed. Hence in the

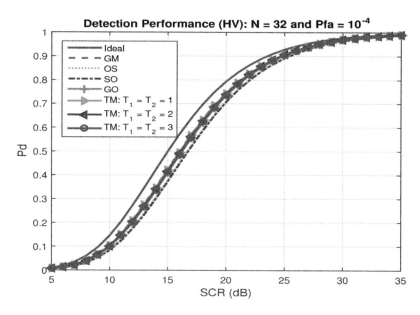

Figure 5.3: Performance of the five decision rules in cross-polarisation.

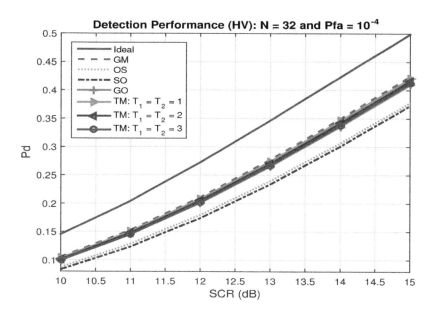

Figure 5.4: Enlargement of the plot in Figure 5.3.

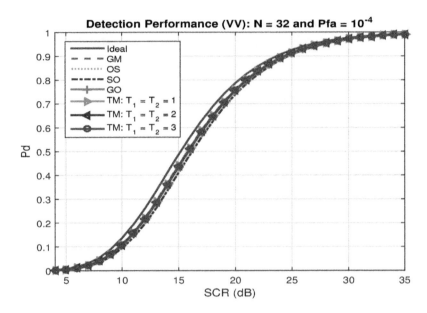

Figure 5.5: Detector performance in the scenario of vertically polarised homogeneous clutter, for the same series of decision rules.

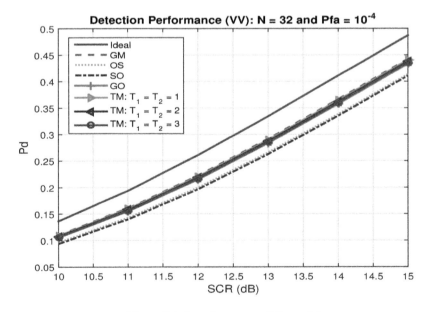

Figure 5.6: An enlargement of Figure 5.5.

Weibull case *a priori* knowledge of its shape parameter is required, while the decision rules are independent of the scale parameter. It is hence of interest to understand this more deeply, and one can ask what clutter environments will result in decision rules that are completely CFAR under the transformation approach. This has been addressed in (Weinberg, 2016b), which is considered in this section.

Toward this objective, it is insightful to reformulate (5.6) in terms of the clutter environment of interest. Suppose that the desired clutter model Y has distribution function of the form

$$F_Y^c(t) = e^{-\mu h(t)} \tag{5.20}$$

where $F_Y^c(t) = 1 - F_Y(t)$ is the complementary distribution function, $\mu > 0$ is a distribution parameter and $h(t)$ is a function that satisfies the conditions that $h(0) = 0$ and $h(t)$ is increasing. The latter conditions are imposed so that (5.20) has the properties of a distribution function. The function $h(t)$ may have dependence on unknown clutter parameters. This general form of a clutter distribution will be referred to as the *h*-clutter model.

The function h is closely related to the function H. In particular, using the definition of H and its inverse, it can be shown that $H^{-1}(t) = \mu h(t)$ and consequently $H(t) = h^{-1}(t/\mu)$. The following result expresses the transformed decision rule (5.6) in terms of the function h:

Lemma 5.2
Suppose that Y_0 is the CUT statistic and $\{Y_1, Y_2, \ldots, Y_N\}$ is the CRP for a sliding window detection process. Then the detector

$$Y_0 \underset{H_0}{\overset{H_1}{\gtrless}} h^{-1}\left(\left\{\frac{\tau}{\mu}\right\} g\left(\mu\left[h(Y_1), \ldots, h(Y_N)\right]\right)\right) \tag{5.21}$$

is equivalent to (5.6), where τ is set with (5.2), and under the clutter model specified by (5.20).

To prove this result, observe that since distribution functions are monotonic, one can apply F_Y to both sides of (5.6) to re-express it in the form

$$F_Y(Y_0) \underset{H_0}{\overset{H_1}{\gtrless}} 1 - e^{-\tau g(H^{-1}(Y_1), \ldots, H^{-1}(Y_N))}, \tag{5.22}$$

since $F_Y(H(t)) = F_Z(t)$ by construction, and applying the distribution function of Z. By some simple analysis, it is not difficult to show that (5.22) is equivalent to

$$-\log F_Y^c(Y_0) \underset{H_0}{\overset{H_1}{\gtrless}} \tau g\left(-\log F_Y^c(Y_1), \ldots, -\log F_Y^c(Y_N)\right). \tag{5.23}$$

The final result now follows with an application of (5.20) to (5.23), as required.

Recalling the assumed scale invariance property of g, the following is immediate:

Corollary 5.1

In the case where the function g satisfies the scale invariance property that $g(\lambda(y_1,\ldots,y_N)) = \lambda g(y_1,\ldots,y_N)$, for fixed $\lambda > 0$, the detector (5.21) reduces to

$$Y_0 \underset{H_0}{\overset{H_1}{\gtrless}} h^{-1}\left(\tau g\left(h(Y_1),\ldots,h(Y_N)\right)\right), \qquad (5.24)$$

implying the decision rule is CFAR with respect to μ.

The results of Lemma 5.2 and Corollary 5.1 provide an give an insight into when a transformed detection process will inherit the CFAR property. Clearly to produce a CFAR process, the function $h(t)$ must have no dependence on unknown clutter parameters. It is also clear that, for the class of transformed detectors, the CFAR property will be only achievable with respect to μ. Hence transformation to a family of distributions which have a single shape parameter, as in the model (5.20), will yield decision rules that are CFAR with respect to μ.

With reference to the Pareto Type I and II distributions it follows that unless the clutter is modelled by a one parameter model it will not be possible to produce a fully CFAR decision rule via the transformation approach alone. As a consequence of this, (Weinberg, 2016b) investigated the application of a one-parameter Lomax model for X-band clutter. This is examined in the next section.

5.5 Lomax-Distributed Clutter and Detector Performance

The validity of a single parameter Pareto-type distribution, as a model for X-band maritime surveillance radar clutter, is now investigated. This is examined by considering the fit of a one-parameter Lomax model with distribution function

$$F_Y(t) = 1 - \left(\frac{1}{t+1}\right)^{\mu}, \qquad (5.25)$$

where $t \geq 0$ and $\mu > 0$ is a distributional shape parameter. Throughout the following (5.25) will be referred to as a Lomax distribution for brevity. Maximum likelihood estimation has been used to estimate the relevant Lomax shape parameter based upon the data used as a basis for this book. For comparison purposes the K-distributional fit is also included. Figure 5.7 shows the fits of distributions in the case of horizontal polarisation, in the lower tail region, while Figure 5.8 shows the same plot except in the upper tail region. Here one observes that the Lomax fit is not as accurate as that of a two parameter Pareto Type II model. Figure 5.9 shows the fits for the cross-polarisation case, while Figure 5.10 corresponds to the vertically polarised case. It is clear that the Lomax model is completely inappropriate for the latter two polarisations. In the horizontally polarised case, it is inaccurate but can be used as an approximation for the fit.

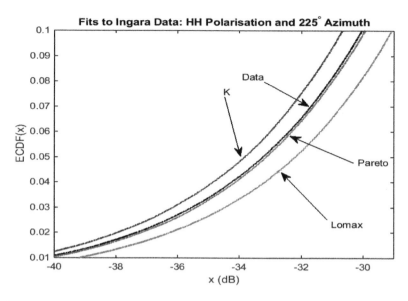

Figure 5.7: Comparison of distribution function fits to the Ingara data set, and in the case of horizontal polarisation, showing the lower tail region. The Lomax one-parameter model is an insufficient fit.

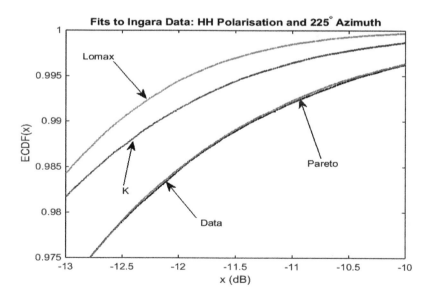

Figure 5.8: Distribution function fits to the Ingara data set, again in horizontal polarisation and showing the upper tail region. The one-parameter Lomax model provides a worse fit than that of the K-distribution. The Pareto Type II model provides the best fit.

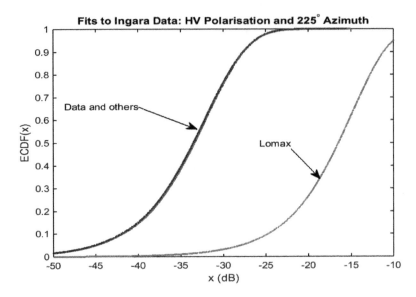

Figure 5.9: Fits to the Ingara data in the case of cross polarisation. Here it is clear that a one-parameter Lomax model is very inaccurate.

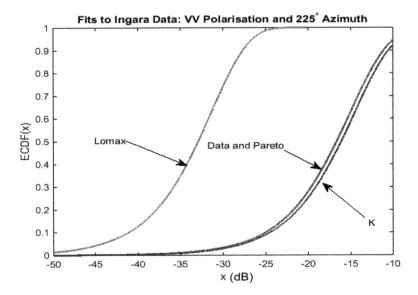

Figure 5.10: Fits for the vertically polarised clutter case, showing that the one-parameter Lomax model is an inappropriate fit to the data.

The fitting of the Lomax model, using maximum likelihood estimation for its shape parameter, resulted in an estimate of 84.8173 for the case of horizontal polarisation, 1.3541×10^3 for cross polarisation and 31.2739 for vertical polarisation.

Although these results discount the suitability of the one-parameter Lomax model for X-band maritime surveillance radar clutter, one can still explore detection performance under such a clutter model assumption. However, attention is limited to the case of horizontal polarisation in terms of performance analysis. Towards this end it is not difficult to produce transformed CA- and OS-detectors under this clutter model assumption. In particular, the transformed CA-detector can be shown to be

$$Y_0 \underset{H_0}{\overset{H_1}{\gtrless}} \prod_{j=1}^{N} (Y_j + 1)^{\tau} - 1, \tag{5.26}$$

with τ set via inversion of (1.7), while the OS-equivalent is

$$Y_0 \underset{H_0}{\overset{H_1}{\gtrless}} \left(Y_{(k)} + 1\right)^{v} - 1, \tag{5.27}$$

with v set through (1.9).

Performance analysis is again with a Pfa of 10^{-4} and $N = 32$, with the order statistic index set to $k = N - 2 = 30$. Figure 5.11 shows the performance of the two decision rules (5.26) and (5.27) in simulated Lomax clutter, with shape parameter $\mu = 84.8173$. Performance is compared with an ideal detector, and also shows the result of two independent interfering targets in the clutter range profile. Both these targets have ICR of 20 dB. The curves show that the detectors work very well in the absence of interference, while the decision rule (5.26) suffers in the presence of increasing interference. Figure 5.12 repeats the same scenario when the interfering targets both have SCR of 80 dB. Here it is clear that (5.27) is better at managing interference than (5.26).

To complete the performance analysis of the decision rules (5.26) and (5.27), Figures 5.13 and 5.14 demonstrate false alarm regulation, with Figure 5.13 for the case of a 0.5 dB power increase and Figure 5.14 corresponding to a 5 dB power increase. Both these Lomax-based decision rules show reasonable management of the false alarm during power increases, with a greater deviation from the design Pfa resulting from a larger power increase.

In summary, it was found that the one-parameter Lomax model did not fit the Ingara clutter as well as a two parameter Pareto Type II model. However, in the case of horizontal polarisation, it could be used as an approximation. Simulated detector performance showed reasonable performance, especially for the CFAR detector based upon an order statistic. This was explored in (Weinberg, 2016b) and shown to improve on the Pareto based detectors. Additional issues with the transformed detectors in the Pareto setting can be illustrated through a closer examination of (5.10) and (5.11). As discussed previously, it has been found that

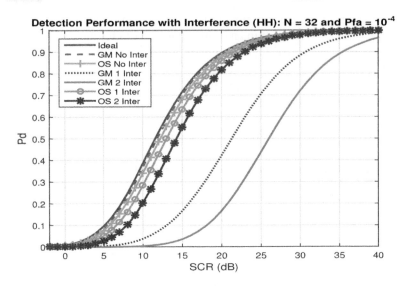

Figure 5.11: Detection performance under a Lomax clutter model assumption. In this simulation, one-parameter Lomax clutter model is simulated, with shape parameter matched to the Ingara fit for horizontal polarisation. The plot shows the two transformed detectors when subjected to up to two interfering targets, with ICR of 20 dB in each case.

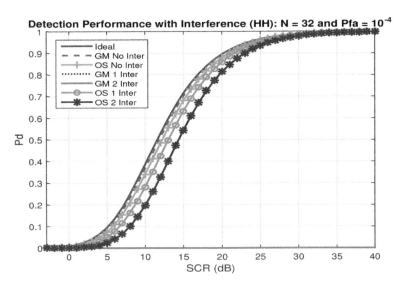

Figure 5.12: Second example of detector performance under a Lomax clutter model assumption. The scenario for Figure 5.11 is replicated with the exception that the ICR has been increased to 80 dB.

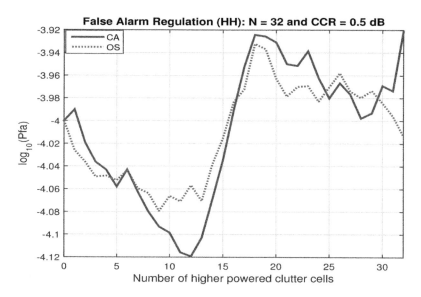

Figure 5.13: False alarm regulation in Lomax distributed clutter, coinciding with horizontal polarisation.

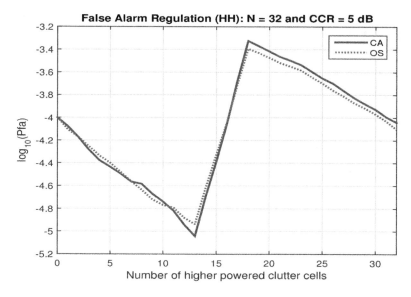

Figure 5.14: Second example of false alarm regulation in Lomax distributed clutter, also for horizontal polarisation but with a larger increase in the clutter power level than that in Figure 5.13.

when the Pareto model is fitted to the Ingara data, the estimate of the Pareto scale parameter was always less than unity. When β is very small, as in the case of horizontal polarisation, the coefficients $\beta^{1-N\tau}$ and $\beta^{1-\tau}$ have the potential to be quite large in practice. This will result in the right hand size of the decision rules (5.10) and (5.11) being larger than the CUT statistic, and so it is quite possible this will result in a significant decline in performance. Hence, given the solution to β-dependence postulated in (Weinberg, 2016b) is not sufficient, it is important to investigate whether the problem of scale parameter dependency in the Pareto-based decision rules can be addressed using a different approach.

5.6 Modification of the General Transformed Detector

The problems outlined in the previous sections can be rectified with the application of a complete sufficient statistic for β in the transformed decision rules. This has been documented in (Weinberg, 2017b), which is discussed in this section. As pointed out in Chapter 3 the minimum OS is a complete sufficient statistic for the Pareto scale parameter. Hence this section examines whether this can be used to produce decision rules without dependency on the Pareto scale parameter, while retaining the CFAR property. In order to facilitate the analysis, it turns out to be advantageous to examine the decision process (5.6) under the assumption of a slightly more general clutter model than the Pareto. Hence suppose that the clutter is modelled by the random variable

$$Y = \Phi_1 W^{\Phi_2},\tag{5.28}$$

where Φ_1 and Φ_2 are nonnegative constants and $W > 0$ is a random variable with no unknown parameters. Clearly the Pareto Type I model fits into such a class of random variables, as does the Weibull distributional model. The K-distribution, with a known shape parameter, can also be formulated in such a form. It is relatively straightforward to show that the distribution function of Y can be expressed in terms of that of W as

$$F_Y(t) = F_W\left(\left(\frac{t}{\Phi_1}\right)^{\frac{1}{\Phi_2}}\right).\tag{5.29}$$

Additionally, the inverse of (5.29) can be shown to be

$$F_Y^{-1}(t) = \Phi_1\left[F_W^{-1}(t)\right]^{\Phi_2}.\tag{5.30}$$

Hence the function (5.4) is

$$H(t) = \Phi_1\left[F_W^{-1}\left(1 - e^{-t}\right)\right]^{\Phi_2},\tag{5.31}$$

with inverse

$$H^{-1}(t) = -\log F_W^c\left(\left(\frac{t}{\Phi_1}\right)^{\frac{1}{\Phi_2}}\right),$$ (5.32)

where $F_W^c(t) = 1 - F_W(t)$ is the complementary distribution function of W. Both (5.31) and (5.32) can be applied to (5.6) to yield the detector

$$Y_0 \underset{H_0}{\overset{H_1}{\gtrless}} \Phi_1\left[F_W^{-1}\left(1 - e^{-\tau g(H^{-1}(Y_1),\dots,H^{-1}(Y_N))}\right)\right]^{\Phi_2}.$$ (5.33)

Based upon the clutter model defined via (5.28), it is clear that Φ_1 acts as a distribution scale parameter, and more importantly, it is essentially the minimum of the model in the case of Pareto Type I clutter. Hence one could examine the cost of applying a minimum order statistic as a replacement for Φ_1 in (5.33). This results in the alternative decision rule

$$Y_0 \underset{H_0}{\overset{H_1}{\gtrless}} Y_{(1)}\left[F_W^{-1}\left(1 - e^{-\tau g(K(Y_1),\dots,K(Y_N))}\right)\right]^{\Phi_2}$$ (5.34)

where

$$K(Y_j) = -\log F_W^c\left(\left(\frac{Y_j}{Y_{(1)}}\right)^{\frac{1}{\Phi_2}}\right)$$ (5.35)

and $Y_{(1)}$ is the minimum of the clutter range profile. Since Φ_1 and Φ_2 are non-negative and $W > 0$ it follows that $Y_{(1)} = \Phi_1 W_{(1)}^{\Phi_2}$ so that under H_0

$$\left(\frac{Y_j}{Y_{(1)}}\right)^{\frac{1}{\Phi_2}} = \frac{W_j}{W_{(1)}},$$ (5.36)

where $W_{(1)}$ is the minimum of the $W_j \overset{d}{=} W$. Hence the argument of F_W^{-1} in (5.34) is independent of both Φ_1 and Φ_2. Thus, by an application of (5.28), the Pfa of (5.34) is equivalent to

$$\begin{aligned}
\text{Pfa} &= \mathbf{P}\left(Y_0 > Y_{(1)}\left[F_W^{-1}\left(1 - e^{-\tau g(K(Y_1),\dots,K(Y_N))}\right)\right]^{\Phi_2} \middle| H_0\right) \\
&= \mathbf{P}\left(W > W_{(1)} F_W^{-1}\left(1 - e^{\tau g\left(\log F_W^c\left(\frac{W_1}{W_{(1)}}\right),\dots,\log F_W^c\left(\frac{W_N}{W_{(1)}}\right)\right)}\right)\right),
\end{aligned}$$ (5.37)

using the assumed scale invariance of g. Consequently (5.37) is independent of Φ_1 and Φ_2, which implies the detector (5.34) is certainly CFAR with respect to Φ_1. It is interesting to note that the choice of $X_{(1)}$ in (5.34) can be replaced by selection of another order statistic, and similar conclusion can be reached.

However, from a statistical analysis perspective, it is advantageous to condition with respect to the minimum, as will become apparent in the following section. If the decision rule (5.34) does not depend on Φ_2 then it will also be CFAR with respect to this parameter. To investigate this further requires specific choices for W, which will be considered next in the context of the Pareto Type I model.

5.7   Specialisation to the Pareto Clutter Case

When the above is specialised to the Pareto Type I case, the choices of $\Phi_1 = \beta$, $\Phi_2 = \alpha^{-1}$ are appropriate, together with $\log(W) \overset{d}{=} \mathrm{Exp}(1)$. Hence $F_W(t) = 1 - \frac{1}{t}$ and $F_W^{-1}(t) = \frac{1}{1-t}$. Observing that (5.35) reduces to $K(Y_j) = \frac{1}{\Phi_2}\log\frac{Y_j}{Y_{(1)}}$, applying these results to (5.34) yields the generic decision rule

$$Y_0 \underset{H_0}{\overset{H_1}{\gtrless}} Y_{(1)} e^{\tau g\left(\log\left(\frac{Y_1}{Y_{(1)}}\right),...,\log\left(\frac{Y_N}{Y_{(1)}}\right)\right)}, \tag{5.38}$$

showing that the detector does not depend on Φ_2, or equivalently α. Hence (5.38) will have the CFAR property with respect to both Pareto parameters. Additionally, this property is also independent of the choice of g, provided the latter possesses the assumed scale invariance property.

In view of Lemma 3.4 the minimum $Y_{(1)} \overset{d}{=} \mathrm{Pareto}(\alpha N, \beta)$, so that the Pareto-exponential duality property (3.21) implies that the ratio $\frac{Y_j}{Y_{(1)}}$ does not depend on β, for all j. Additionally, since Corollary 3.1 states that $Y_{(1)}$ is a complete sufficient statistic for β, it can be concluded that the minimum is independent of the second term on the right hand side of (5.38) (see (Beaumont, 1980), Section 2.6, page 36). Such a result parallels those in Lemmas 3.8 and 3.9. Suppose that X_0 and X_j, for $j \in \{1,2,\ldots,N\}$ are the associated exponential random variables defined through the Pareto-exponential duality property (3.21). Then the Pfa of (5.38) can be reduced to

$$\mathrm{Pfa} = \mathrm{P}(X_0 > X_{(1)} + \tau g(X_1 - X_{(1)},\ldots,X_N - X_{(1)}). \tag{5.39}$$

These considerations imply that the minimum is independent of each of the $X_j - X_{(1)}$ and so by conditioning on $X_{(1)}$ (5.39) becomes

$$\mathrm{Pfa} = \int_0^\infty Ne^{-Nt}\mathrm{P}(X_0 > t +$$

$$\tau g(X_1 - X_{(1)},\ldots,X_N - X_{(1)})|\{X_{(1)} = t\})dt, \tag{5.40}$$

using the fact that the minimum of a series of exponential random variables is also exponentially distributed but with parameter equal to the number of random variables over which the minimum is taken. Hence if one identifies a random variable $\Xi = g(X_1 - X_{(1)}, \ldots, X_N - X_{(1)}) | \{X_{(1)} = t\}$ with density f_Ξ then since this is independent of X_0, (5.40) reduces to

$$
\begin{aligned}
\text{Pfa} &= \int_0^\infty N e^{-Nt} \int_0^\infty f_\Xi(s) P(X_0 > t + \tau s) ds \, dt \\
&= \int_0^\infty N e^{-(N+1)t} dt \int_0^\infty e^{-\tau s} f_\Xi(s) ds \\
&= \frac{N}{N+1} M_\Xi(-\tau),
\end{aligned}
\tag{5.41}
$$

where M_Ξ is the moment generating function of Ξ. Result (5.41) is quite significant, in view of the fact that the moment generating function of Ξ, evaluated at $-\tau$ is essentially the Pfa of a detector, based upon the function g, but operating in exponentially distributed clutter, and conditioned on the minimum. To investigate this further, it is necessary to understand the behaviour of the exponentially distributed random variables involved in the definition of Ξ, together with their interaction with the function g. Two specific examples of g, complementing those on which the decision rules (5.10) and (5.11) are based, will be considered.

With an application of conditional probability, it can be shown that for each $j \in \{1, 2, \ldots, N\}$

$$
\begin{aligned}
P(X_j \leq x | X_{(1)} = t) \\
&= P(X_j \leq x | X_1 \geq t, \ldots, X_N \geq t) \\
&= P(X_j \leq x | X_j \geq t) = 1 - e^{-(x-t)},
\end{aligned}
\tag{5.42}
$$

provided $x \geq t$. Hence if one defines $\aleph_j = X_j | \{X_{(1)} = t\}$, then it follows that $\aleph_j - t \overset{d}{=} \text{Exp}(1)$. Since the X_j are independent, $\aleph_j$ will inherit this property.

In the situation where the function g is a sum of its arguments, it will follow from this that $M_\Xi(-\tau)$ will reduce to the Pfa (1.7), since the sum of a series of independent and identically exponential random variables is gamma distributed, with shape parameter N and scale parameter unity in this case. Hence the decision rule

$$
Y_0 \underset{H_0}{\overset{H_1}{\gtrless}} Y_{(1)}^{1-N\tau} \prod_{j=1}^N Y_j^\tau
\tag{5.43}
$$

has its Pfa given by

$$
\text{Pfa} = \frac{N}{N+1}(1+\tau)^{-N}.
\tag{5.44}
$$

The scenario where g is an order statistic requires more careful considerations. Suppose that $g(x_1,\ldots,x_N) = x_{(j)}$, the jth order statistic for some $1 < j \le N$. With this choice for g it is relatively simple to show that (5.38) reduces to

$$Y_0 \underset{H_0}{\overset{H_1}{\gtrless}} Y_{(1)}^{1-v} Y_{(j)}^{v}, \tag{5.45}$$

with the proviso that $j \neq 1$, so that (5.45) is meaningful. In this situation, the variable $\Xi = X_{(j)} - X_{(1)} | \{X_{(1)} = t\}$, and since the sequence $\{\aleph_j - t, j \in \{1, 2, \ldots, N\}\}$ defined above consists of independent and identically distributed exponential random variables, it is evident that Ξ will be the $(j-1)$th order statistic of a series of $N-1$ independent and identically distributed exponential random variables with unit mean. Hence with an application of (1.9) with such parameters, in conjunction with (5.41), the Pfa of (5.45) is

$$\text{Pfa} = \frac{N!}{(N+1)(N-j)!} \frac{\Gamma(v+N-j+1)}{\Gamma(v+N)}. \tag{5.46}$$

Consequently, based upon the analysis presented in this section, one can conclude that with the application of the complete sufficient statistic $Y_{(1)}$ as an approximation for β in the original decision rules (5.10) and (5.11), the cost in terms of the resultant Pfa is a multiplicative factor of $\frac{N}{N+1}$ applied to the original Pfa expression of the pre-transformed detector, operating in exponentially distributed clutter, but conditioned on the minimum. This provides a very convenient technique, to be coupled with the transformation approach, to produce CFAR detectors for operation in Pareto Type I clutter. Next the performance of (5.43) and (5.45) will be examined. As a final word on the development outlined here, it is worth observing that the same approach can be applied to the SO, GO and TM detectors, to produce CFAR variants of these partial CFAR processes.

As an example, one can consider the SO detector (5.13) and examine the variant

$$Y_0 \underset{H_0}{\overset{H_1}{\gtrless}} Y_{(1)}^{1-M\tau} \min\left(\prod_{j=1}^{M} Y_j, \prod_{j=M+1}^{N} Y_j\right)^{\tau}, \tag{5.47}$$

where the scale parameter has been replaced by the CRP minimum. It can be shown that the Pfa of this detector is given by

$$\text{Pfa} = \frac{2N}{N+1} \sum_{j=0}^{M-1} \binom{M+j-1}{j} (2+\tau)^{-(M+j)}, \tag{5.48}$$

demonstrating that the detector (5.47) attains the full CFAR property. The cost in achieving this is a multiplicative factor of $\frac{N}{N+1}$ as in the case of the GM-CFAR. Similar results can be obtained for the GO and TM detectors, which is left as an exercise for the interested reader.

5.8   Performance of the New CFAR Processes

To conclude the analysis presented in this chapter, it is of interest to investigate the performance of the new CFAR decision rules (5.43) and (5.45). These are compared with the detectors on which they are based, namely (5.10) and (5.11). In addition to this, the performance of the order statistic detector (4.8) is included for comparison purposes, as well as the upper bound provided by the ideal detector. Throughout the target in the CUT is modelled by a Swerling I fluctuation law.

5.8.1   Performance in Homogeneous Clutter

Analysis of the new decision rules begins with the case of homogeneous clutter, and Figure 5.15 is for the case where $N = 32$ and Pfa $= 10^{-4}$. The order statistic-based detectors utilise $k = N - 2$ as previously employed, and the figure is for the horizontally polarised case. The legend used here indicates which parameter is assumed known, so for example, the detector (5.10) is marked *GM with β known*. The new decision rules are marked as CFAR as shown. Here one observes that the new detectors (5.43) and (5.45) match very closely the performance of the detector on which they are based. This is excellent because they have the CFAR property with respect to both Pareto clutter parameters. The order statistic detector (4.8) has the best performance due to the fact it assumes *a priori* knowledge of the Pareto shape parameter.

It is of interest to consider the effect of reducing the size of the clutter range profile. Figure 5.16 repeats the same experiment as in Figure 5.15 with the exception that $N = 16$. There is a slight improvement in the performance of the new CFAR decision rule (5.43) relative to (5.10). To explore this further, Figure 5.17 further reduces N to 10. It appears that the decision rule (5.43) improves on (5.10) when N is smaller, while the situation is reversed for the case of (5.45).

Figure 5.18 examines the case where $N = 32$ but the Pfa is reduced to 10^{-6}, again in horizontally polarised homogeneous clutter. Here we observe that the new detector (5.43) has a slight gain on (5.10), while the order statistic variant (5.45) matches the performance of (5.11).

Next performance in the case of cross polarisation is considered, with Figure 5.19 for the scenario where $N = 32$ and Pfa $= 10^{-4}$, and Figure 5.20 corresponding to $N = 16$ with the same Pfa. These results show similar performance as for the horizontally polarised case, although the detector performance is closer to ideal in this scenario.

Finally, performance in vertically polarised clutter is investigated in Figures 5.21 and 5.22. These two figures parallel the scenario illustrated in Figures 5.19 and 5.20 respectively. Here one observes the performance of the decision rules becomes closer to that of the ideal detector.

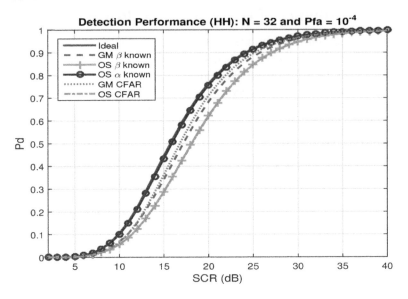

Figure 5.15: Performance of the two new CFAR detectors in the case of horizontal polarisation. These are compared with relevant alternatives, where the plot indicates the assumptions made for each alternative detector. Included is the ideal decision rule.

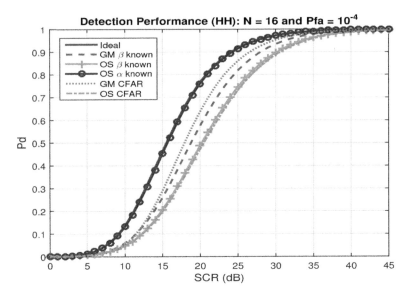

Figure 5.16: Second example of performance in homogeneous clutter, with a reduced CRP length.

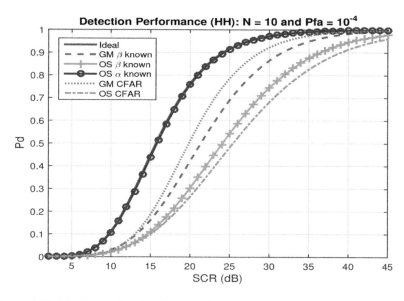

Figure 5.17: Third example of performance in horizontally polarised clutter, with another reduction in the size of the CRP.

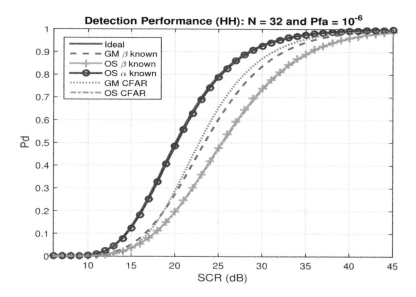

Figure 5.18: Final example of performance in horizontally polarised clutter. In this simulation the CRP has been increased to 32, while the Pfa has been reduced to 10^{-6}.

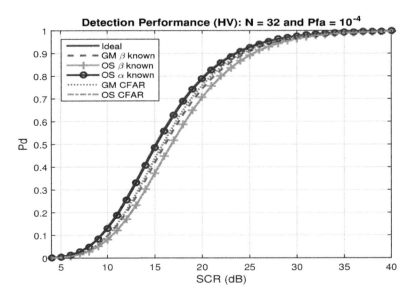

Figure 5.19: Detector performance in the cross polarisation case, with N increased to 32 and the Pfa set to 10^{-4}.

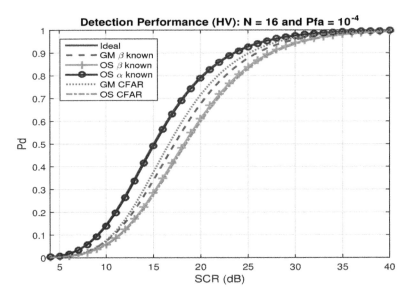

Figure 5.20: Second example of performance in cross polarised clutter, where N has been reduced to 16.

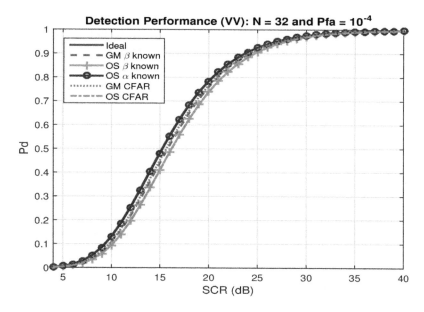

Figure 5.21: Detector performance comparison in the case of vertical polarisation.

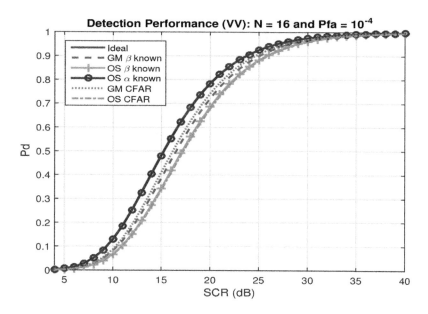

Figure 5.22: Second example of detector performance in vertically polarised clutter, where *N* has been reduced to 16.

These results are quite promising, since they show that the new CFAR detectors (5.43) and (5.45) perform very well in homogeneous clutter, with a smaller detection loss incurred as the polarisation is shifted from horizontal to vertical. It is especially interesting to further examine (5.43) since this decision rule not only has the capacity to improve on (5.10) for smaller N, but also has the CFAR property.

5.8.2   Performance in the Presence of Interference

Some examples of the effect of an interfering target are now considered, with Figures 5.23 and 5.24 for the case of horizontal polarisation. Both these figures are for the case where $N = 32$ and Pfa $= 10^{-4}$. All detectors, with the exception of the ideal, are subjected to one interfering target, where the latter is modelled by an independent Swerling I fluctuation. Figure 5.23 is for the case where the SCR of the interfering target is 20 dB, while it is increased to 80 dB in Figure 5.24. These figures show that each of the detectors, with the exception of (4.8), experience a degradation in performance. The decision rule (5.43) suffers more severely in the presence of strong interference.

In the case of cross and vertical polarisation, the performance is similar. Figure 5.25 is for the case of cross polarisation and (5.26) is for the vertically polarised case. Both scenarios are for $N = 32$ with Pfa $= 10^{-4}$ and with a 40 dB interfering target. The key observations is that the GM-based detectors suffer more in the presence of interference, while the detection loss of the OS-based detectors is reduced in vertical polarisation.

5.8.3   Clutter Transitions

False alarm regulation is now investigated, and Figures 5.27 and 5.28 demonstrate the effect on the resultant Pfa, in horizontal polarisation, for a 0.5 and 5 dB power increase respectively. In all cases to be examined, $N = 32$ and the design Pfa is 10^{-4}. These two figures show that the new OS CFAR detector performs relatively well in managing the Pfa. The detector (4.8), which requires *a priori* knowledge of α, manages to regulate the Pfa very well until the CUT is affected with higher power clutter. Thereafter it tends to increase the number of false alarms considerably, and more severely than the other decision rules. The GM-CFAR tends to increase the number of false alarms throughout the clutter power transitions, while the other OS-based detectors, and the original GM detector (5.10), are reasonably robust to the clutter power transition.

Next the cross polarisation case is examined, with Figure 5.29 for the case of a 0.5 dB power increase, and Figure 5.30 corresponding to a 2 dB power increase. The conclusions are similar to that for horizontal polarisation. Here a 2 dB power level increase was selected because the Monte Carlo estimators used in this case required more than 10^{6} runs to produce an accurate estimate of the Pfa.

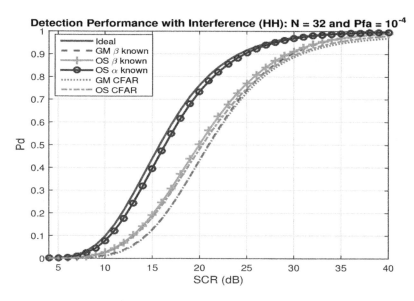

Figure 5.23: Interference in the horizontally polarised case, with a 20 dB interfering target. Only performance when subjected to the interference is shown.

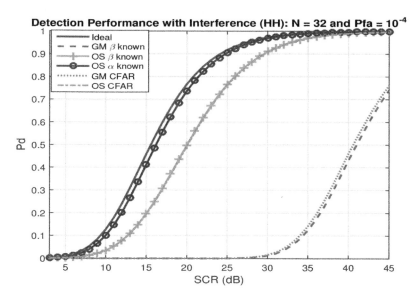

Figure 5.24: Second example of detector performance with interference, horizontal polarisation, with ICR of 80 dB. As in the previous plot the cases where the decision rules are not subjected to interference are not shown for brevity.

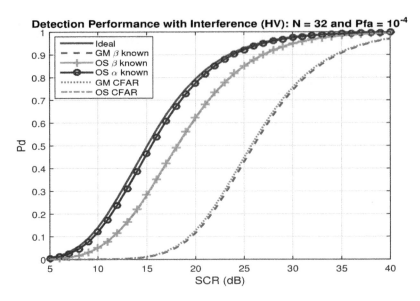

Figure 5.25: Cross polarisation and interference with a 40 dB interfering target.

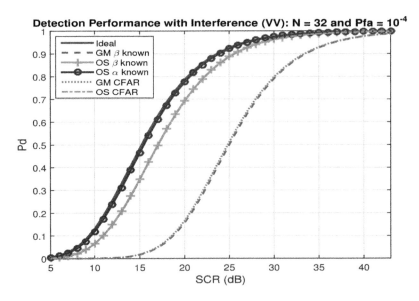

Figure 5.26: Another example of performance with interference, except in vertical polarisation. The ICR is again 40 dB.

Figures 5.31 and 5.32 examine the case of vertical polarisation, with Figure 5.31 for a CCR of 0.5 dB, and Figure 5.32 corresponding to a CCR of 1 dB.

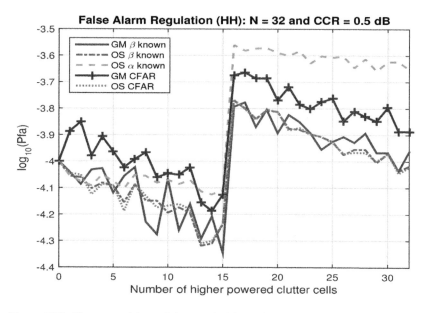

Figure 5.27: Clutter transitions of the new decision rules, together with the alternatives analysed previously, in horizontal polarisation.

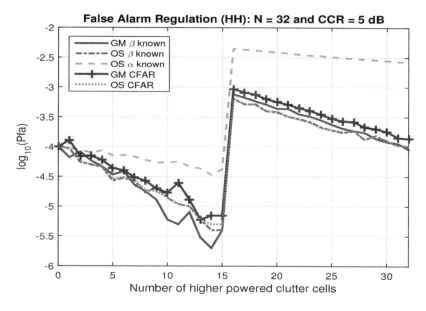

Figure 5.28: Second example of clutter transitions in horizontally polarised clutter.

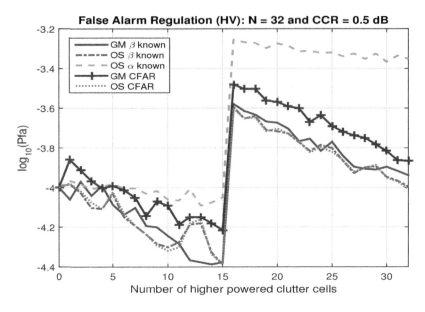

Figure 5.29: False alarm regulation in cross polarisation.

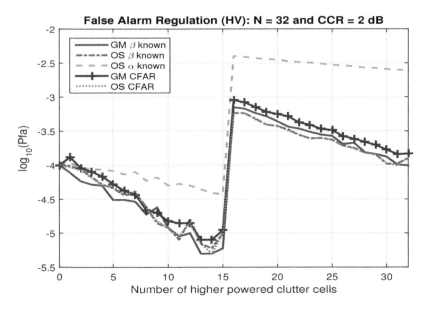

Figure 5.30: A second example of false alarm regulation in cross polarisation.

The conclusions are similar to the previous cases, with the exception of a larger deviation from the design Pfa for this polarisation.

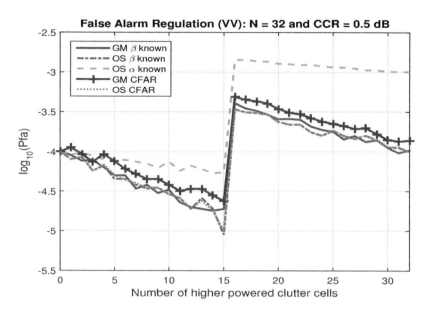

Figure 5.31: False alarm regulation in the vertically polarised case.

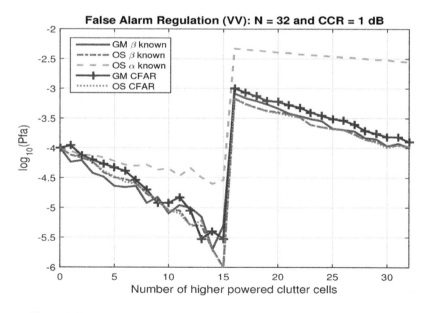

Figure 5.32: Second example of false alarm regulation in vertically polarised clutter.

5.9 Conclusions

This chapter overviewed the transformation approach for detector design, focusing on the evolution of results in the Pareto Type I clutter case. The original work of (Weinberg, 2013d), and its generalisation in (Weinberg, 2014b), showed that the CFAR detectors for operation in the exponentially distributed clutter case could be transformed to the Pareto realm, while retaining the original Pfa and threshold multiplier relationship. This approach resulted in the transformed detectors inheriting dependency on the Pareto scale parameter. The best solution to this issue is application of a complete sufficient statistic for this unknown parameter, with the choice of the minimum of the CRP being the most convenient solution. As a consequence of this the resultant decision rules acquired the CFAR property with respect to both Pareto clutter parameters. It was shown that the GM-based detector could improve significantly on the transformed detector on which it was based, for smaller N. As the latter increased, the new decision rules performed comparably to the variants upon which they are structured. The OS-based detectors had the greatest resilience to interfering targets, while during clutter transitions the OS-CFAR proved to be a more reliable solution.

Chapter 6

Modified Minimum Order Statistic Detector

6.1 Introduction

A minimum-based decision rule has the potential to manage interfering targets very well. Additionally, such a detection process has a strong capability to regulate the Pfa during clutter power transitions, especially in the event of power level increases. Hence the purpose of this chapter is to explore the minimum-based detector, for the Pareto Type I clutter case, in more detail. This analysis has been presented in (Weinberg, 2015a), where a modification of the original minimum detector in (Weinberg, 2013d) has been investigated. This has been, in turn, motivated by the investigation of coherent multilook detection processes in (Weinberg, 2013c), where a scale-type factor is introduced in order to enhance detection performance.

The minimum-based order statistic detector has been shown to perform very well in cases where its threshold multiplier is not too large. Saturation issues tend to occur when the Pfa is smaller than 10^{-2} and the size of the CRP exceeds 2. Detector saturation is a term used, in the context of sliding window decision rules, to describe the situation where the normalised measurement of clutter is so large that the CUT never exceeds it. This means that in simulated performance, the detector returns a probability of detection of zero for every SCR.

As will be observed, the modification of the minimum-based detector produces a decision rule that requires *a priori* knowledge of both Pareto clutter parameters. In such cases one could justify using the Pareto ideal decision rule. However, such a detector is unable to regulate the Pfa during clutter transitions.

Hence it is of interest to explore this modified minimum since it may be useful in some detection scenarios, such as when a clutter map is available to produce updated measurements of the clutter parameters. Hence throughout this chapter it will be assumed that one has *a priori* knowledge of the Pareto clutter parameters. The cost, in terms of the resultant Pfa, in cases where an approximation for these parameters is applied, is discussed in detail in (Weinberg, 2015a).

6.2 Transformed Order Statistic Detectors

The original minimum-based detector introduced in (Weinberg, 2013c) is given by

$$Z_0 \underset{H_0}{\overset{H_1}{\gtrless}} \beta^{1-\tau} Z_{(1)}^{\tau}, \tag{6.1}$$

where Z_0 is the CUT statistic, $Z_{(1)}$ is the minimum of the CRP $Z_1, Z_2, \ldots, Z_N$ and the threshold multiplier τ is given by

$$\tau = N(\text{Pfa}^{-1} - 1). \tag{6.2}$$

It is clear from (6.2) that in most situations of practical usefulness, the threshold multiplier will be quite large, resulting in (6.1) saturating. To illustrate this, suppose that Pfa $= 10^{-6}$ and $N = 16$, then $\tau = 15999984$. Similarly, if $N = 32$ and Pfa $= 10^{-4}$ then $\tau = 319968$. In practice, such a large threshold will result in no detections being made, and so the detector (6.1) will saturate. The situation is not improved even with a larger Pfa. For instance, if Pfa $= 10^{-3}$ and $N = 8$ are selected, then $\tau = 7992$, which can also result in the same detector issues. In order to rectify this problem, one can consider scaling the detector (6.1) using a fixed constant. For purposes of generality, one can examine the OS-detector (5.11) and consider the modification

$$Z_0 \underset{H_0}{\overset{H_1}{\gtrless}} \beta^{1-v} (M Z_{(k)})^v \tag{6.3}$$

for some $M > 0$ that is independent of both Pareto clutter parameters, for a given $1 \leq k \leq N$. This detector introduces a scale transformation, with the motivation being that for the case where $k = 1$, one is trying to produce a decision rule that is useful in practice. The following result shows how the threshold multiplier v is related to the Pfa:

Lemma 6.1
The Pfa of the transformed OS detector (6.3) is given by

$$\text{Pfa} = \frac{N!}{(N-k)!} M^{-\alpha v} \frac{\Gamma(N-k+v+1)}{\Gamma(N+v+1)}. \tag{6.4}$$

To prove this result, observe that, under H_0, the kth OS for the Pareto case can be written $Z_{(k)} = \beta W_k^{-\frac{1}{\alpha}}$, where W_k has a beta distribution with parameters $N-k+1$ and k respectively (Weinberg, 2013e). Suppose that the density of this is f_{W_k}. Then observing that under H_0 the CUT can be decomposed to $Z_0 = \beta e^{\alpha^{-1}X_0}$, where X_0 is exponentially distributed with unity mean, it follows that

$$
\begin{aligned}
\text{Pfa} \quad &= \quad \text{P}(e^{\alpha^{-1}X_0} > M^\nu W_k^{-\frac{\nu}{\alpha}}) \\[2mm]
&= \quad \int_0^1 f_{W_k}(t)\text{P}(X_0 > \alpha \log(M^\nu t^{-\frac{\nu}{\alpha}}))dt \\[2mm]
&= \quad M^{-\alpha\nu} \int_0^1 t^\nu f_{W_k}(t)dt,
\end{aligned}
\tag{6.5}
$$

where it has been assumed, without loss of generality, that $M > 1$. Applying the beta distribution density (2.96), with the appropriate parameters, to (6.5) it follows that

$$
\begin{aligned}
\text{Pfa} \quad &= \quad M^{-\alpha\nu} k \binom{N}{k} \int_0^1 t^{N-k+\nu}(1-t)^{k-1}dt \\[2mm]
&= \quad M^{-\alpha\nu} k \binom{N}{k} \mathcal{B}(N-k+\nu+1,k),
\end{aligned}
\tag{6.6}
$$

where $\mathcal{B}(\cdot,\cdot)$ is the beta function. The final result follows by expanding the latter function in terms of gamma functions.

As discussed previously the motivation in this chapter is to produce a modified version of the minimum based decision rule. Hence, one can consider the detector

$$
Z_0 \underset{H_0}{\overset{H_1}{\gtrless}} \beta^{1-\nu}(MZ_{(1)})^\nu
\tag{6.7}
$$

whose threshold is determined by inversion of

$$
\text{Pfa} = \frac{N}{N+\nu}M^{-\alpha\nu},
\tag{6.8}
$$

which results from setting $k=1$ in Lemma 6.1. Throughout the rest of the chapter this modified minimum detector (6.8) will be examined. The interested reader may apply a similar analysis to the general transformed OS detector (6.3).

6.3 Detector Comparison

It is now relevant to provide a mathematical investigation as to when the decision rule (6.7) provides improved detection performance relative to the standard minimum (6.1). In addition to this, it is also informative to apply the same mathematical analysis to compare (6.7) with the other decision rules introduced in Chapter

5. Since it is being assumed that one has *a priori* knowledge of the Pareto shape and scale parameters, the detectors (5.10) and (5.11) are only considered.

The key to providing a mathematical analysis of performance is to utilise a detector comparison lemma, developed in the context of coherent multilook detection, in (Weinberg, 2013b). This result can be adapted to provide a quantitative assessment of the performance of pairs of sliding window decision rules. Suppose that we have a pair of detectors $T_1 \underset{H_0}{\overset{H_1}{\gtrless}} \tau_1$ and $T_2 \underset{H_0}{\overset{H_1}{\gtrless}} \tau_2$, where T_1 and T_2 are functions of the CUT and CRP, and τ_1 and τ_2 are the corresponding threshold multipliers. It is assumed that both these detectors operate with identical probability of false alarm. Then if we define the statistic

$$\eta := \frac{\tau_1 \, T_2}{\tau_2 \, T_1} \tag{6.9}$$

then it is shown in (Weinberg, 2013b) that the detector based upon T_2 will have better performance than that based upon T_1 if and only if $\eta > 1$. This result can be used to determine environments or scenarios in which a particular decision rule will have better performance. The proof is included for completeness and is adapted from that in (Weinberg, 2013b).

Beginning with sufficiency, suppose that $\eta > 1$, and let Pd_1 and Pd_2 be the probabilities of detection for T_1 and T_2 respectively. Then

$$Pd_2 = P(T_2 > \tau_2 | H_1) = P(\eta T_1 > \tau_1 | H_1), \tag{6.10}$$

using the definition of η. Observe that if $T_1 > \tau_1$, then $\eta T_1 > \tau_1$, due to the assumption on η. Then it follows the event $\{T_1 > \tau_1\}$ will imply $\{\eta T_1 > \tau_1\}$ and so $P(T_1 > \tau_1 | H_1) \leq P(\eta T_1 > \tau_1)$, implying $Pd_1 \leq Pd_2$, using (6.10), completing the proof of sufficiency.

The simplest way in which necessity can be demonstrated is with a proof by contradiction. Towards this end, assume that $Pd_2 \geq Pd_1$ and suppose $\eta < 1$. Then $\eta T_1 < T_1$, and so the event $\{\eta T_1 > \tau_1\}$ will imply $\{T_1 > \tau_1\}$ and thus $P(\eta T_1 > \tau_1 | H_1) \leq P(T_1 > \tau_1 | H_1)$, which is equivalent to stating that $Pd_2 \leq Pd_1$, which is a contradiction. Hence it follows that $\eta > 1$, which completes the proof of necessity.

Note that the situation where $\eta = 1$ implies the two detection schemes are identical, and hence is ignored.

To see how the decision rules developed in this book can be expressed in a form suitable for analysis based upon (6.9), observe that the standard minimum detector (6.1) can be written

$$\log\left(\frac{Z_0}{\beta}\right) \underset{H_0}{\overset{H_1}{\gtrless}} \tau_1 \log\left(\frac{Z_{(1)}}{\beta}\right). \tag{6.11}$$

Hence we can select

$$T_1 := \log\left(\frac{Z_0}{\beta}\right) \bigg/ \log\left(\frac{Z_{(1)}}{\beta}\right), \tag{6.12}$$

observing that since $Z_{(1)} > \beta$ the logarithmic term is nonnegative and so the decision rule is $T_1 \underset{H_0}{\overset{H_1}{\gtrless}} \tau_1$. The next section provides some applications of the theory developed here to investigate when the new decision rule (6.7) provides detection gains.

6.4   Mathematical Analysis of Detectors

A mathematical analysis, based upon the detector comparison lemma of the previous section, is now provided, to determine whether there is any performance improvements achieved via (6.7). This is done by determining (6.9) for pairs of detectors, and then determining the probability that it exceeds unity. The reason probabilities are used to assess improvements is due to the fact that η is a random variable. If it can be shown that it is likely that $\eta > 1$ then it is plausible to conclude that the second decision rule improves on the first. For the first case, the detectors (6.1) and (6.7) are compared. Using a similar approach in the derivation of (6.12), one can select

$$T_2 = \log\left(\frac{Z_0}{\beta}\right) \Big/ \log\left(\frac{MZ_{(1)}}{\beta}\right). \tag{6.13}$$

For convenience the threshold multipler corresponding to (6.1) is τ_1, while that for (6.7) is τ_2. Recalling that the minimum of a series of Pareto distributed returns is also Pareto distributed, we can write $Z_{(1)} = \beta e^W$, where $W \overset{d}{=} \mathrm{Exp}(\alpha N)$, then it is not difficult to show

$$\eta = \frac{\tau_1}{\tau_2} \frac{W}{\log(M) + W}. \tag{6.14}$$

Since M is arbitrary, suppose that $M > 1$. Then for the same Pfa for the two detectors T_1 and T_2,

$$\frac{N}{N + \tau_1} = \frac{N}{N + \tau_2} M^{-\alpha\tau_2} < \frac{N}{N + \tau_2}, \tag{6.15}$$

since $\alpha\tau_2 > 0$. Therefore it follows from (6.15) that $\tau_2 < \tau_1$, resulting in a reduction in the size of the threshold multiplier, for the detector (6.7). As a result of this, one can determine the probability that the new minimum (6.7) provides an improvement on (6.1), for the same Pfa. Observe that

$$P(\eta > 1) = P\left(W > \frac{\tau_2}{\tau_1}[\log(M) + W]\right) = M^{-\frac{\alpha N \tau_2}{\tau_1 - \tau_2}}, \tag{6.16}$$

where the distribution function of W has been applied. Therefore (6.16) provides a statistical measure of improvement of the new detector (6.7) when compared to the standard minimum (6.1). Given τ_1 tends to be large for scenarios of interest, it therefore follows that this probability is likely to be very close to 1 in practice. To illustrate this, Table 6.1, which has been taken from (Weinberg, 2015a), computes (6.16) for a range of parameter values, and shows that decision rule (6.7) almost certainly always outperforms the standard minimum (6.1).

α	N	Pfa	$P(\eta > 1)$	τ_1	τ_2
3	6	10^{-2}	0.9516	594	0.3553
3	10	10^{-3}	0.9927	9990	0.5308
3	16	10^{-4}	0.9990	159984	0.7053
15	8	10^{-2}	0.9539	792	0.9539
15	12	10^{-3}	0.9930	11988	0.1014
15	16	10^{-6}	1.000	15999984	0.2027

Table 6.1: A comparison of the standard minimum (6.1) and the transformed version (6.7), where $M = 100$ throughout. The case $\alpha = 3$ corresponds to very spiky clutter, while $\alpha = 15$ is typical in the vertical polarisation case. These probabilities indicate (6.7) has better performance in general.

The results of the previous analysis are not really surprising, since the modified minimum is designed so that its threshold multiplier is not excessive, and the standard minimum will tend to saturate in scenarios of interest. Hence a somewhat fairer comparison involves analysis of the performance of (6.7) relative to the GM-detector (5.10). To this end, the GM decision rule can be written in the equivalent form

$$T_3 = \log\left(\frac{Z_0}{\beta}\right) \Bigg/ \log\left(\sum_{j=1}^{N}\left(\frac{Z_j}{\beta}\right)\right), \qquad (6.17)$$

and suppose its threshold multiplier is τ_3, which is given by inversion of (1.7). Using the fact that the sum of N independent exponentially distributed random variables with parameter α has a gamma distribution with parameters N and α, it follows that

$$\eta = \frac{\tau_3}{\tau_2} \frac{Q}{\log(M) + W}, \qquad (6.18)$$

where Q is the gamma distribution and $W \stackrel{d}{=} \text{Exp}(\alpha N)$ as before. Suppose that f is the density of Q, given by

$$f(t) = \frac{\alpha(\alpha t)^{N-1}e^{-\alpha t}}{\Gamma(N)}. \qquad (6.19)$$

Then

$$P(\eta > 1) \quad = \quad P\left(Q > \frac{\tau_2}{\tau_3}(\log(M) + W)\right)$$

$$= \quad \int_0^\infty f(t) P\left(W < \frac{\tau_3}{\tau_2}t - \log(M)\right) dt. \qquad (6.20)$$

Based upon (6.20), it is clear that the distribution function of W will be nonzero only when $t > \frac{\tau_2}{\tau_3}\log(M) := t^*$, and hence,

$$P(\eta > 1) \quad = \quad \int_{t^*}^\infty f(t)\left(1 - M^{\alpha N}e^{-\frac{\alpha N \tau_3}{\tau_2}t}\right) dt$$

$$= \quad 1 - F(t^*) - M^{\alpha N}\int_{t^*}^\infty e^{-\frac{\alpha N \tau_3}{\tau_2}t}f(t)dt$$

$$= \quad 1 - F(t^*) - \frac{M^{\alpha N}}{\Gamma(N)}\mathcal{G}_I\left(N, \alpha\left[\frac{\tau_2}{\tau_3} + N\right]\log(M)\right), \qquad (6.21)$$

where F is the distribution function of W, Γ is the gamma function and $\mathcal{G}_I(\cdot, \cdot)$ is the upper incomplete gamma function. This expression can be evaluated numerically; examples of it are provided in Table 6.2, which has been presented in (Weinberg, 2015a). The probabilities in Table 6.2 are not as close to 1 as those in Table 6.1, but show that the detector (6.7) can still provide performance improvements when $N \geq 16$ and Pfa $\leq 10^{-6}$.

α	N	Pfa	$P(\eta > 1)$	τ_3	τ_2
3	8	10^{-2}	0.7233	0.7783	0.1724
3	16	10^{-4}	0.8272	0.7783	0.3440
3	16	10^{-6}	0.9358	1.3714	0.5150
15	8	10^{-3}	0.8586	1.3714	0.0504
15	16	10^{-4}	0.8497	0.7783	0.0671
15	16	10^{-6}	0.9455	1.3714	0.1007

Table 6.2: Comparison of (6.7) and the GM detector, where $M = 10^4$ throughout. In cases of practical interest, where $N \geq 16$ and Pfa $\leq 10^{-6}$, the new detector (6.7) is very likely to introduce detection gains.

As a final comparison, the kth order statistic detector (5.11) is compared with (6.7). This detector can be expressed in the form

$$T_4 = \log\left(\frac{Z_0}{\beta}\right)\Big/\log\left(\frac{Z_{(k)}}{\beta}\right), \qquad (6.22)$$

and the corresponding threshold multiplier τ_4 is given by numerical inversion of (1.9), with ν replaced by τ_4. In order to derive the probability comparing (6.22) and (6.7), recall that Lemma 3.5 states that the kth OS for a Pareto distributed random sample is related to beta distributions. In particular, $Z_{(k)} = \beta W_k^{-1/\alpha}$, where W_k has a beta distribution with parameters $N - k + 1$ and k. If we let $g(t)$ be the density of this beta distribution, it can be shown as before that

$$P(\eta > 1) = \int_0^1 P\left(W_k < \frac{\tau_4}{\tau_2}\log\left(t^{-1/\alpha}\right) - \log(M)\right)g(t)dt. \qquad (6.23)$$

The distribution function of W_k in (6.23) requires the integral to be restricted to $t < M^{-\frac{\alpha\tau_2}{\tau_4}} := t^{**}$, and by the restriction placed on M, it can be demonstrated that $0 < t^{**} < 1$. Defining F_{W_k} to be this distribution function, it is not difficult to shown that the probability becomes

$$P(\eta > 1) = \int_0^{t^{**}} \left(1 - M^{\alpha N} t^{N\tau_4/\tau_2}\right)g(t)dt. \qquad (6.24)$$

The expression (6.24) can be further reduced to

$$P(\eta > 1) = 1 - F_{W_k}(t^{**}) - M^{\alpha N}k\binom{N}{k} \times$$

$$\mathcal{B}_I\left(t^{**}, N\left[1 + \frac{\tau_4}{\tau_2}\right] - k + 1, k\right), \qquad (6.25)$$

where $\mathcal{B}_I(\cdot, \cdot, \cdot)$ is the incomplete beta function. Table 6.3 shows some typical resultant probabilities, and it is clear that the new detector (6.7) generally has much better performance than the OS detector. This table first appeared in (Weinberg, 2015a).

α	N	Pfa	k	$P(\eta > 1)$	τ_4	τ_2
3	5	10^{-2}	3	0.8497	14.1895	0.2323
3	10	10^{-4}	5	0.9630	41.6927	0.4628
3	16	10^{-6}	15	0.9522	10.8951	0.6920
15	8	10^{-3}	4	0.9334	29.5194	0.0673
15	12	10^{-4}	10	0.9115	10.0348	0.0897
15	16	10^{-6}	9	0.9838	42.4307	0.1346

Table 6.3: Probabilities measuring the gain in using (6.7) relative to an OS detector with given k as shown. Here $M = 10^3$ throughout.

6.5   Selection of Parameter M

The previous section indicated that the choice of $M > 1$ is appropriate, but no further guidelines have been established for appropriate selection of M in practice. Hence, before examining detector performance in detail, an analysis of the scaling parameter M is undertaken. In the first instance, it is useful to consider the effect that M has on the threshold multiplier v through (6.8). Suppose that the probability of false alarm, clutter range profile length and Pareto clutter parameters are fixed. Then by viewing the threshold multiplier as a function of M, it is not difficult to show, using implicit differentiation, that its derivative is given by

$$\frac{dv}{dM} = \frac{-\alpha v}{M\left(\alpha\log(M) + (\mathrm{Pfa}/N)M^{\alpha v}\right)},\tag{6.26}$$

from which it follows that $v(M)$ is a decreasing function. Since it is bounded from below by zero, it follows by monotonic convergence (Rudin, 1987) that $v(M)$ converges as $M \to \infty$. In addition to this, it is clear from (6.26) that

$$\lim_{M\to\infty}\frac{dv}{dM} = 0,$$

implying that the threshold function $v(M)$ becomes flatter as $M \to \infty$. Hence it is plausible to select an M sufficiently large to attain the limiting threshold in practice.

To investigate this further, consider the probability of detection of scheme (6.7) with a general target model. Suppose that the signal plus clutter in the CUT has intensity distribution Z_0, with probability density function f_{Z_0}. In addition, suppose that Z_0 has support $[\gamma, \infty)$, where $\gamma > 0$ is independent of M. The probability of detection is

$$
\begin{aligned}
\mathrm{Pd} \quad &= \quad \mathbf{P}(Z_0 > \beta^{1-v}M^\tau Z_{(1)}^v | H_1)\\[2mm]
&= \quad \int_\gamma^\infty f_{Z_0}(t)\mathbf{P}\left(W < v^{-1}\log\left(\frac{t}{\beta M^v}\right)\right) dt\\[2mm]
&= \quad 1 - \left(\beta^{1/v}M\right)^{\alpha N}\int_\gamma^\infty t^{-\frac{\alpha N}{v}} f_{Z_0}(t)\,dt,\tag{6.27}
\end{aligned}
$$

where statistical conditioning has been applied, and W is an exponential random variable with parameter αN. An important observation from (6.27) is that it indicates in order to maximise the probability of detection, one must reduce the detection scaling factor $\left(\beta^{1/v}M\right)^{\alpha N}$ if possible. Lebesgue's Dominated Convergence Theorem (Rudin, 1987) demonstrates the integral in (6.27) remains finite as $M \to \infty$. For the DST Group's Ingara data, it has been found that $\beta < 1$, and so since v is decreasing as a function of M, its reciprocal is increasing. Thus the factor $\beta^{1/\nu}$ may decrease faster than M increases. Consequently, it is suitable to select an M sufficiently large to ensure the detection probability is maximised.

Based upon these observations, one can select $M \gg 1$ in practical implementation. Observe that in view of (6.8), since $\nu(M)$ converges,

$$\lim_{M \to \infty} M^{\nu(M)} = \left(\frac{N \text{Pfa}^{-1}}{N + \nu(\infty)} \right)^{1/\alpha}, \tag{6.28}$$

where $\nu(\infty)$ is the limit of $\nu(M)$. Hence, since the limit (6.28) exists, $M^{\nu(M)}$ is bounded for all M, and so the detector specified in (6.7) is well defined, for any $M \gg 1$.

6.6 Performance in Homogeneous Clutter

Analysis of the transformed minimum (6.7) is now undertaken, where the design Pfa is again set to 10^{-4} with CRP length of $N = 32$, and with clutter simulated with a Pareto Type I distribution whose parameters are matched to the same Ingara data set as before. The reader is referred to Table 3.1 for clarification of these parameters. The transformed minimum (6.7) is compared with an ideal decision rule as well as the GM-detector (5.10) and the OS detector (5.11), where the latter uses $k = N - 2 = 30$. For the transformed minimum, $M = 10$ has been selected and applied throughout. Additionally a Swerling I target model is used, both for the CUT as well as for interfering targets, although in the latter case this Swerling I target is generated independently.

Figures 6.1–6.3 show simulated performance in homogeneous clutter, for the cases of horizontal, cross- and vertical polarisation respectively. In these three cases the new decision rule matches the ideal detector almost exactly and clearly outperforms the GM- and OS detectors.

6.7 Examples of Management of Interference

To provide examples of the resilience of (6.7) to interfering targets, Figures 6.4–6.6 show the same scenarios as for Figures 6.1–6.3, except with the insertion of a 30 dB interfering target in the CRP. The decision rule (6.7) is completely unaffected by this, while the GM and OS detectors tend to experience a detection loss. Subjecting the detector (6.7) to further interfering targets results in very little variation to the results, until the number of interfering targets almost matches the size of the CRP N.

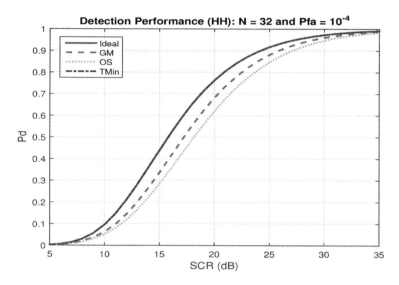

Figure 6.1: An example of the performance of the transformed minimum detector in horizontal polarisation. Its performance is gauged against that of the GM- and OS detectors, together with the ideal decision rule. The transformed minimum is indistinguishable from the ideal detector.

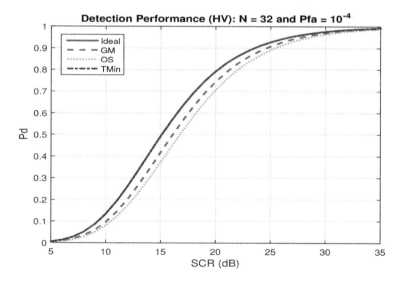

Figure 6.2: Performance analysis in cross-polarisation. In this experiment the transformed minimum again matches the performance of the ideal detector.

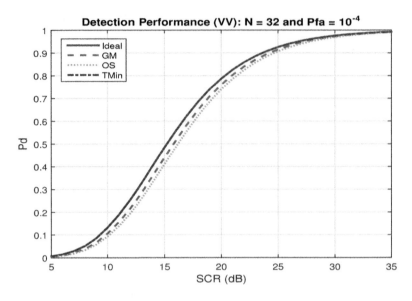

Figure 6.3: Performance of the transformed minimum in vertical polarisation. As for the two previous examples this decision rule matches the ideal detector in terms of performance.

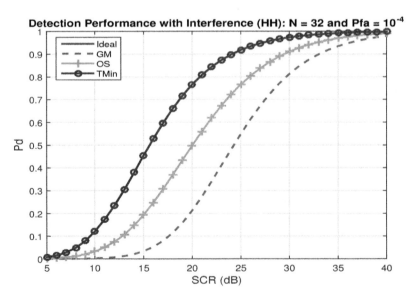

Figure 6.4: Effects of an interfering target in horizontal polarisation, with ICR of 30 dB. The transformed minimum shows no detection loss relative to the ideal decision rule, while the other two detectors experience a detection loss.

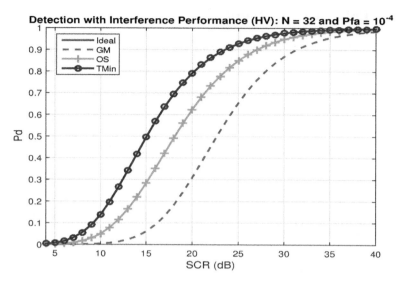

Figure 6.5: Replication of the scenario of Figure 6.4, except in cross-polarisation. The interfering target has ICR of 30 dB, and the transformed minimum experiences no discernable detection loss.

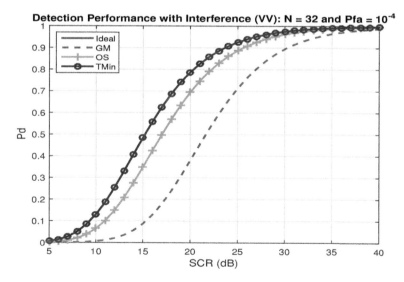

Figure 6.6: A repeat of the previous two numerical experiments, with the same level of interference with ICR of 30 dB, except in vertical polarisation. As in the two previous cases the transformed decision rule experiences no detection loss relative to the ideal detector.

6.8 False Alarm Regulation

False alarm regulation analysis of the new minimum (6.7) can be performed very simply since it is possible to derive a simple closed form expression for the resultant Pfa during clutter power transitions. This can be related to clutter power transitions for exponentially distributed clutter as follows. By an application of logarithms to the decision rule (6.7) the detector is equivalent to

$$\log(Z_0) \underset{H_0}{\overset{H_1}{\gtrless}} (1-v)\log(\beta) + v\log(M) + v\log(Z_{(1)}). \qquad (6.29)$$

Since the interest is in clutter power transitions, one can set $Z_0 = \beta e^{Y_0}$ and also $Z_{(1)} = \beta e^{Y_{(1)}}$, where Y_0 is exponentially distributed with parameter μ and $Y_{(1)}$ is the minimum of a series of exponentially distributed clutter returns. An application of these to (6.29) results in the simplification

$$Y_0 \underset{H_0}{\overset{H_1}{\gtrless}} v\left[\log(M) + Y_{(1)}\right]. \qquad (6.30)$$

The idea is that the parameter μ is either α or the corresponding Pareto shape parameter resulting from a given clutter power level increase, and $Y_{(1)}$ is the minimum of two sets of exponentially distributed random variables, coinciding with the partition of the statistics in the CRP into lower and higher powered returns. If we suppose that there are H higher powered clutter returns and $N - H$ lower powered returns, then the random variable $Y_{(1)}$ is the minimum of the minimum of the two sets of variables in the partition of the CRP. If it is assumed that the Pareto shape parameter for higher powered returns is α_I then it is not difficult to show that $Y_{(1)} \overset{d}{=} \mathrm{Exp}(H\alpha_I + (N-H)\alpha)$, under the assumption that the CRP statistics are independent. Thus the Pfa during clutter transitions is given by

$$
\begin{aligned}
\mathrm{Pfa} \;=\;& \mathbf{P}(Y_0 > v\left[\log(M) + Y_{(1)}\right]) \\[2mm]
=\;& \int_0^\infty f_{Y_{(1)}}(w)\mathbf{P}(Y_0 > v\left[\log(M) + w\right])dw \\[2mm]
=\;& M^{-\mu v}\int_0^\infty (H\alpha_I + (N-H)\alpha)e^{-(H\alpha_I + (N-H)\alpha)w - v\mu w}dw \\[2mm]
=\;& \frac{H\alpha_I + (N-H)\alpha}{H\alpha_I + (N-H)\alpha + v\mu}M^{-\mu v}, \qquad (6.31)
\end{aligned}
$$

where $f_{Y_{(1)}}$ is the density of $Y_{(1)}$. Hence (6.31) provides a closed form expression for the resultant Pfa during clutter power increases, where $\mu = \alpha$ until the CUT is considered affected by higher power returns, resulting in $\mu = \alpha_I$. To illustrate (6.31) Figures 6.7–6.9 plot the resultant Pfa during clutter power transitions for the simulation scenario adopted throughout. Figure 6.7 is for horizontal polarisation and with a clutter power level increase of 0.5 dB. It is clear the decision rule (6.7) manages to maintain the design Pfa very well until the CUT is affected, resulting in an increase in the number of false alarms. However, the resultant Pfa is increased to $10^{-3.5}$, which is not too extreme an increase. Figure 6.8 shows the same situation except with a 2 dB power increase. Once the CUT is affected, the resultant Pfa is roughly $10^{-2.7}$; the latter Pfa further increases as the CCR increases.

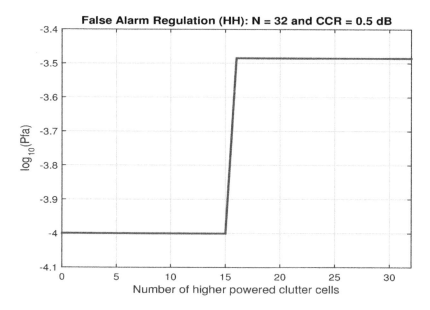

Figure 6.7: False alarm regulation of the transformed minimum in horizontal polarisation.

In the case of cross and vertical polarisation the situation is repeated, except the false alarm regulation after the CUT is affected worsens as the polarisation is changed from the horizontal case. Figure 6.9 provides an illustration of this for the vertically polarised case, with a 0.5 dB power increase. Clearly once the CUT is affected there is a substantial increase in the estimated Pfa.

Observe that one can apply (6.8) to (6.31) to express the design Pfa in terms of the resultant Pfa; if we let Pfa_D and Pfa_R be these two respectively, then

$$\text{Pfa}_R = \frac{H\alpha_I + (N-H)\alpha}{H\alpha_I + (N-H)\alpha + \nu\mu} \left(\frac{N}{\nu} + 1\right)^{\frac{\nu}{\alpha}} \text{Pfa}_D^{\frac{\nu}{\alpha}}. \tag{6.32}$$

Based upon (6.32) it is clear that the larger the deviation ν has from α, the more severe the increase in the resultant Pfa.

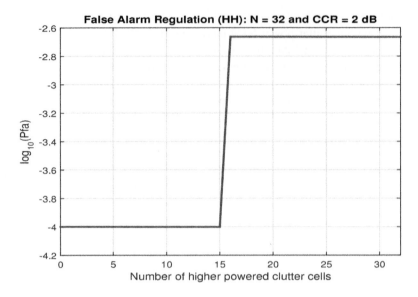

Figure 6.8: A second example of false alarm regulation of the transformed minimum in horizontal polarisation.

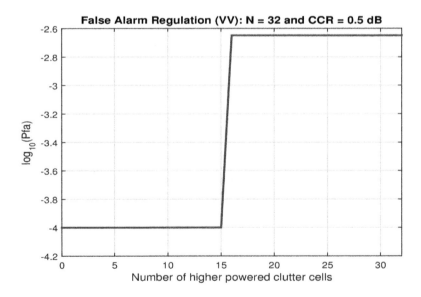

Figure 6.9: Example of false alarm regulation of the transformed minimum in vertical polarisation.

6.9   Conclusions

A modification of the standard minimum detector was shown to produce a new decision rule whose performance in homogeneous clutter matched that of an ideal decision rule. This detector required *a priori* knowledge of the Pareto clutter parameters. It was shown, using a detector comparison lemma, that the level of improvement could be quantified mathematically. Numerical experiments demonstrated the success of such a decision rule in managing interfering targets and regulating the Pfa during clutter transitions.

Chapter 7

Dual Order Statistic CFAR Detectors

7.1 Introduction

Dual order statistic detection processes have appeared within the signal processing literature in several different clutter model contexts. The idea of such a detection process has been to extend Rohling's orginal OS-CFAR process to allow two order statistics of the CRP to provide a measurement of the clutter level. Such an approach, in the case of clutter modelled by a Pareto Type I distribution, has been examined in (Weinberg and Alexopoulos, 2016), and so this is overviewed in this chapter. This complements the development of CFAR decision rules via complete sufficient statistics in Chapter 5, since the CFAR detector (5.45) is a special case of a dual order statistic CFAR process.

7.2 Motivation and Definition of Detection Process

The application of pairs of order statistics as a measurement of the level of clutter was first examined in (Weber and Haykin, 1985), in the context of Weibull distributed clutter. In the latter a heuristic motivation was provided to justify the approach. A modified version, in a setting relevant to the focus of this book, is now examined. Suppose that a series of clutter measurements of the form $Z_j = \Phi_1 W_j^{\Phi_2}$ are available, where the constants Φ_1 and Φ_2 are nonnegative and unknown distributional parameters, and for each j the nonnegative random variable W_j has a known statistical structure. By this it is being assumed that these random variables are distributed according to some known model, where there are no un-known model parameters. Since the function $g(t) = \Phi_1 t^{\Phi_2}$ is increasing on the interval $[0, \infty)$ it follows that each order statistic is given by $Z_{(j)} = \Phi_1 W_{(j)}^{\Phi_2}$. To illustrate this Figure 7.1 plots an example of this function together with a series of order statistics. Suppose that two such order statistics are available, namely $Z_{(i)}$ and $Z_{(j)}$ for indices i and j, and let v_1 and v_2 be two constants. Then it is immediate that

$$Z_{(i)}^{v_1} Z_{(j)}^{v_2} = \Phi_1^{v_1 + v_2} \left(W_{(i)}^{v_1} W_{(j)}^{v_2} \right)^{\Phi_2}. \tag{7.1}$$

Recall that in a sliding window detection process, if Z_0 is the CUT, then under H_0 this statistic will have the same distribution as each of the Z_j. Hence

$$Z_0 | H_0 \overset{d}{=} \Phi_1 W_0^{\Phi_2}, \tag{7.2}$$

where W_0 has the same distribution as each of the W_j. Therefore, if one is to de-sign a detection process based upon (7.1), then it follows that in view of (7.2), for the corresponding Pfa expression to be independent of Φ_1 and Φ_2 it is neces-sary to select v_1 and v_2 so that their sum is unity. To illustrate this, suppose one selects $v_2 = \tau$ then with the choice of $v_1 = 1 - \tau$, the corresponding decision rule is given by

$$Z_0 \underset{H_0}{\overset{H_1}{\gtrless}} Z_{(i)}^{1-\tau} Z_{(j)}^{\tau}. \tag{7.3}$$

It can be assumed that $i < j$ without loss of generality. The detector (7.3) is known as a dual order statistic, and it is clear that the case where $i = 1$ is equivalent to the CFAR detector (5.45).

A decision rule of the form (7.3) was first considered in (Weber and Haykin, 1985), who showed that in the case of Weibull clutter returns it has the CFAR property with respect to both Weibull distributional parameters. This provided a significant advance on Rohling's original OS-detection process, since when this is applied directly to the Weibull setting, as in the approach in Chapter 4, the resultant detector requires *a priori* knowledge of the Weibull shape parameter. However, the detector (7.3) is reported to experience a larger detection loss than a single OS variant, which can be interpreted as the cost in achieving CFAR. A second analysis of the detector (7.3) can be found in (Levanon and Shor, 1990), who also examine it in Weibull distributed clutter. In addition to this, it is shown that a variant of (7.3) can be constructed which uses more than two order statistics to measure the level of clutter. The major finding in (Levanon and Shor, 1990) was that the latter provided no significant gains on a dual order statistic detector.

A third example of analysis of (7.3) can be found in (Al-Hussaini, 1988), who examined (7.3) in log normal distributed clutter, showing that it also achieves the CFAR property.

The discussion above explains why (7.3) achieves the CFAR property in a number of clutter model assumptions. The following Lemma clarifies this:

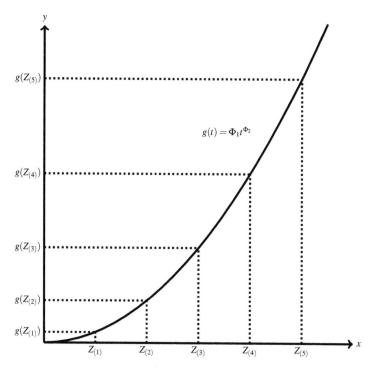

Figure 7.1: An illustration of the fact that for increasing functions of the form $g(t) = \Phi_1 t^{\Phi_2}$ processed order statistics form a series of increasing functions.

Lemma 7.1

The detector (7.3) has the CFAR property if the independent and identically distributed clutter returns can be represented by random variables of the form $Z_k = \Phi_1 W_k^{\Phi_2}$, where Φ_1 and Φ_2 are (possibly) unknown clutter parameters, and the random variables W_k have no unknown parameters ($k \in \{1, 2, \ldots, N\}$). Additionally it is understood that $Z_0 | H_0 \stackrel{d}{=} \Phi_1 W_0^{\Phi_2}$, where $W_0 \stackrel{d}{=} W_k$.

The proof of this lemma has been outlined above, but is included below for completeness. Let $Z_0 = \Phi_1 W_0^{\Phi_2}$ and observe that the OS can be written $Y_{(i)} = \Phi_1 W_{(i)}^{\Phi_2}$ and $Y_{(j)} = \Phi_1 W_{(j)}^{\Phi_2}$. Then the Pfa of (7.3) is

$$
\begin{aligned}
\text{Pfa} &= P\left(\Phi_1 W_0^{\Phi_2} > \Phi_1 W_{(i)}^{(1-\tau)\Phi_2} W_{(j)}^{\tau\Phi_2}\right) \\
&= P\left(\log(W_0) > (1-\tau)\log(W_{(i)}) + \tau \log(W_{(j)})\right),
\end{aligned}
$$

$$(7.4)$$

and since the probability (7.4) does not depend on Φ_1 and Φ_2, it follows that the detector has the claimed CFAR property, completing the proof.

Observe that Lemma 7.1 applies trivially to the case of exponentially distributed clutter, although it appears that the detector (7.3) has never been explored in such a clutter model environment. Recall that if $Z \stackrel{d}{=} \text{Weib}(k, \lambda)$ then $Z = \lambda W^{\frac{1}{k}}$, where $W \stackrel{d}{=} \text{Exp}(1)$, demonstrating the reason (7.3) is also CFAR in Weibull distributed clutter. Since if Z has a log normal distribution, $\log(Z) \stackrel{d}{=} \mathcal{N}(\mu, \sigma^2)$ so that $Z = e^{\mu} \left(e^X \right)^{\sigma}$, where $X \stackrel{d}{=} \mathcal{N}(0, 1)$, also explaining the CFAR property of (7.3) in log normal clutter. Finally, an appeal to the Pareto-exponential duality property (3.21) establishes the fact that (7.3) is CFAR in Pareto Type I clutter.

The detector (7.3) has a generalisation in terms of an arbitrary number of order statistics from its clutter range profile, as discussed in (Levanon and Shor, 1990). To see this, suppose that $S \subset \{1, 2, \ldots, N\}$ and consider the detector

$$Z_0 \underset{H_0}{\overset{H_1}{\gtrless}} Z_{(i)} \prod_{(k,l) \subset S \times S} \left(\frac{Z_{(l)}}{Z_{(k)}} \right)^{\tau}, \tag{7.5}$$

where $1 \leq i \leq N$. The dual order statistic detector (7.3) arises as a special case of (7.5) with the selection of $S = (i, j)$. The following demonstrates that it is also CFAR:

Lemma 7.2
Under the conditions of Lemma 7.1 the generalised order statistic detector (7.5) has the CFAR property with respect to Φ_1 and Φ_2.

The proof is similar to that of the proof of Lemma 7.1, and so is omitted. The key component in the proof is showing that the Pfa of (7.5) reduces to

$$\text{Pfa} = \text{P}\left(W_0 > W_{(i)} \prod_{(k,l) \subset S \times S} \left(\frac{W_{(l)}}{W_{(k)}} \right)^{\tau} \right). \tag{7.6}$$

The reason that (7.5) works as an extension of the dual order statistic CFAR (7.3) can be understood from its Pfa expression (7.6). With the constraints imposed on the clutter model, one order statistic term (which is $Z_{(i)}$ in the formulation) is necessary to eliminate Φ_1 dependence that arises from the CUT under H_0. It is then necessary to have ratios of order statistics, since these will not depend on Φ_1 and in virtue of the structure of the clutter model, will be raised to the power of Φ_2. The fact that the CUT and the initial order statistic $Z_{(i)}$ are also raised to this power will result in Φ_2 being eliminated from the entire expression.

Since the focus in this book is on detection processes operating in Pareto Type I clutter, the following sections focus on (7.3) in such clutter. It is expected that the multiple order statistic version (7.5) may not provide significant benefits over (7.3) in Pareto distributed clutter. The exploriation of this will be left to the interested reader.

7.3   Specialisation to the Pareto Type I Case

Here the focus is specialised to (7.3) operating in Pareto Type I distributed clutter. Instead of deriving a closed form expression for the Pfa of (7.3) immediately, it is useful to derive a general expression for the Pd, which is encapsulated in the following result:

Lemma 7.3
The detector (7.3), operating in independent Pareto Type I distributed clutter, has probability of detection given by

$$\text{Pd} \; = \; \frac{N!}{(i-1)!(j-i-1)!(N-j)!} \times$$

$$\int_{\phi=0}^{1} \int_{\psi=0}^{1} [1-\phi]^{i-1} \phi^{N-i} F_{Z_0}^c \left(\beta \phi^{-1/\alpha} \psi^{-\tau/\alpha} \right) \times$$

$$[1-\psi]^{j-i-1} \psi^{N-j} d\phi d\psi, \tag{7.7}$$

where $F_{Z_0}^c(t) = \mathbf{P}(Z_0 \geq t)$ is the complementary cumulative distribution function of the CUT under H_1.

In order to prove Lemma 7.3, the Pareto-exponential duality property (3.21) can be applied to the two order statistics so that $Z_{(i)} = \beta e^{W_{(i)}}$ and $Z_{(j)} = \beta e^{W_{(j)}}$ where $W_{(.)}$ is the corresponding order statistic for a series of exponentially distributed random variables. Then, when applied to the expression for the Pd, one obtains

$$\text{Pd} = \mathbf{P} \left(\log \left(\frac{Z_0}{\beta} \right) > (1-\tau) W_{(i)} + \tau W_{(j)} \Big| H_1 \right). \tag{7.8}$$

Now the joint density of the order statistics $W_{(i)}$ and $W_{(j)}$, for a sample of independent and identically distributed exponential random variables with mean $1/\alpha$ can be determined from (2.98), and shown to be

$$f_{W_{(i)}, W_{(j)}}(x,y) \; = \; \frac{N!}{(i-1)!(j-i-1)!(N-j)!} \times$$

$$\alpha^2 e^{-\alpha x - \alpha y} (1 - e^{-\alpha x})^{i-1} \times$$

$$(e^{-\alpha x} - e^{-\alpha y})^{j-i-1} e^{-\alpha y (N-j)}. $$

$$\tag{7.9}$$

Hence by conditioning on this joint distribution, the probability of detection becomes

$$
\text{Pd} = \int\int_{\mathcal{D}} f_{W_{(i)},W_{(j)}}(x,y) \times
$$

$$
P\left(\log\left(\frac{Z_0}{\beta}\right) > (1-\tau)W_{(i)} + \tau W_{(j)} \,\Big|\,
$$

$$
W_{(i)} = x, W_{(j)} = y, H_1 \right) dxdy, \tag{7.10}
$$

where $\mathcal{D} \subset \mathbf{R}^+ \times \mathbf{R}^+$ is the support of the joint random variables. By observing that the order statistics satisfy the condition that $W_{(i)} < W_{(j)}$, it can be shown that (7.10) reduces to

$$
\text{Pd} = \frac{N!}{(i-1)!(j-i-1)!(N-j)!} \times
$$

$$
\int_{x=0}^{\infty}\int_{y=x}^{\infty} P\left(Z_0 > \beta e^{(1-\tau)x+\tau y} \right) \times
$$

$$
\alpha^2 e^{-\alpha(x+y)} \left[1 - e^{-\alpha x} \right]^{i-1}
$$

$$
\left[e^{-\alpha x} - e^{-\alpha y} \right]^{j-i-1} e^{-\alpha y(N-j)} dydx, \tag{7.11}
$$

where (7.9) has been applied. Observe that by a transformation of $\theta = e^{-\alpha y}$, the corresponding integral in (7.11), with x fixed, becomes

$$
I: = \alpha e^{-\alpha x} \left[1 - e^{-\alpha x} \right]^{i-1} \times
$$

$$
\int_{\theta=0}^{\theta=e^{-\alpha x}} P\left(Z_0 > \beta e^{(1-\tau)x}\theta^{-\frac{\tau}{\alpha}} \right) \left[e^{-\alpha x} - \theta \right] \theta^{N-j} d\theta. \tag{7.12}
$$

With a second transformation of $\phi = e^{-\alpha x}$ applied to (7.11), together with (7.12), the probability of detection reduces to

$$
\text{Pd} = \frac{N!}{(i-1)!(j-i-1)!(N-j)!} \times
$$

$$
\int_{\phi=0}^{\phi=1}\int_{\theta=0}^{\theta=\phi} P\left(Z_0 > \beta\phi^{-\left(\frac{1-\tau}{\alpha}\right)}\theta^{-\frac{\tau}{\alpha}} \right) \times
$$

$$
[1-\phi]^{i-1} [\phi - \theta]^{j-i-1} \theta^{N-j} d\theta d\phi. \tag{7.13}
$$

Finally, a change of variables of $\psi = \frac{\theta}{\phi}$ within the second integral in (7.13) reduces the latter to (7.7) after simplification, completing the proof of the lemma.

Expression (7.8) provides an interesting representation for the probability of detection of (7.3) operating in Pareto Type I distributed clutter, in terms of the CUT statistic's complementary distribution function. Based upon this one can derive the Pfa as a corollary:

Corollary 7.1
The probability of false alarm of (7.3), operating in Pareto Type I distributed clutter, is given by

$$\text{Pfa} = \frac{(N-i+1)!}{(N+1)(N-j)!} \frac{\Gamma(\tau+N-j+1)}{\Gamma(\tau+N-i+1)}. \tag{7.14}$$

The proof of Corollary 7.1 follows by noting under H_0 the CUT is Pareto distributed and so its distribution function can be substituted into the expression in Lemma 7.3 to show that $F_{Z_0}^c \left(\beta\phi^{-1/\alpha}\psi^{-\tau/\alpha} \right) = \phi\psi^\tau$, and the final result follows by simplification.

Corollary 7.1 confirms the fact that the detection process (7.3) is CFAR with respect to both the Pareto shape and scale parameters. It is also interesting to note that if $i = 0$ in (7.14), then it reduces to that of the original transformed order statistic detector, namely (1.9). Hence one can interpret the order statistic $Z_{(0)}$ as β for the Pareto Type I model.

In order to apply the dual order statistic detection process (7.3) in practice, it is necessary to determine appropriate choices for the order statistic indices i and j. To facilitate this, the following lemma is a useful tool:

Lemma 7.4
For fixed j, N and Pfa, the threshold multiplier τ, given by inversion of (7.14), is an increasing discrete function of $i \in \{1, \ldots, j-1\}$, and additionally $\tau \geq 1$ for all i.

To prove Lemma 7.4 define a function $g_i(\tau)$ to be the right hand side of the expression in (7.14), so that the threshold τ is determined via $g_i(\tau) = \text{Pfa}$. Then it is not difficult to show that

$$\frac{g_{i+1}(\tau)}{g_i(\tau)} = 1 + \frac{\tau-1}{N-i+1}, \tag{7.15}$$

where properties of the gamma function have been applied. Note that when $\tau > 1$ the right hand side of (7.15) exceeds unity, so that $g_{i+1}(\tau) > g_i(\tau)$, implying for fixed $\tau > 1$, $g_i(\tau)$ is an increasing function of i. Additionally, at $\tau = 1$, it is evident from (7.15) that $g_{i+1}(1) = g_i(1) = 1 - \frac{j}{N+1}$, where the latter results from evaluation of the right hand side of (7.14) at $\tau = 1$.

By the definition of the gamma function it can be shown that since $i < j$,

$$\Gamma(\tau+N-i+1) = \Gamma(\tau+N-j) \prod_{k=i}^{j} (\tau+N-k), \tag{7.16}$$

which when applied to the definition of $g_i(\tau)$ yields

$$g_i(\tau) = \frac{(N-i+1)!}{(N+1)(N-j)!} \prod_{k=i}^{j-1} \frac{1}{\tau+N-k}, \qquad (7.17)$$

from which we conclude that for fixed i, $g_i(\tau)$ is a decreasing function of τ.

Based upon these observations it follows that since thresholds are determined via the solution to $g_i(\tau) = $ Pfa, the sequence of thresholds must be increasing with i. Noting that $1 - \frac{j}{N+1} > \frac{1}{N+1} > $ Pfa when $N < $ Pfa$^{-1} - 1$, it follows also that these thresholds must exceed unity in scenarios of interest. To illustrate this, Figure 7.2 plots a series of $g_i(\tau)$ in the case where $N = 32$ and $j = 29$. It is evident from the figure that the thresholds are increasing and exceed unity.

Another interesting result is that it turns out that the probability of detection of (7.3) is a decreasing discrete function of the first index i:

Lemma 7.5
For fixed j, N and Pfa, the detection probability

$$\mathrm{Pd}_i = \mathbf{P}\left(Z_0 > Z_{(i)}^{1-\tau_i} Z_{(j)}^{\tau_i} \,\Big|\, H_1 \right) \qquad (7.18)$$

is a decreasing discrete function of i, for $i \in \{1, 2, \ldots, j-1\}$, where τ_i is given by inversion of (7.14).

The proof of Lemma 7.5 now follows. Note that the probability of detection can be written in the form

$$\mathrm{Pd}_i = \mathbf{P}\left(Z_0 > Z_{(i)} \left[\frac{Z_{(j)}}{Z_{(i)}}\right]^{\tau_i} \,\Big|\, H_1 \right) \qquad (7.19)$$

Observe that by definition of order statistics that $Z_{(i+1)} > Z_{(i)}$ and $Z_{(j)} > Z_{(i)}$, where the latter is valid for all $i < j$. Hence the ratio $\frac{Z_{(j)}}{Z_{(i)}} > 1$, and consequently the sequence $\{a_i, i \in \{1, 2, \ldots, j-1\}\}$ defined by

$$a_i = \left[\frac{Z_{(j)}}{Z_{(i)}}\right]^{\tau_i} \qquad (7.20)$$

is increasing, due to the fact that τ_i is an increasing sequence, based upon Lemma 7.4. Thus it follows that $Z_{(i+1)}a_{i+1} > Z_{(i)}a_i$ and so the event $\{Z_0 > Z_{(i+1)}a_{i+1}\}$ will imply $\{Z_0 > Z_{(i)}a_i\}$ and consequently

$$\mathrm{Pd}_{i+1} = \mathbf{P}\left(Z_0 > Z_{(i+1)}^{1-\tau_{i+1}} Z_{(j)}^{\tau_{i+1}} \,\Big|\, H_1 \right) \leq \mathbf{P}\left(Z_0 > Z_{(i)}^{1-\tau_i} Z_{(j)}^{\tau_i} \,\Big|\, H_1 \right) = \mathrm{Pd}_i, \quad (7.21)$$

from which it follows that $\{\mathrm{Pd}_i, i \in \{1, 2, \ldots, j-1\}\}$ is a decreasing sequence, as required.

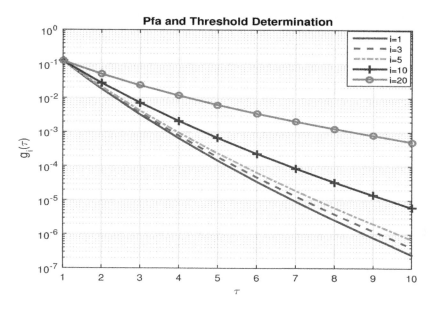

Figure 7.2: Thresholds can be determined by examining which τ satisfies $g_i(\tau) = \text{Pfa}$. These plots show that $g_i(\tau)$ is increasing with i, while decreasing with τ.

Hence, in homogeneous clutter, one expects that the best performance of (7.3) will be achieved with $i = 1$. Typically, it is useful to select the upper order statistic index j to account for a number of expected interfering targets, as in previous scenarios where the decision rule is based upon an order statistic. Thus, for example, in the case where $N = 32$, one could select $j = 30$ in an attempt to filter out up to two interfering targets with large signal to clutter ratios. In view of Lemma 7.4, the fact that the threshold multiplier is increasing with the lower order statistic index i and greater than unity, suggests that detector saturation could occur if i is selected too large. Furthermore, if i is selected to be too small, then interfering targets with smaller SCR may have a detrimental effect on performance. These issues need to be considered by the radar practitioner with the application of (7.3). Next the performance of the dual order statistic detector is analysed.

7.4 Performance in Homogeneous Clutter

As in previous chapters performance analysis begins with the homogeneous clutter case. For the numerical analysis, the two CFAR decision rules (5.43) and (5.45) are included for comparison purposes, as well as the ideal detector. Since (5.45) coincides with (7.3) in the case where the latter utilises $i = 1$, only the cases where $i > 1$ are considered for (7.3). To provide an analysis consistent with previous chapters, the Pfa is set to 10^{-4} while the length of the CRP is $N = 32$, while the upper order statistic is set to $j = N - 2 = 30$ for all decision rules. Pareto clutter parameters are again based upon the estimates in Table 3.1, and the figures indicate the underlying polarisation mode employed to acquire the parameter estimates.

Figure 7.3 plots performance of the decision rules in homogeneous horizontally polarised clutter, while Figure 7.4 shows an enlargement to clarify the close band of decision rules. The GM CFAR (5.43) has the best performance, followed by (5.45). As the index i is increased in the DOS detector, the performance decreases as shown. The tight cluster of performance curves coincide with the cases where $i \in \{1, 2, 3, 4, 5\}$. It was found that once i exceeded around 15 the performance degraded sharply, as illustrated, and expected based upon the theoretical analysis.

Figures 7.5 and 7.6 repeat the same scenario, except in cross polarisation. The same results are repeated, with the exception that performance has improved in comparison to the horizontally polarised case.

Figures 7.7 and 7.8 show the performance of (7.3) in vertically polarised clutter, and the conclusions are similar to as before. Comparing performance across polarisations one concludes that as the polarisation is changed from horizontal to vertical, the performance improves as shown. Next analysis of (7.3) in the presence of interference is examined.

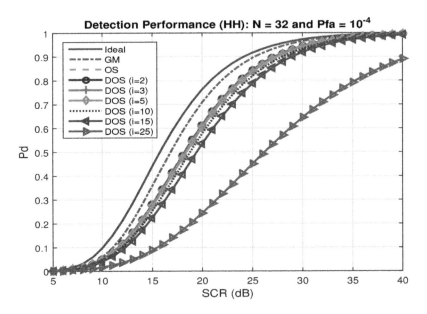

Figure 7.3: Performance of several dual order statistic CFARs, compared with the GM and OS-CFARs, in the case of homogeneous clutter in horizontal polarisation.

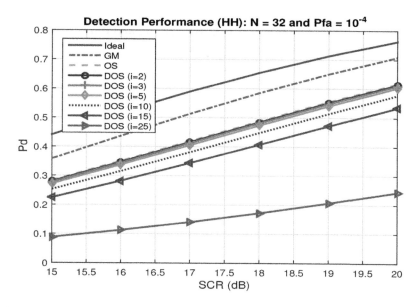

Figure 7.4: Enlargement of Figure 7.3, showing that when $i \in \{1,2,3,4,5\}$ the detector (7.3) tends to have very little variation.

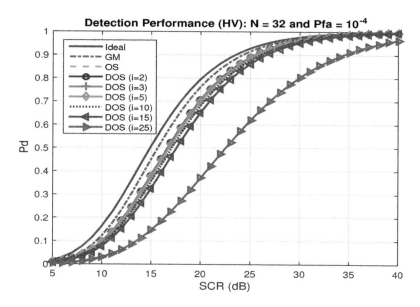

Figure 7.5: Detector performance comparisons in cross-polarisation, for the same set of decision rules examined in the horizontally polarised case.

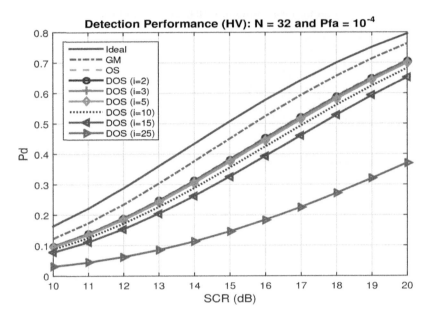

Figure 7.6: Magnification of Figure 7.5 to clarify performance.

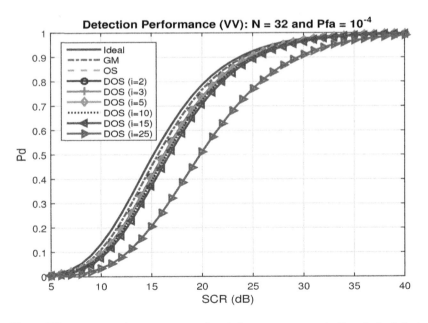

Figure 7.7: Detector performance comparison in homogeneous vertically polarised clutter.

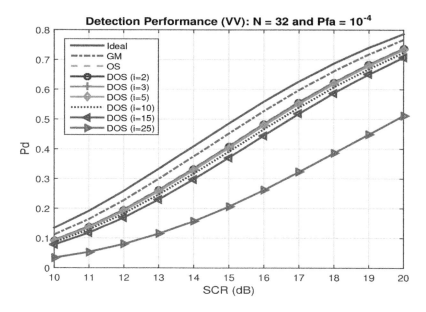

Figure 7.8: Snapshot of part of the plot of Figure 7.7.

7.5 Management of Interfering Targets

As in the preceding chapters the effect of an interfering target in the CRP is examined. This interfering target is modelled by an independent Swerling I fluctuation with a fixed SCR, as in previous chapters. Figure 7.9 shows the effect that a 20 dB interfering target has on the detection processes, while Figure 7.10 provides an enlargement to clarify the behaviour of the decision processes that are closely banded together. This is for the case of horizontal polarisation. The figures show that the GM CFAR (5.43) has degraded in performance considerably, while the DOS detectors with $i \in \{2, 3, 5\}$ tend to perform comparably to (5.45).

Figure 7.11 shows the consequence of doubling the interference SCR to 40 dB, while Figure 7.12 is for the case of an 80 dB interfering target. Again, these simulations are in the case of horizontal polarisation. The main conclusion is that (5.43) does not manage higher levels of interference very well, while (5.45) and the DOS CFAR (7.3) do this well, with the proviso that i is not to large for (7.3).

Next the case of interference in cross polarisation is examined, and Figure 7.13 shows the result of a 20 dB interfering target, with Figure 7.14 providing a magnification. Here is it interesting to observe that there is a slight gain in applying the DOS detector (7.3) with $i \in \{2, 3, 5\}$ over the decision rule (5.45). As in the case of horizontal polarisation, the detector (5.43) experiences a significant performance degradation.

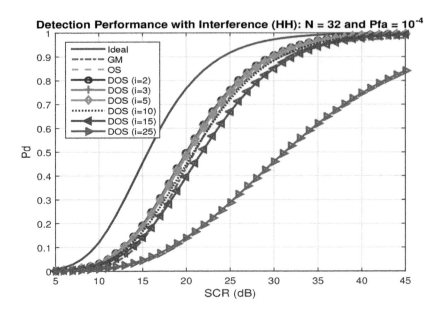

Figure 7.9: Performance with interference in horizontal polarisation with an independent Swerling I target model with ICR of 20 dB.

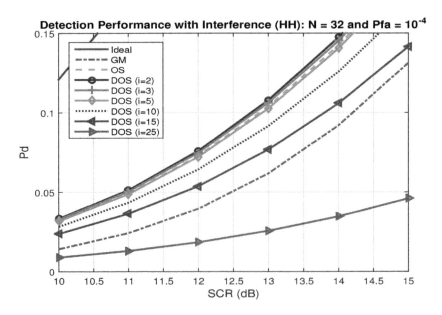

Figure 7.10: Closer examination of the results in Figure 7.9.

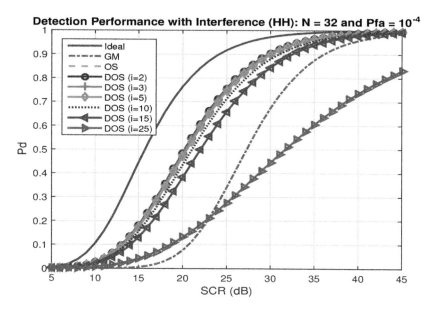

Figure 7.11: Second example of interference in horizontal polarisation, with ICR of 40 dB.

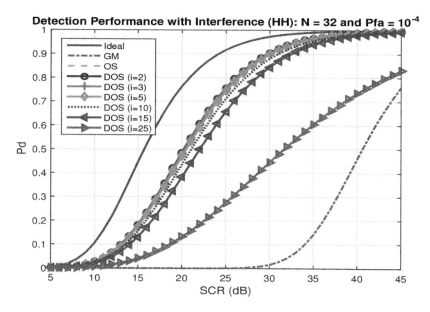

Figure 7.12: Third example of interference in the horizontally polarised case. In this scenario the ICR is 80 dB.

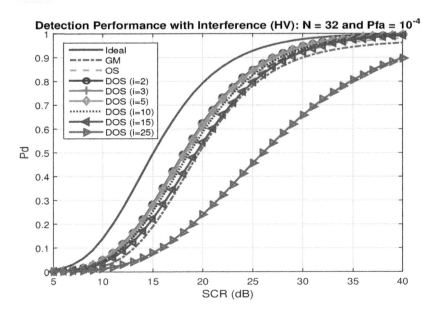

Figure 7.13: Detection performance in the present of interference in the cross polarisation case, with ICR of 20 dB.

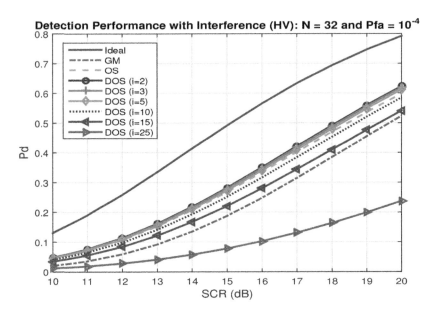

Figure 7.14: A closer look at Figure 7.13.

As a final example for this polarisation, Figure 7.15 shows the effect of increasing the interference SCR to 60 dB. As expected, the GM CFAR (5.43) has very poor performance, while the OS-CFAR (5.45) and the DOS CFAR (7.3) perform reasonably well. As before the DOS detector requires $i < 10$ to have reasonable performance.

The case of vertical polarisation is considered next, with Figures 7.16 and 7.17 for the case of a 40 dB interfering target, while Figure 7.18 corresponding to 60 dB interference. The conclusions are similar to those reached before for the other two polarisations, with the difference being slightly better performance in this polarisation, in general.

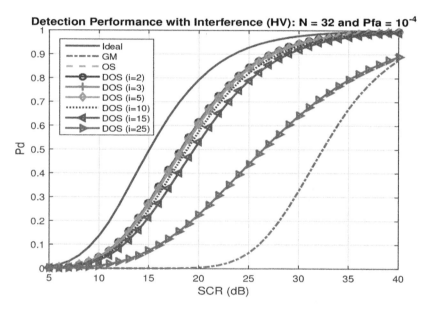

Figure 7.15: Second example of detector performance in cross polarisation with interference, where the ICR is 60 dB.

7.6 False Alarm Regulation

The final stage of performance analysis of the DOS CFAR involves analysis of its capability to manage variations in clutter, via an analysis of its ability to manage robustly its resultant Pfa. Here this is not only analysed by considering the effect of slowly increasing the clutter power level in the clutter range profile, but the effect on the resultant Pfa is analysed when there is an interfering target also present. Any sliding window detection process will always be affected by variations in the underlying clutter power, and the hope is that the effects of

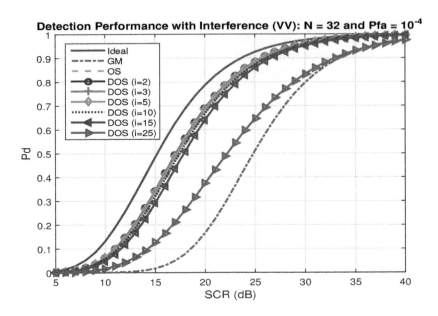

Figure 7.16: Performance in vertical polarisation with interference, where the latter is provided via an independent Swerling I target with ICR of 40 dB.

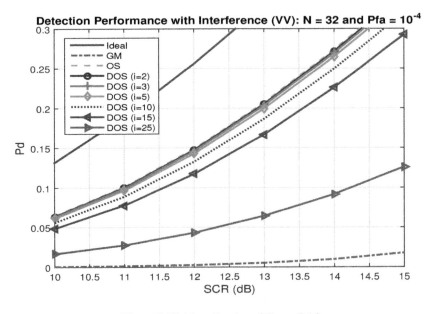

Figure 7.17: Magnification of Figure 7.16.

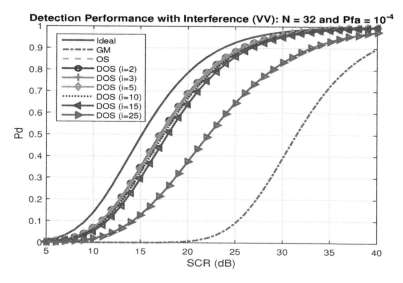

Figure 7.18: Final example of detection performance with interference, for vertical polarisation, with ICR of 60 dB.

an interfering target will be mitigated by the underlying function used to measure the clutter level. Given the detectors (5.43), (5.45) and (7.3) all achieve the CFAR property with respect to Pareto Type I clutter, this analysis is now deemed appropriate. This interfering target is generated as an independent Swerling I fluctuation model with a fixed SCR relative to the lower powered clutter returns.

Figure 7.19 shows the regulation of the Pfa in horizontal polarisation. The DOS CFAR (7.3) is shown for $i \in \{2, 3, 25\}$, with the performance of (5.43) and (5.45) also shown. In this scenario the clutter power is increased by 0.5 dB. One observes that the GM-CFAR has a tendency to increase the Pfa, while the OS-CFAR and the DOS-CFARs tend to manage the Pfa similarly. It is interesting to note that for $i = 25$ the DOS-CFAR tends to manage the Pfa slightly better, although this could be attributed to variation due to the Monte Carlo sample size (10^6 runs in this case).

Figure 7.20 shows the same situation with an interfering target inserted into the lower powered clutter. The latter has SCR of 30 dB, representing strong interference. The SCR in such cases is known as the interference to clutter ratio (ICR). The GM-CFAR tends to reduce significantly its resultant Pfa, while the other decision rules are also affected but to a lesser degree as shown.

Figure 7.21 shows the behaviour when the clutter power is increased by 2 dB, for the same scenario as for Figure 7.19. The situation is similar as in the latter figure, with the GM-CFAR tending to increase the resultant Pfa. Figure 7.22 includes a 15 dB interfering target. Here the same phenomenon is observed as before.

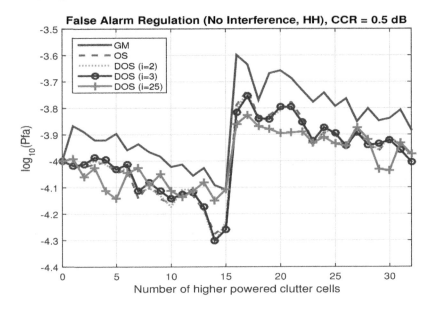

Figure 7.19: False alarm regulation in horizontally polarised clutter.

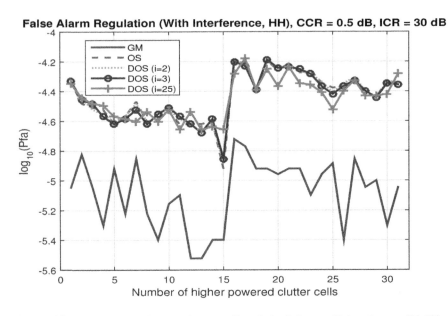

Figure 7.20: False alarm regulation in horizontally polarised clutter with interference (30 dB).

Next the cross polarisation case is investigated, with Figure 7.23 for the case of a 0.5 dB power increase, and Figure 7.24 showing the effects of a 30 dB inter-

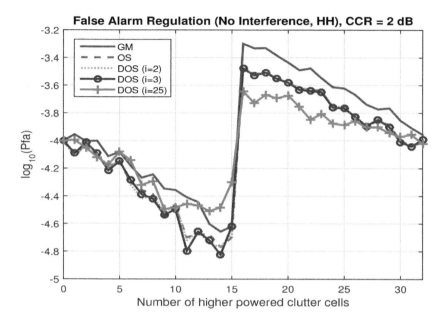

Figure 7.21: Second example of false alarm regulation in horizontally polarised clutter.

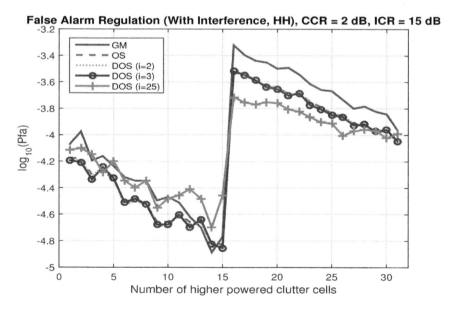

Figure 7.22: The scenario of Figure 7.21 but with interference, with ICR of 15 dB.

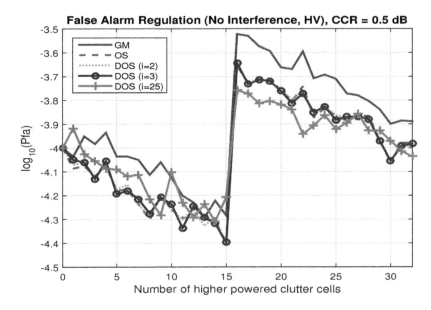

Figure 7.23: False alarm regulation in cross polarisation.

fering target inserted into the lower powered clutter. The performance illustrated in Figure 7.23 is consistent with that in the horizontal polarisation case, although the behaviour shown in Figure 7.24 demonstrates that each of the detection processes manages the Pfa in much the same way, although the GM-CFAR appears to increase the resultant Pfa more significantly.

Figures 7.25 and 7.26 are for the case of a 2 dB power increase, while the latter figure includes a 30 dB interfering target. Again the performance is consistent with previous results, although there is a sharper deviation away from the design Pfa for this polarisation.

The final set of figures correspond to vertical polarisation, with Figure 7.27 for a 0.5 dB power increase, Figure 7.28 for the same level of power increase with the addition of a 15 dB interfering target, Figure 7.29 for a 1 dB power increase and Figure 7.30 including a 5 dB interfering target. These results are consistent with those for the previous polarisations.

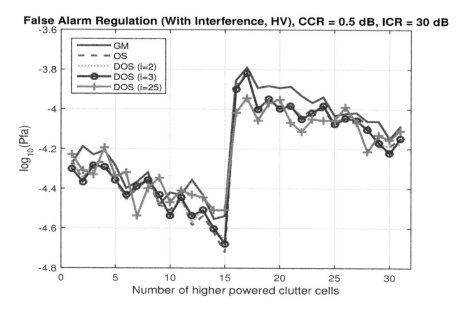

Figure 7.24: False alarm regulation, with interference, in cross-polarisation. In this situation the ICR is 30 dB.

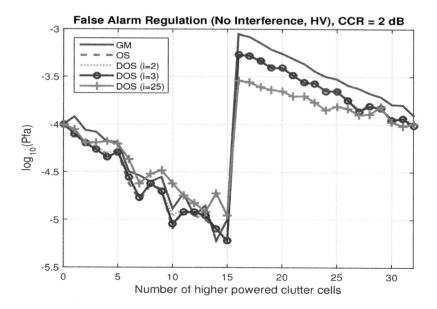

Figure 7.25: Second example of false alarm regulation in cross polarisation.

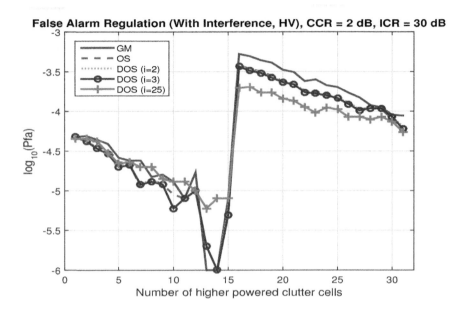

Figure 7.26: Another example of false alarm regulation with interference in cross-polarisation (30 dB).

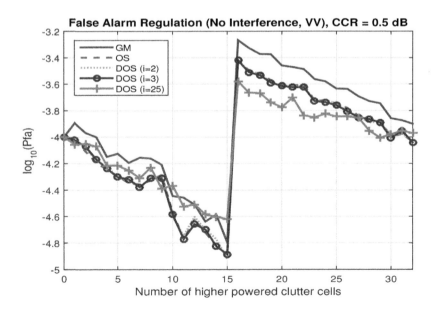

Figure 7.27: False alarm regulation in vertical polarisation.

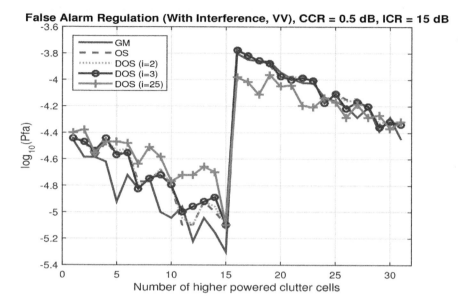

Figure 7.28: The same scenario as in Figure 7.27 but with interference (15 dB).

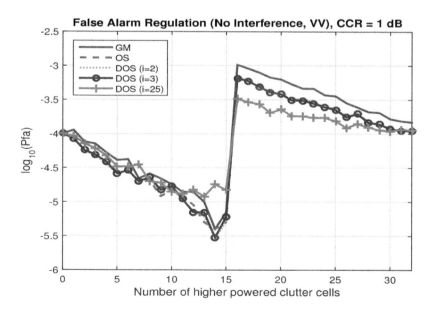

Figure 7.29: Second example of false alarm regulation in vertically polarised clutter.

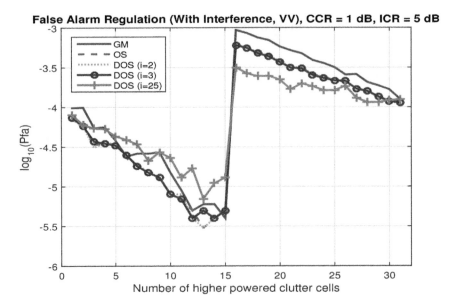

Figure 7.30: 5 dB interference added to the scenario of Figure 7.29.

7.7 Conclusions

This chapter investigated a dual order statistic detector that has been of interest to signal processing researchers for many years. It was demonstrated that this detector has the CFAR property with respect to a large class of clutter models. Additionally it was shown that it had a natural generalisation to a set of order statistics of the clutter range profile. When specialised to the Pareto Type I case it was proven that its probability of detection could be expressed elegantly in terms of beta-type functions, which then could be used to extract its probability of false alarm in terms of its threshold multiplier. The latter was demonstrated to increase with the first index in the dual order statistic detector.

From an application perspective it was demonstrated that one could select the first and last order statistics to manage interference, noting that as the first index increased, the probability of detection tended to reduce significantly. During clutter transitions the dual order statistic was shown to be robust, as well as able to manage an interfering target. However, the conclusion reached from this analysis is that the OS-CFAR (5.45), which is a special case of the DOS CFAR, is the best option for managing interference in a sliding window detection process.

Chapter 8

On Goldstein's Log-t Detector

8.1 Introduction

This chapter examines a well-known detection process in radar signal processing known as the log-t detector. This was introduced into the signal processing literature by Goldstein in 1973 (Goldstein, 1973). Motivated by analysis of decision rules for target detection in log normal clutter, Goldstein formulated a detector which was shown to possess an invariance property with respect to certain classes of clutter models. Goldstein then analysed this detector for application to log normal and Weibull clutter environments. A deficiency in Goldstein's original work was that the decision rule was not examined in terms of its performance relative to interfering targets appearing in the clutter range profile. Additionally, its performance during clutter transitions was also neglected. The focus in (Goldstein, 1973) was to consider the problem of threshold approximation, since in the Weibull case it is very difficult to determine the distribution of the cell under test statistic under the null hypothesis.

Several studies followed subsequently where the log-t detector was analysed in terms of its detection performance, and where its threshold multiplier was determined via simulation. Its performance in Weibull distributed clutter can be found in (Sekine and Mao, 2009), (Pourmottaghi et al., 2012) and (Detouche and Laroussi, 2012), while a brief analysis of it in K-distributed clutter can be found in (Jakubiak, 1982). The latter demonstrates that the log-t detector loses the CFAR property in K-distributed clutter. This fact explains the lack of application of it to X-band maritime surveillance radar in the literature.

It is interesting to observe that the Pareto Type I model fits into the class of distributions originally postulated by Goldstein to result in the log-t detector possessing the CFAR property. Hence it is of relevance to re-examine this decision rule to determine whether it provides any gain for target detection in X-band maritime surveillance radar. This topic has been examined in (Weinberg and Glenny, 2017) and the results of this analysis are reported in this chapter. This includes a new version of the log-t detector based upon order statistics.

8.2   The Log-t Detector

The original formulation of the log-t detector is based upon considerations of envelope detectors whose inputs are passed through an ideal log-amplifier. The reason this is important in this context is that it explains the distribution of the log-t statistic under the null hypothesis is exactly the *t*-distribution. To clarify this, and following the development in (Goldstein, 1973), suppose that the output of the linear envelope detector matched filter is the series $\{X_0, X_1, \ldots, X_N\}$, where the cell under statistic is X_0 and the other statistics form the CRP. Goldstein then shows that if the clutter is modelled as log normal with density function

$$f_{X_i}(t) = \frac{1}{\sqrt{2\pi}\sigma_c t} e^{\left\{-\frac{1}{2\sigma_c^2}\left[\log^2\left(\frac{t}{\mu_2}\right)\right]\right\}},$$

(8.1)

where μ_c and σ_c are clutter parameters, then the ideal logarithmic amplifier output given by the transformation $Y_i = \log(X_i)$ results in the density

$$f_{Y_i}(t) = \frac{1}{\sqrt{2\pi}\sigma_c} e^{\left[-\frac{1}{2\sigma_c^2}(t - \log(\mu_c))^2\right]},$$

(8.2)

showing that the Y_i are Gaussian distributed, as expected. This fact suggests that a *t*-statistic formulation could be used to produce a CFAR detector for log-normal distributed clutter. Based upon this realisation, (Goldstein, 1973) proposed the test statistic

$$T_1 = \frac{Y_0 - \frac{1}{N}\sum_{j=1}^{N} Y_j}{\sqrt{\frac{1}{N}\sum_{i=1}^{N}\left(Y_i - \frac{1}{N}\sum_{j=1}^{N} Y_j\right)^2}},$$

(8.3)

where the CUT statistic is Y_0 and the CRP consists of the statistics $Y_1, Y_2, \ldots, Y_N$. Goldstein points out that when the clutter is log normal, (8.3) has a *t*-distribution with $N-1$ degrees of freedom. This realisation enables the setting of an appropriate threshold multiplier for (8.3), and consequently the statistic (8.3) is CFAR with respect to the underlying log normal clutter distribution. Additionally, it is explained in (Goldstein, 1973) that the log-t detector is essentially the optimal envelope detector with minimum mean squared estimates applied for the unknown clutter parameters. Hence in the class of CFAR detection processes, it is often described as the optimal decision rule in log normal clutter.

The innovation of Goldstein's analysis is the realisation that the detector based upon the statistic (8.3) can be extended to produce a CFAR decision rule for a large class of underlying clutter models. This can be shown by reformulating the statistic (8.3) in terms of the matched filter outputs to yield the statistic

$$T_1 = \frac{\ln \left[\prod_{j=1}^{N} \left(\frac{Z_0}{Z_j} \right) \right]}{\sqrt{\frac{1}{N} \sum_{k=1}^{N} \left\{ \ln \left[\prod_{l=1}^{N} \left(\frac{Z_k}{Z_l} \right) \right] \right\}^2}} \tag{8.4}$$

and the corresponding decision rule

$$T_1 \underset{H_0}{\overset{H_1}{\gtrless}} \tau_1, \tag{8.5}$$

where $\tau_1 > 0$ is the detection threshold multiplier. Goldstein argues that if each of the Z_i were replaced by $\Phi_1 Z_i^{\Phi_2}$, then the statistic (8.4) remains invariant. This implies that the detector (8.5) will have the CFAR property for classes of random variables including the Weibull. The fact that the test (8.5) has the CFAR property with respect to the Pareto Type I model follows from the Pareto-exponential duality property (3.21), since this implies that under H_0, $Z_j = \beta e^{\alpha^{-1} W_j}$, where each W_j has an exponential distribution with unity mean, and applying this to the detection statistic (8.5), noting under H_0 the CUT has the same statistical distribution, and showing that it does not depend on α and β.

It will become apparent, once the reader consults Chapter 12, that the detector (8.5) is in fact based upon an invariant statistic with respect to the group of transformations outlined in the aforementioned chapter. This fact, not realised by Goldstein, explains why (8.5) achieves the CFAR property for a large number of clutter distributions.

8.3   An Order Statistic Based Log-t Detector

A significant issue with the log-t detector is that it can be adversely affected by interfering targets in the clutter range profile (Pourmottaghi et al., 2012). Additionally, the detector can perform suboptimally in the presence of clutter transitions. Hence, some attempts have been made to rectify its performance. The most notable contribution is repoted in (Pourmottaghi et al., 2012), who examine the application of a supplementary clutter edge detection algorithm to enhance its performance. However it is possible to reformulate the detector so that it can be designed to manage interfering targets with the application of order statistics. The key to this is to note that in the statistic (8.4) one can also replace the elements of the clutter range profile with their order statistic equivalents, and Goldstein's original observation of invariance should also apply.

To see this, suppose S is a subset of the indices $\{1, 2, \ldots, N\}$ and consider the statistic

$$
T_2 = \frac{\ln\left[\prod_{j \in S}\left(\frac{Z_0}{Z_{(j)}}\right)\right]}{\sqrt{\frac{1}{|S|}\sum_{k \in S}\left\{\ln\left[\prod_{l \in S}\left(\frac{Z_{(k)}}{Z_{(l)}}\right)\right]\right\}^2}}, \tag{8.6}
$$

where Z_0 is the CUT as before, $Z_{(i)}$ is the ith order statistic of the CRP and $|S|$ is the size of the set S. One can then define the corresponding detector

$$
T_2 \underset{H_0}{\overset{H_1}{\gtrless}} \tau_2, \tag{8.7}
$$

where the detection threshold is $\tau_2 > 0$. The detector (8.7) not only attains the CFAR property with respect to Pareto Type I distributed clutter, but also attains it for a large class of clutter model distributions, complementing the original discovery by Goldstein:

Lemma 8.1
Suppose that the clutter returns are of the form $Z_i = \Phi_1 W_i^{\Phi_2}$, where Φ_1 and Φ_2 are unknown constants, and W_i is a random variable with a known distribution but with no unknown parameters. Additionally we assume that under H_0 the CUT statistic can be represented in the same form, namely $Z_0 = \Phi_1 W_0^{\Phi_2}$. Then the detector (8.7) attains the CFAR property with respect to Φ_1 and Φ_2.

To see this, observe that $Z_{(i)} = \Phi_1 W_{(i)}^{\Phi_2}$ and so applying this to (8.6) yields

$$
\begin{aligned}
T_2 &= \frac{\ln\left[\prod_{j \in S}\left(\Phi_1 W_0^{\Phi_2}/\Phi_1 W_{(j)}^{\Phi_2}\right)\right]}{\sqrt{\frac{1}{|S|}\sum_{k \in S}\left\{\ln\left[\prod_{l \in S}\left(\Phi_1 W_{(k)}^{\Phi_2}/\Phi_1 W_{(l)}^{\Phi_2}\right)\right]\right\}^2}} \\[2em]
&= \frac{\ln\left[\prod_{j \in S}\left(W_0/W_{(j)}\right)\right]}{\sqrt{\frac{1}{|S|}\sum_{k \in S}\left\{\ln\left[\prod_{l \in S}\left(W_{(k)}/W_{(l)}\right)\right]\right\}^2}}
\end{aligned} \tag{8.8}
$$

which does not depend on parameters Φ_1 or Φ_2, and thus the claimed CFAR property holds, as required. Hence the new decision rule (8.7) achieves the CFAR property and will be referred to as the log-t OS detector.

The fact that (8.7) is CFAR in Pareto Type I clutter follows immediately from the Pareto-exponential duality property (3.21). With reference to the analysis presented in Chapter 12 one can deduce that (8.6) is an invariant statistic with respect to the same group of transformations as specified in the formulation of the original log-t detector (8.5).

As a more practical formulation of the decision rule (8.7), it is possible to re-write the test as the following shows:

Corollary 8.1

The detector (8.7) is equivalent to

$$Z_0 \underset{H_0}{\overset{H_1}{\gtrless}} \prod_{j \in S} X_{(j)}^{\frac{1}{|S|}} \exp \left\{ \frac{\tau_2}{|S|} \sqrt{\frac{1}{|S|} \sum_{k \in S} \left\{ \ln \prod_{l \in S} \left(\frac{Z_{(k)}}{Z_{(l)}} \right) \right\}^2} \right\}. \tag{8.9}$$

The proof is omitted and left as an exercise. Before proceeding to the analysis of the performance of (8.5) and (8.7) it is interesting to observe that the latter is a generalisation of the dual order statistic decision rule (7.3):

Lemma 8.2

When specialised to the case where $|S| = 2$, the detector (8.7) reduces to the dual order statistic detector (7.3).

To prove this result, let $S = \{i, j\}$ where $1 \leq i < j \leq N$, then the detector (8.9) given in Corollary 8.1 reduces to

$$Z_0 \underset{H_0}{\overset{H_1}{\gtrless}} Z_{(i)}^{\frac{1}{2}} Z_{(j)}^{\frac{1}{2}} e^{\frac{\tau_2}{2} \log \left(\frac{Z_{(j)}}{Z_{(i)}} \right)}. \tag{8.10}$$

The final result follows by simplification of (8.10) and by defining $\kappa = \frac{1}{2}(1 - \tau_2)$, showing that (8.10) is equivalent to

$$Z_0 \underset{H_0}{\overset{H_1}{\gtrless}} Z_{(i)}^{1-\kappa} Z_{(j)}^{\kappa}, \tag{8.11}$$

which completes the proof. Consequently the detector (8.7) is a generalisation of (7.3). As has been remarked in (Goldstein, 1973), the difficulty with the log-t detector is determination of the threshold multiplier τ_1. Under the null hypothesis, and in Pareto Type I distributed clutter, the distribution of the statistic (8.3) is difficult to determine exactly, and unfortunately this problem is inherited by (8.7). Hence it is necessary to determine the threshold multipliers via simulation of the appropriate statistic under the null hypothesis. In the numerical analysis to follow, this has been applied, with 10^6 Monte Carlo runs used to determine the appropriate threshold multiplier. Since the Pfa will be set to 10^{-4} in the following, this is an acceptable practice.

8.4 Performance in Homogeneous Clutter

Several examples of detection performance in homogeneous clutter are now considered, with the usual practice of applying $N = 32$ with the Pfa set to 10^{-4}. For comparison purposes, the ideal detector is again included, together with the GM-CFAR (5.43) and the OS-CFAR (5.45). The latter has been applied with $k = N - 2$ throughout. Clutter parameters have again been sourced from Table 3.1. As in previous analyses the target in the CUT is modelled by a Swerling I fluctuation law.

Figures 8.1 and 8.2 are for the case of horizontal polarisation, with the latter figure providing an enlargement of the former. The figure shows the log-t detector (8.5) together with three examples of (8.7). These correspond to the selection of $S = \{2, 3, \ldots, N - 1\}$ for Case 1, $S = \{3, 4, \ldots, N - 2\}$ for Case 2 and $S = \{4, 5, \ldots, N - 3\}$ for Case 3. These two figures show that the GM-CFAR has the best performance, followed by the log-t detector and then the OS-CFAR. The new variation of the log-t detector is slighly suboptimal to the OS-CFAR for Case 1, and then as the number of elements of the set S is reduced, the performance decreases as shown.

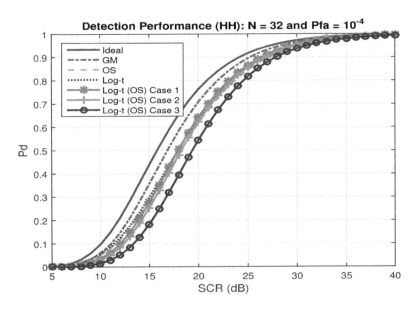

Figure 8.1: Performance analysis of the log-t and log-t OS detectors in homogeneous horizontally polarised clutter. These detectors are compared with the GM- and OS-CFARs, with the ideal decision rule also included.

The same phenomenon is observed in the case of cross polarisation, as shown in Figures 8.3 and 8.4, and in vertical polarisation (Figures 8.5 and 8.6). The log-

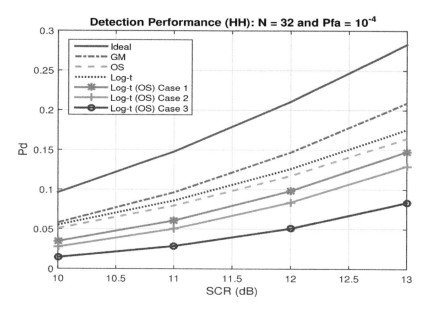

Figure 8.2: Magnification of the results in Figure 8.1.

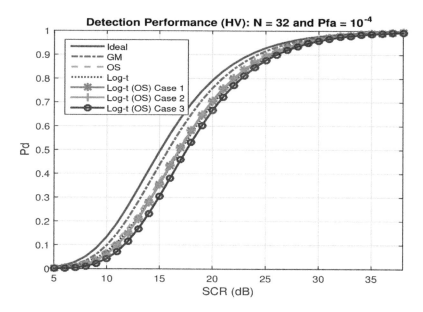

Figure 8.3: Performance of the same set of detectors in the homogeneous clutter in the cross-polarisation case.

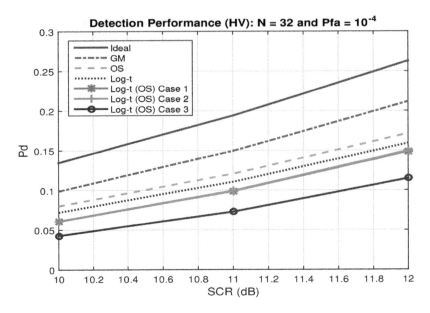

Figure 8.4: Enlargement of Figure 8.3.

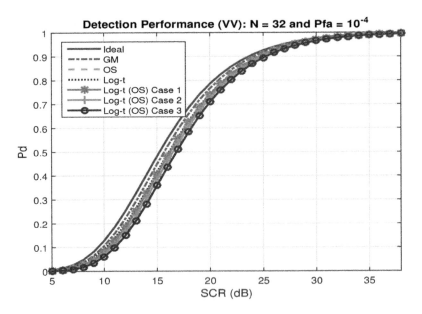

Figure 8.5: Detector performance in the vertically polarised case, comparing the log-t and log-t OS detectors with the GM and OS-CFARs.

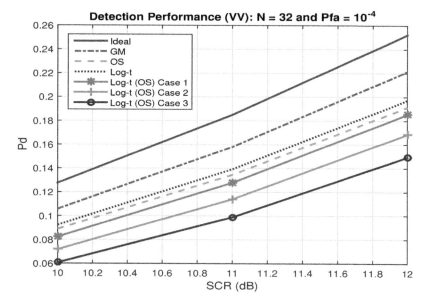

Figure 8.6: Different perspective on Figure 8.5.

ical conclusion is that the new decision rule (8.7) does not provide a significant gain on the standard log-t detector, nor the two comparison CFARs, to warrant application in homogeneous clutter.

8.5 Interference

Analysis of the effects of interference in the clutter range profile is now undertaken, and for simplicity the detector (8.7) is applied with $S = \{2, 3, \ldots, N-2\}$, and as in previous cases an independent Swerling I target model is utilised for interference. Three examples in horizontal polarisation are considered, with Figure 8.7 for a 40 dB interfering target, Figure 8.8 for 60 dB and Figure 8.9 corresponding to 80 dB interference.

These three examples show that the log-t detector completely saturates and performs suboptimally in the presence of strong interference. By contrast, the new decision rule (8.7) is robust to strong interference. It is interesting to observe that in some cases the test (8.7) performs slightly better than the OS-CFAR, while in all cases the GM-CFAR also has very poor performance.

These results are replicated for the other polarisations. To illustrate this, Figures 8.10 and 8.11 are for the case of cross- and vertical polarisation respectively, where the interference is 40 dB in strength.

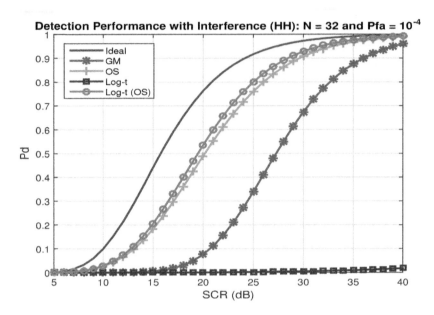

Figure 8.7: First example of interference in horizontally polarised clutter, where the ICR is 40 dB. The failure of the log-t detector is apparent, while the log-t OS has the best performance.

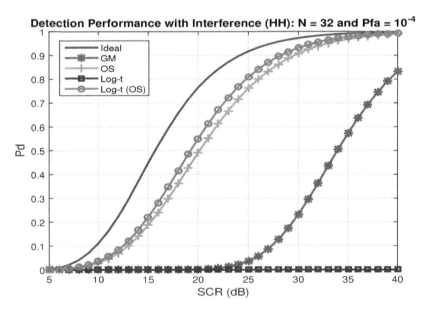

Figure 8.8: Second example of interference in horizontally polarised clutter, with ICR equal to 60 dB. As for Figure 8.7 the log-t OS detector has the best performance.

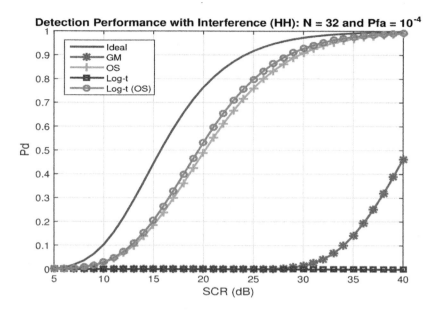

Figure 8.9: Final example of performance with interference in horizontally polarised clutter (80 dB). This simulation demonstrates the robust nature of the log-t OS detector.

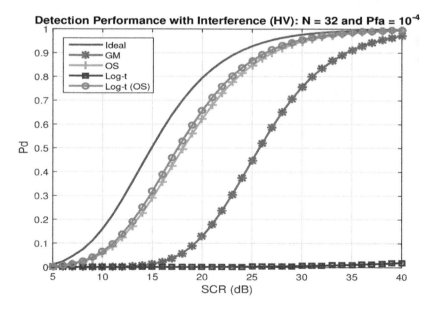

Figure 8.10: Cross polarisation interference example, with ICR of 40 dB.

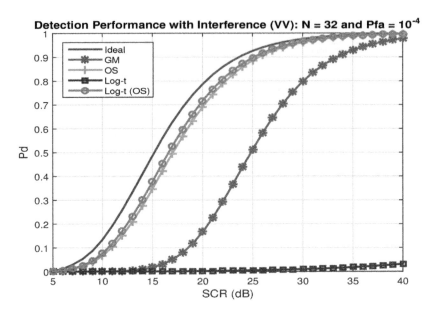

Figure 8.11: Vertical polarisation interference example, where the ICR is 40 dB.

8.6 False Alarm Regulation

False alarm regulation of the log-t detector and its order statistic variant is now investigated, where they are both compared with the GM-CFAR (5.43) and the OS-CFAR (5.45). This is again undertaken with a design Pfa of 10^{-4} and $N = 32$. The log-t OS detector is based upon $S = \{2, 3, \ldots, N - 2\}$ in all cases, while the OS-CFAR applies $k = N - 2$. As in the previous chapter clutter transitions will also be investigated in the presence of an interfering target in the CRP.

Beginning with the horizontally polarised case, Figure 8.12 shows the estimated Pfa for a clutter power increase of 1 dB. As can be observed, the GM-CFAR tends to increase the Pfa, while the log-t detector also does the same but to a much smaller degree. The log-t OS detector tends to match the performance of the OS-CFAR.

Figure 8.13 includes a 20 dB interfering target into the clutter range profile, embedded within the lower powered clutter. The figure shows an overall decrease in the resultant Pfa, with the log-t detector experiencing a sharper decrease before the CUT is affected with higher powered clutter. The GM-CFAR still has a tendency to increase the resultant Pfa, while the log-t OS and OS-CFAR detectors perform comparably.

As a second example of false alarm regulation in the horizontally polarised case, Figure 8.14 shows the effect of reducing the CCR to 0.5 dB, while Figure 8.15 introduces a 10 dB interfering target into the scenario. The conclusions are similar to the previous case.

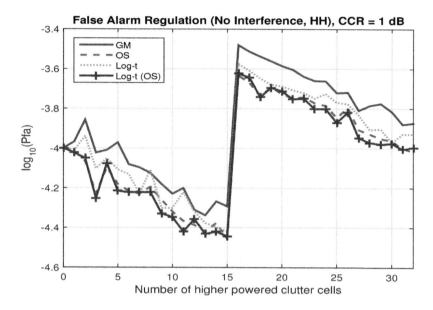

Figure 8.12: False alarm regulation in horizontally polarised clutter.

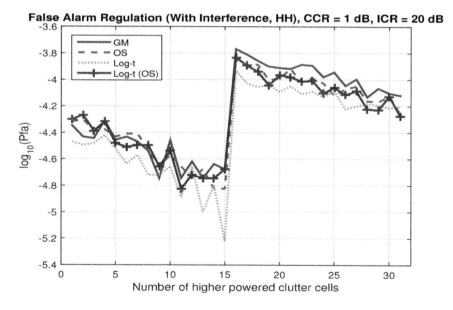

Figure 8.13: False alarm regulation with interference, horizontal polarisation, where the ICR is 20 dB.

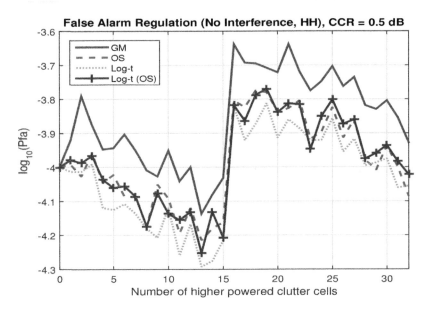

Figure 8.14: Second example of false alarm regulation in horizontal polarisation.

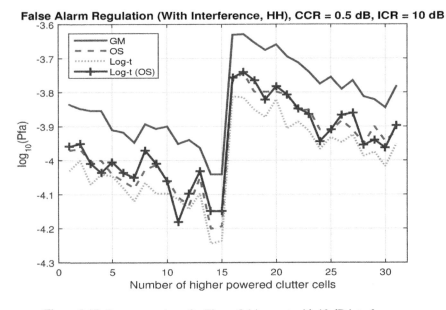

Figure 8.15: Same scenario as for Figure 8.14 except with 10 dB interference.

Next the case of cross polarisation is considered, with Figure 8.16 for the situation when the clutter power is increased by 1 dB, and with Figure 8.17 including a 20 dB interfering target. In these two examples the Monte Carlo estimation has less variation, showing less fluctuation in the estimated Pfa. The GM-CFAR has a tendency to increase the number of false alarms while the other decision rules tend to perform comparably.

Figures 8.18 and 8.19 repeat the experiment with the CCR reduced to 0.5 dB and with a 10 dB interfering target. The conclusions are similar to before although it is interesting to observe that the log-t OS detector tends to reduce the number of false alarms.

Finally the case of vertical polarisation is examined, beginning with the situation where the clutter power is increased by 1 dB (Figure 8.20) and when a 10 dB interfering target is included (Figure 8.21). The performance is comparable to that in previous polarisations with an exception being a larger overall deviation from the design Pfa in this case.

Figures 8.22 and 8.23 are final examples for vertical polarisation, where the clutter power is increased by 0.5 dB with a 10 dB interfering target included in Figure 8.23. The reduced CCR results in a smaller overall deviation from the design Pfa, but performance is similar to previous cases.

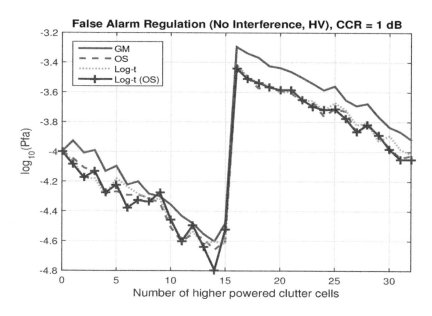

Figure 8.16: False alarm regulation in the cross polarisation case.

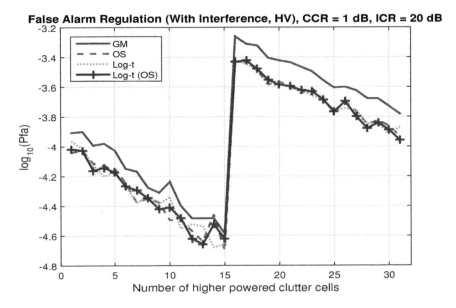

Figure 8.17: False alarm regulation with interference in cross polarisation, with ICR of 20 dB.

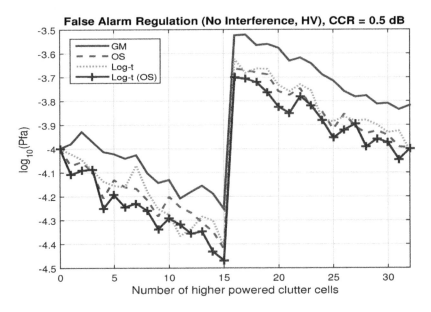

Figure 8.18: A second example of false alarm regulation in cross polarisation.

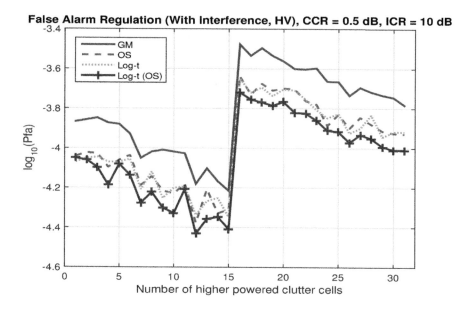

Figure 8.19: Same scenario as for Figure 8.18 with the addition of a 10 dB interfering target.

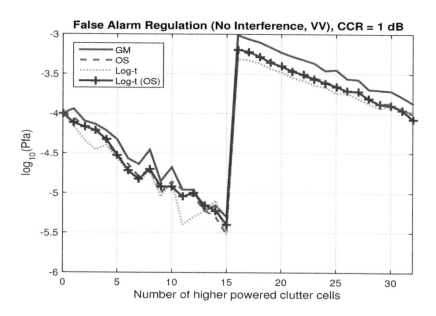

Figure 8.20: False alarm regulation in vertical polarisation.

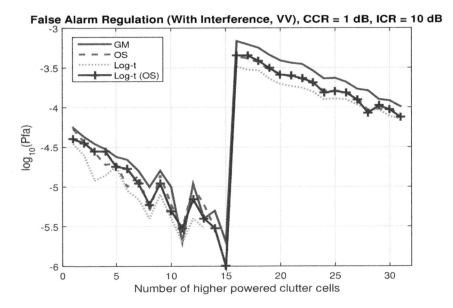

Figure 8.21: 10 dB interference added to the scenario of Figure 8.20.

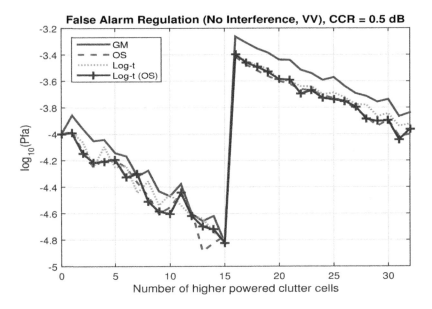

Figure 8.22: Final example of false alarm regulation in vertically polarised clutter.

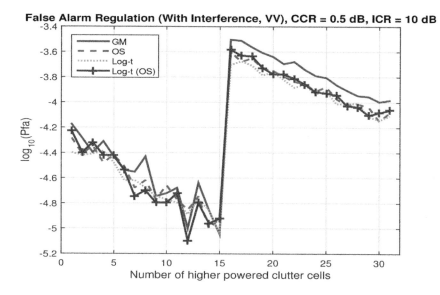

Figure 8.23: The scenario of Figure 8.22 with 10 dB interference.

8.7 Conclusions

This chapter re-examined Goldstein's log-t detector, demonstrating why it is also CFAR when applied to Pareto Type I distributed clutter. A new version of it, based upon order statistics was presented, which was shown to also possess the CFAR property with respect to a large class of clutter models. The advantage of the log-t OS detector was that it could manage interfering targets better than a conventional log-t detector.

In homogeneous clutter, and during clutter transitions, the log-t OS detector tended to perform no better than the OS-CFAR. Given the complexity of determining its threshold multiplier (as well as that of the conventional log-t detector), it is difficult to justify its application, within the signal processing architecture of a radar system, as a detection algorithm when the OS-CFAR is a simpler alternative. Nonetheless, the log-t detector, and its OS variant, are of interest and hence merited examination.

This chapter has highlighted a number of open problems that the interested researcher could pursue. It would be of relevance to determine whether there exists an optimal choice for the index set S in the formulation of (8.7), and to also examine whether, in the Pareto Type I case, a more efficient means of determining threshold multipliers could be produced, without the need for a simulation-based approach. Since the Weibull distribution is still of relevance in radar signal processing, it would be useful to examine the performance of (8.7) in such a clutter model environment.

SPECIALISED DEVELOPMENTS

Chapter 9

Switching Based Detectors

9.1 Introduction

Many of the decision rules introduced thus far experience difficulty in the balance between the management of false alarm regulation and eliminating the effects of interfering targets in the clutter range profile. As has been demonstrated, the GM-based detectors tended to have better performance than OS-based decision rules, when the clutter is homogeneous. However, in the presence of interference, the GM decision rules have a tendency to degrade in performance as an interfering target's SCR increases. In such situations, the OS-based detectors tend to perform better, as demonstrated previously. During false alarm regulation, the GM and OS based detectors perform somewhat similarly, although the GM-based detectors tended to deviate more significantly from the design Pfa in the presence of interference in the clutter range profile.

One of the solutions to this problem is to apply the idea of a switching-based detector. Such decision rules are based upon a given detection process, such as the GM-detector (5.10), but with the added complexity of filtering the detector's clutter range profile prior to the calculation of the measurement of the clutter level. Specifically, the clutter returns are partitioned into two disjoint sets, and if there are too many unusual observations, then the clutter range profile is reduced by eliminating excessive returns. The original decision rule is then applied with this reduced clutter range profile. Otherwise, if there are not too many unusual observations then the detector is run on the complete clutter range profile. Thus the detection process switches between two possible clutter range profiles. Such a process requires adjustment to the threshold multiplier, and an analysis of how to partition the clutter range profile successfully. The hope in this approach is that the performance of detectors which operate well in homogeneous clutter may be preserved while making them more robust during variations from the homogeneous clutter scenario.

Switching based detectors were first examined in (Van Cao, 2004a), (Van Cao, 2004b), and later a more systematic analysis appeared in (Van Cao, 2008). The interest in these publications was to enhance the performance of the cell-averaging CFAR, designed to operate in homogeneous exponentially distributed clutter, but when subjected to heterogeneous clutter. The analysis of (Van Cao, 2008) was improved subsequently in (Meng, 2009). The main conclusion of (Van Cao, 2008) was that a switched version of the cell averaging detector could manage interfering targets very well in comparison to a standard CA-CFAR. Additionally, the switched detector could be designed so that it regulated the false alarm probability very well.

Consequently, further developments and analysis of switching-based detectors have been reported in (Moazen and Akhavan-Sarraf, 2007), (Erfanian and Faramarzi, 2008), (Tom and Viswanathan, 2008), (Erfanian and Vakili, 2008), (Song et al., 2009), (Erfanian and Vakili, 2009) and (Zhang et al., 2013). Additionally, the extension of the switching-based detector design has been examined in the scenario of K-distributed clutter: see (Erfanian and Faramarzi, 2008) and (Erfanian and Vakili, 2009) for example. More recently, (Weinberg, 2014c) presented a development of the switching detector design for the case of Pareto Type I distributed clutter. The mathematical analysis in (Weinberg, 2014c) showed that there is a more elegant approach to the detector design than that presented in the original approach of (Van Cao, 2008). Subsequently, a generalisation of the methodology of (Weinberg, 2014c) appeared in (Weinberg, 2015c). This chapter will overview these developments with a particular focus on the Pareto Type I clutter scenario. The first two sections will overview the development and generalisation of switching detectors, while the latter two will show how a switching CFAR detector can be constructed from (5.43).

9.2   Development of a Switching Detector

This section now explores the developments of switching-based detectors based upon that in (Weinberg, 2014c), with the focus on the GM-detector (5.10). This will require the assumption of *a priori* knowledge of β.

Suppose, as before, that the clutter range profile is $\{Z_1, Z_2, \ldots, Z_N\}$, and the CUT statistic is Z_0. Partition the clutter range profile through the set

$$S_0 = \left\{ Z_j : Z_j < \beta^{1-a} Z_0^{\,a} \right\}, \tag{9.1}$$

for some fixed $a > 0$. The union of (9.1) and its complement form a partition of the clutter range profile, based upon a function of the CUT statistic. The motivation for (9.1) will be provided subsequently. Suppose that the size of S_0 is n_0; hence it is clear that $0 \le n_0 \le N$. Since the elements of the set S_0 will depend on the underlying statistical nature of the elements of the clutter range profile and the CUT, it follows that n_0 is a discrete random variable.

The motivation for the definition of (9.1) is that this form of partitioning is equivalent to that in (Van Cao, 2008), which can be seen when one applies the Pareto-exponential duality property to (9.1). To see this, suppose that $Z_j = \beta e^{\alpha^{-1} X_j}$, where $X_j \overset{d}{=} \text{Exp}(1)$ for all $j \in \{0, 1, \dots, N\}$. Then it follows that

$$S_0 | H_0 \equiv \{X_j : \beta e^{\alpha^{-1} X_j} < \beta^{1-a} \beta^a e^{\alpha^{-1} a X_0}\} = \{X_j : X_j < a X_0\}, \qquad (9.2)$$

which is exactly the partitioning employed in (Van Cao, 2008).

It is possible to alter the definition of (9.1) to eliminate β from the definition of S_0, and sort the clutter range profile, for example, using the set $\{Z_j : Z_j < \kappa Z\}$, for some $\kappa > 0$. As is explained in (Weinberg, 2014c), application of such a partitioning will introduce Pareto shape parameter dependence into the detector design. Thus it is preferential to use (9.1).

To specify a switching detector based upon the set S_0, and the GM detector (5.10), a target is declared present in the CUT if one of the two conditions are met:

$$Z_0 > \beta^{1-n_0 \tau} \prod_{Z_j \in S_0} Z_j^\tau \text{ when } n_0 > N_T$$

$$Z_0 > \beta^{1-N\tau} \prod_{j=1}^{N} Z_j^\tau \quad \text{when } n_0 \le N_T, \qquad (9.3)$$

where $\tau > 0$ is a threshold multiplier and $N_T \in \{1, 2, \dots, N\}$ is an integer threshold constant. Guidelines are established in (Van Cao, 2008) on appropriate choices for N_T in practical implementation of a switching scheme. In order to manage N_I expected interfering targets, the choice of $N_T = N - N_I - 1$ is justified in (Van Cao, 2008), which will be adopted in the numerical analysis to follow. Appropriate choices for the parameter a will be discussed subsequently.

Figure 9.1 provides a schematic of the implementation of a switching-based detector. It is assumed that the CUT and CRP have been pre-processed as in previous schematics. The CUT and CRP are then applied to produce the set S_0, after which a test is applied to the magnitude of S_0. The function ζ, used in the definition of S_0, indicates that the set S_0 is determined by comparing each clutter measurement with a specified function of the CUT. This then allows selection of the appropriate statistics for the measurement of clutter, as indicated by the set over which the clutter measurement function g is applied. Normalisation follows

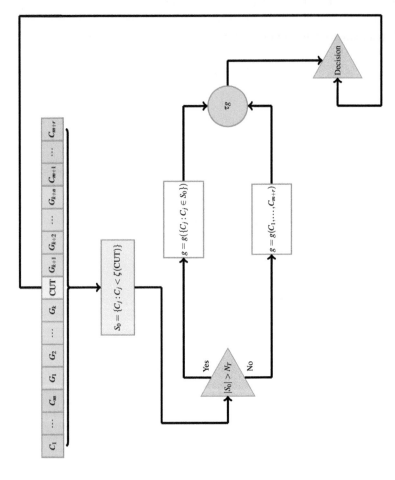

Figure 9.1: Schematic of a switching-based detector, where the pre-processing functions have been omitted for brevity.

and the result is compared with the CUT to decide on the presence or absence of a target.

In order to apply (9.3) it is necessary to determine an appropriate τ, based upon a specific Pfa. The following result provides an appropriate expression:

Lemma 9.1

The probability of false alarm of the detector (9.3) is given by

$$\text{Pfa} \; = \; (1+\tau)^{-N}\text{P}(n_0 \le N_T|H_0)$$

$$+ \sum_{k=N_T+1}^{N} (1+\tau)^{-k}\text{P}(n_0 = k|H_0). \qquad (9.4)$$

The proof of Lemma 9.1 proceeds by conditioning on the random variable n_0, and then using the probability of false alarm and threshold multiplier relationship of the GM-CFAR for the two separate components of the switching detector. To see this, observe that

$$\text{Pfa} \; = \; \sum_{k=0}^{N_T}\text{P}\left(Z_0 > \beta^{1-N\tau}\prod_{j=1}^{N}Z_j^\tau \,\middle|\, H_0, n_0 = k\right)\text{P}(n_0 = k|H_0)$$

$$+ \sum_{k=N_T+1}^{N}\text{P}\left(Z_0 > \beta^{1-n_0\tau}\prod_{Z_j\in S_0}Z_j^\tau \,\middle|\, H_0, n_0 = k\right)\text{P}(n_0 = k|H_0)$$

$$= \; \sum_{k=0}^{N_T}\text{P}\left(Z_0 > \beta^{1-N\tau}\prod_{j=1}^{N}Z_j^\tau \,\middle|\, H_0\right)\text{P}(n_0 = k|H_0)$$

$$+ \sum_{k=N_T+1}^{N}\text{P}\left(Z_0 > \beta^{1-k\tau}\prod_{Z_j\in S_0}Z_j^\tau \,\middle|\, H_0, n_0 = k\right)\text{P}(n_0 = k|H_0)$$

$$(9.5)$$

where the fact that in the case where $n_0 \le N_T$, the switching detector employs the full clutter range profile, and so the conditional false alarm probability reduces to that of the GM detector. In the case where $n_0 \ge N_T$, the switching detector uses the subset S_0 of clutter returns, but since these are independent and identically distributed, the conditional probability of false alarm and threshold multiplier relationship is also based upon that of a GM detector with k clutter cells. This completes the proof.

The threshold multiplier τ is set, for a desired Pfa, by numerical inversion of (9.4) in practice. Hence it is clear that one requires knowledge of the distribution of n_0 under H_0. The following shows that its point probabilities can be expressed in terms of a beta function:

Lemma 9.2
The discrete random variable n_0, under H_0, has probability mass function given by

$$\mathrm{P}(n_0 = k|H_0) = \binom{N}{k} a^{-1} \mathcal{B}(N - k + a^{-1}, k + 1), \tag{9.6}$$

where $k \in \{0, 1, \ldots, N\}$ and $\mathcal{B}(\cdot, \cdot)$ is the beta function.

To prove Lemma 9.2, observe that the event $\{n_0 = k|H_0\}$ is equivalent to the event that k of the Z_j in S_0 are less than $\beta^{1-a} Z_0^a$, and the remaining $N - k$ of the Z_j exceed $\beta^{1-a} Z_0^a$. The number of ways in which this can occur can be expressed as a binomial coefficient. Hence we focus on the probability of one such sequence occurring. Recalling that under H_0, we can write $Z_0 = \beta e^{X_0}$ and $Z_j = \beta e^{X_j}$, where X_0 and each X_j are independent and identically distributed exponential random variables with parameter α, it is not difficult to show that the event specified above can be re-expressed as requiring k of the $X_j < aX_0$ and the remaining $N - k$ of the $X_j > aX_0$. Hence, by conditioning on the random variable X_0, and applying independence,

$$\mathrm{P}(\text{one sequence}) = \int_0^\infty \alpha e^{-\alpha t} \mathrm{P}(k \text{ of the } X_j < at, N - k \text{ of the } X_j > at) dt$$

$$= \int_0^\infty \alpha e^{-\alpha t} \mathrm{P}(X_1 < at)^k \mathrm{P}(X_1 > at)^{N-k} dt$$

$$= \int_0^\infty \alpha e^{-\alpha t} \left(1 - e^{-\alpha a t}\right)^k e^{-\alpha a t(N-k)} dt$$

$$= \int_0^1 a^{-1} y^{N-k-1+1/a} (1 - y)^k dy, \tag{9.7}$$

where the exponential distribution function has been used, and the change of variables $y = e^{-\alpha a t}$ has been applied to produce the final integral in (9.7). Observing that the latter is a beta function, the proof of the Lemma is completed by multiplying (9.7) by the binomial coefficient giving the total number of possible sequences.

The probability of false alarm (9.4) can be written explicitly as a function of the switching detector parameter a. To see this, observe that by applying the definition of the beta function to (9.6), one can derive the following result:

Lemma 9.3

The point probabilities given by Lemma 9.2 are also given by

$$\mathbf{P}(n_0 = k|H_0) = \frac{P_k^N a^k}{(aN+1)(a(N-1)+1)\ldots(a(N-k)+1)}, \tag{9.8}$$

where $P_k^N = N(N-1)(N-2)\ldots(N-k+1)$ is the permutation coefficient.

It is a straightforward exercise to apply (9.8) to (9.4). The interesting fact that (9.8) demonstrates is that the point probabilities given by it are roughly of the order of the reciprocal of a.

To examine this Figure 9.2 plots (9.8) for three values of a. In this case N is set to 16 and a is varied from 0.1, 1 to 10. This example, and other cases considered, showed that selecting an $a < 1$ resulted in the point probabilities being concentrated in lower values of k. In the case where $a > 1$, the point probabilities were larger for k near N. When $a = 1$, the distribution (9.8) is uniform. Figure 9.3 plots the Pfa as a function of a, in the scenario where $N = 16$, $N_T = 13$, with τ set to 0.7783, which corresponds to a threshold multiplier for the GM detector with a Pfa of 10^{-4}. The interesting characteristic of the figure is that as a increases, the resultant Pfa limits to 10^{-4}. This suggested that suitable choices for a could be based upon the reciprocal of the Pfa. Hence, for a Pfa of 10^{-4}, $a = 10^4$, for example.

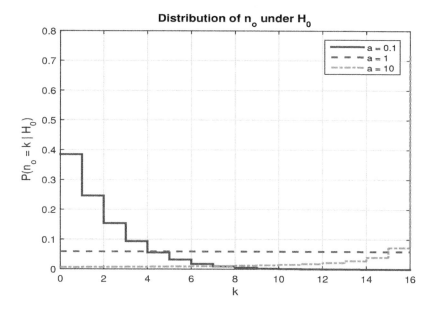

Figure 9.2: The effect of switching parameter a on the distribution of n_0 under H_0.

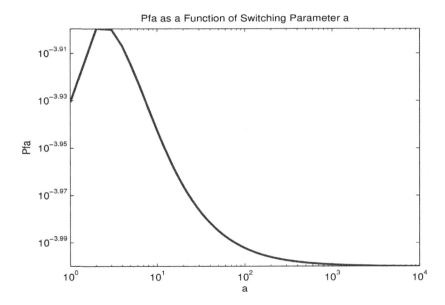

Figure 9.3: The Pfa as a function of a, for the case where $N = 16$, $N_T = 13$ and with τ set for the GM detector with a Pfa of 10^{-4}.

9.3 Generalisation of the Switching Detector

Next the generalisation of the formulation of the switching detector of the previous section is investigated. As remarked previously this has been examined in (Weinberg, 2015c) on which the following is based. It is not difficult to generalise a switching detector by changing the underlying detection process. The complexity is selection of the set S_0. The main objective in this section is to show how S_0 can be selected, for a given clutter environment, and then to illustrate how it can be coupled with a general switching detection process.

Suppose that the underlying detector takes the form

$$Z_0 \underset{H_0}{\overset{H_1}{\gtrless}} g(Z_1, Z_2, \ldots, Z_N; \tau) \tag{9.9}$$

where the dependency on the threshold multiplier τ is expressed explicity in the definition of g. As in previous developments, the CUT statistic is Z_0 and the clutter range profile consists of $Z_1, Z_2, \ldots, Z_N$. It is assumed that these statistics are independent and identically distributed with distribution function F_Z. Let $g_{|S_0}$ be the restriction of the clutter measurement function g to the set S_0, which will be specified subsequently. To illustrate this, suppose g is a sum, with clutter range profile $\{Z_1, Z_2, \ldots, Z_N\}$ and that $S_0 = \{Z_1, Z_3, Z_5\}$. Then $g_{|S_0}(Z_1, Z_2, \ldots, Z_N) = Z_1 + Z_3 + Z_5$. If instead, g was an order statistic, then $g_{|S_0}(Z_1, Z_2, \ldots, Z_N)$ would be the corresponding order statistic on the set $\{Z_1, Z_3, Z_5\}$.

The general form of a switching-based detector, based upon (9.9), can be formulated as follows. A target is declared present in the CUT if one of the two conditions are met:

$$Z_0 > g_{|S_0}(Z_1, Z_2, \ldots, Z_N; \kappa) \text{ when } n_0 > N_T$$

$$Z_0 > g(Z_1, Z_2, \ldots, Z_N; \kappa) \quad \text{when } n_0 \leq N_T, \tag{9.10}$$

where $\kappa > 0$ is a threshold factor, n_0 is the size of the set S_0 and $N_T \in \{1, 2, \ldots, N\}$ is an integer threshold constant. Selection of N_T can be as before, where it is designed to manage an expected number of interfering targets.

Examination of an appropriate selection for S_0 is now undertaken. The way in which this is done is to generalise the approach in (Weinberg, 2014c) using the memoryless nonlinear transformation between random variables discussed in the design of transformed detectors in Chapter 5. Suppose that X_0 and X_j are exponentially distributed. Then recall that based upon the transformation function (5.7) these can be related to the random variables Z_0 and Z_j via $H(X_0) \stackrel{d}{=} Z_0$ and $H(X_j) \stackrel{d}{=} Z_j$. Hence under the null hypothesis, the original definition of S_0 in (Van Cao, 2008) is

$$S_0 := \{X_j : X_j < aX_0\} \equiv \{Z_j : H^{-1}(X_j) < aH^{-1}(Z_0)\}$$

$$= \{Z_j : Z_j < H(aH^{-1}(Z_0))\}, \tag{9.11}$$

where $X_j = H^{-1}(Z_j)$ for all j. Thus, (9.11) suggests an appropriate way in which clutter statistics may be sorted for a switching detector, based upon the clutter model of interest. Observe that if the set S_0 is determined through (9.11), then any properties of S_0 under H_0 that have been established can be applied directly to the clutter environment of interest. This means that the results of Lemma 9.2 can be used, for example.

Finally, it is necessary to specify the Pfa of (9.10), to produce the threshold factor κ. In the detection scheme (9.10), since it is assumed that the clutter statistics are independent and identically distributed, suppose that Pfa(n) denotes the false alarm probability when there are n clutter statistics. Then, since clutter is sorted via (9.11), it follows by conditional probability that

$$
\begin{aligned}
\text{Pfa} \;=\; & \sum_{k=0}^{N_T} P(Z_0 > g_{|S_0}(Z_1, Z_2, \ldots, Z_N; \kappa)|H_0)P(n_0 = k|H_0) \\
& + \sum_{k=N_T+1}^{N} P(Z_0 > g(Z_1, Z_2, \ldots, Z_N; \kappa)|H_0)P(n_0 = k|H_0) \\
=\; & \text{Pfa}(N)P(n_0 \le N_T|H_0) + \sum_{k=N_T+1}^{N} \text{Pfa}(k)P(n_0 = k|H_0). \quad (9.12)
\end{aligned}
$$

Therefore, one can determine κ via numerical inversion of (9.12). Thus (9.10), together with (9.6), (9.11) and (9.12), provide a generalised switching detector. Clearly if the detector on which it is based is CFAR with respect to a certain clutter parameter, then the switching detector will inherit this property. The switching detector parameter a, used in the definition of S_0, can also be selected based upon the guidelines established previously.

9.4 Switching CFAR Detector

This section begins an examination of a switched version of the GM-CFAR (5.43); the results in this section are based upon (Weinberg, 2017a). Since (5.43) is designed to operate in Pareto Type I distributed clutter, the set S_0 defined via (9.1) will depend on the Pareto scale parameter, which is usually assumed for application of the detector (5.10). However, it is of interest to investigate whether the necessity for assuming *a priori* knowledge of the Pareto scale parameter can be eliminated. In order to address this, and motivated by the application of the minimum of the CRP as a statistic for β, one can consider the set

$$
S_0 = \{Z_j : Z_j < Z_{(1)}^{1-a} Z_0^a\} \quad (9.13)
$$

for some $a > 0$. If the distribution of the cardinality of this set, under H_0 is independent of the Pareto clutter parameters, then the corresponding switching detector, based upon (9.13), will be completely CFAR. It turns out that this is the case, as the following result shows:

Lemma 9.4
Under H_0, the distribution of $n_0 = |S_0|$ for the set S_0 defined in (9.13), is given by

$$
P(n_0 = k|H_0) = \begin{cases} \frac{1}{N+1}\left(1 + \frac{N}{aN+1}\right), & \text{if } k = 0 \\[2mm] a^{-1}\binom{N}{k}\frac{N}{N+1}\mathcal{B}\left(N-k+a^{-1}, k+1\right) & \text{if } k \in \{1,2,\ldots,N\}, \end{cases} \quad (9.14)
$$

where $\mathcal{B}$ is the beta function.

In order to prove this result it is necessary to apply the Pareto-exponential duality property (3.21), so that under H_0 one can write $Z_j = \beta e^{\alpha^{-1} X_j}$ where $0 \leq j \leq N$ and X_j is exponentially distributed with mean unity. In addition $Z_{(1)} = \beta e^{\alpha^{-1} X_{(1)}}$ where $X_{(1)}$ is the minimum of a series of independent exponentially distributed random variables with mean 1. Then the set S_0 under H_0 is equivalent to $\{X_j : X_j < (1-a)X_{(1)} + aX_0\}$. Recall that $X_j | \{X_{(1)} = t\} \stackrel{d}{=} t + Y_j$ where Y_j is exponentially distributed with unity mean. Hence

$$
\begin{aligned}
P(n_0 = k | H_0) \quad &= \quad P(k \text{ of the } X_j < (1-a)X_{(1)} + aX_0, \\
&\qquad N - k \text{ of the } X_j \text{ exceed it}).
\end{aligned}
\tag{9.15}
$$

By conditioning on X_0 and $X_{(1)}$, one can express (9.15) in the equivalent form

$$
\begin{aligned}
P(n_0 = k | H_0) \quad = \quad &\int_0^\infty \int_0^\infty P\Big(k \text{ of the } Y_j < a(w-t) \\
&\quad N - k \text{ of the } Y_j > a(w-t)\Big) N e^{-Nt} e^{-w} dt\, dw.
\end{aligned}
\tag{9.16}
$$

For the case where $k \geq 1$ observe that when $w < t$ the probability within the integral in (9.16) is identically zero, so the integral is non-zero over the interval $w > t$ assuming that $a > 0$. Hence

$$
\begin{aligned}
P(n_0 = k | H_0) \quad = \quad &\int_{t=0}^\infty \int_{w=t}^\infty \binom{N}{k} P(Y < a(w-t))^k P(Y > a(w-t))^{N-k} \times \\
&\quad N e^{-Nt} e^{-w} dw\, dt \\
= \quad &\binom{N}{k} \int_{t=0}^\infty \int_{w=t}^\infty (1 - e^{-a(w-t)})^k e^{-a(w-t)(N-k)} \\
&\quad N e^{-Nt} e^{-w} dw\, dt.
\end{aligned}
\tag{9.17}
$$

Through the transformation $z = w - t$ in the second integral in (9.17), followed by a subsequent transformation of $\phi = e^{-z}$ expression (9.16) reduces to

$$
P(n_0 = k | H_0) = \binom{N}{k} \frac{N}{N+1} \int_0^1 \phi^{a(N-k)} (1 - \phi^a)^k d\phi.
\tag{9.18}
$$

With a final transformation of $\psi = \phi^a$ (9.18) reduces to

$$
P(n_0 = k | H_0) = a^{-1} \binom{N}{k} \frac{N}{N+1} \int_0^1 \psi^{N-k+a^{-1}-1} (1 - \psi)^k d\psi
\tag{9.19}
$$

which can be then expressed in terms of the beta function, establishing the result as required.

The case where $k = 0$ results in integrals being non-zero over the full integration region. Proceeding in a fashion similar to the previous case

$$P(n_0 = 0|H_0) = \int_0^\infty \int_0^\infty P\left(0 \text{ of the } Y_j < a(w-t)\right.$$

$$\left. N \text{ of the } Y_j > a(w-t)\right) Ne^{-Nt} e^{-w} dt dw. \quad (9.20)$$

In the region where $w < t$ the probability within the integral is exactly unity. Hence it follows that

$$P(n_0 = 0|H_0) = \int_{t=0}^\infty \int_{w=0}^t Ne^{-Nt} e^{-w} dw dt$$

$$+ \int_{t=0}^\infty \int_{w=t}^\infty P(Y > a(w-t))^N Ne^{-Nt} e^{-w} dw dt$$

$$= \frac{1}{N+1} + \frac{N}{aN+1} \frac{1}{N+1}, \quad (9.21)$$

which has been obtained by evaluating the integrals. Consequently (9.21) establishes the result for $k = 0$, completing the proof of Lemma 9.4, as required.

It is now possible to specify a switching GM-CFAR, based upon this set S_0 given by (9.13) and the detector (5.43). Therefore one can specify the decision rule to reject H_0 if the CUT is such that

$$Z_0 > Z_{(1)}^{1-n_0\tau} \prod_{Z_j \in S_0} Z_j^\tau \text{ when } n_0 > N_T$$

$$Z_0 > Z_{(1)}^{1-N\tau} \prod_{j=1}^N Z_j^\tau \text{ when } n_0 \leq N_T. \quad (9.22)$$

The following Lemma specifies the Pfa of (9.22):

Lemma 9.5
The probability of false alarm of the detector (9.22) is given by

$$\text{Pfa} = \frac{N}{N+1}(1+\tau)^{-N} P(n_0 \leq N_T|H_0)$$

$$+ \sum_{k=N_T+1}^N \frac{k}{k+1}(1+\tau)^{-k} P(n_0 = k|H_0), \quad (9.23)$$

where the distribution of n_0 under H_0 is given by Lemma 9.4.

The proof of the Lemma is omitted since it parallels that of Lemma 9.1, with an application of the detector (5.43) and its Pfa (5.44). Hence it follows that the switching detector (9.22) will attain the CFAR property with respect to the Pareto Type I shape and scale parameters.

Figure 9.4 plots the distribution of n_0 under H_0 for $N = 32$ and for a series of values of a.

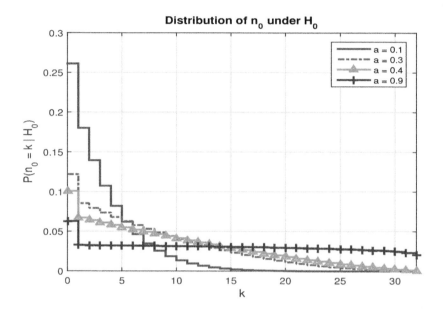

Figure 9.4: Plots of n_0, under H_0 based upon the distribution defined in Lemma 9.4.

In order to determine an appropriate a for practical use in Lemma 9.4 Figure 9.5 plots the Pfa (9.23) as a function of a for the case where $N = 32$ and with $\tau = 0.3322$, where the latter is the corresponding threshold required for a Pfa of 10^{-4} with the GM-CFAR (5.43). Here $N_t = N - 2$ has been selected. It is clear that for $a < 0.3$ the resultant Pfa is roughly of the order 10^{-4}. Hence this suggests that suitable selection of a, for the decision rule (9.22), can be based upon small a.

The reader will observe that numerical analysis of the switching-based detectors has not yet been included. The following section examines the performance of the SW-CFAR constructed in this section, since it is a CFAR process. The interested reader can consult (Weinberg, 2014c) for performance analysis of the detector (9.3), while (Weinberg, 2015c) provides examples of performance of a switching CFAR designed to operate in a one parameter Lomax clutter model environment.

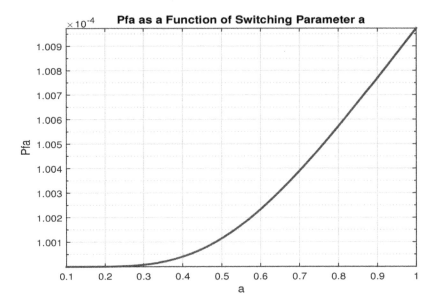

Figure 9.5: The resultant Pfa as a function of a for the SW-CFAR (9.22). In this case the threshold multiplier is set so that the GM-CFAR (5.43) achieves a Pfa of 10^{-4}.

9.5 Performance of the SW-CFAR Detector

Some examples of the performance of the decision rule (9.22) are now considered, with the same numerical analysis procedure adopted as in previous chapters. Hence it will be assumed that $N = 32$, Pfa $= 10^{-4}$ and $N_T = N - 2 = 30$, where the latter choice has been based upon the desire to manage up to two interfering targets. Based upon the analysis of the previous section, the choice of $a = 0.25$ has been adopted. A Swerling I target model is used as previously, and the simulated clutter is based upon the test cases used throughout the numerical investigations in this book. Pareto clutter parameters have also been based upon Table 3.1.

9.5.1 Performance in Homogeneous Clutter

Figures 9.6–9.8 show detector performance in homogeneous clutter for the cases of horizontal, cross and vertical polarisation respectively. The performance of the new switching CFAR is compared with the GM-CFAR (5.43) on which it is based, as well as the ideal decision rule. As can be observed from these three figures the SW-CFAR and the GM-CFAR detectors coincide in homogeneous clutter, which is to be expected.

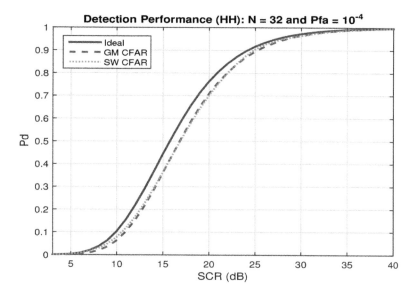

Figure 9.6: Switching CFAR detector performance in horizontal polarisation, and in homogeneous clutter. The SW-CFAR is compared with the GM-CFAR on which it is based. Since the clutter is homogeneous without interference, these two detectors coincide.

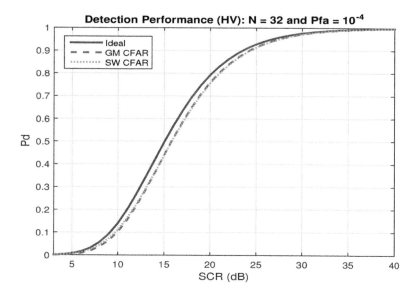

Figure 9.7: Switching detector performance in cross polarisation, showing the latter matches the performance of the GM-CFAR on which it is based.

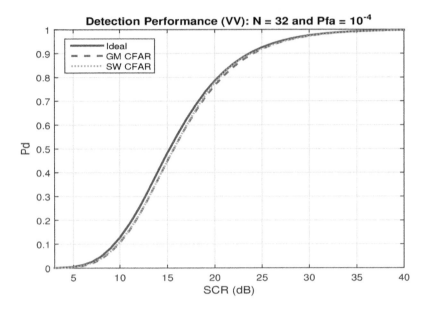

Figure 9.8: Switching detector performance in vertical polarisation, and homogeneous clutter, showing similar performance as for the other two polarisations.

9.5.2   Performance with Interference

Interference is examined as in previous chapters, with the insertion of an independent Swerling I target model into the CRP, with fixed SCR. Due to numerical complexity it is somewhat simpler to use receiver operating characteristic (ROC) curves to examine the performance. These plot the Pd as a function of the Pfa, where the target model in the CUT has fixed SCR. Since switching detectors compare the CRP to measurements of the CUT it is faster to produce ROC curves.

The effects of interference are examined when the CUT SCR is 30 dB, with a 20 dB interfering target. Figures 9.9–9.11 illustrate performance in horizontal, cross and vertical polarisation. Each of these figures show that the switching-based CFAR is robust to strong interference. It is clear that for smaller Pfa the SW-CFAR has superior performance, but for larger Pfa the SW-CFAR and GM-CFAR perform comparably.

9.5.3   False Alarm Regulation

To complete the analysis of the switching CFAR false alarm regulation is examined, where its performance is compared with the GM-CFAR on which it is based. Figures 9.12–9.14 show the estimated Pfa for a 1 dB power level increase for horizontal, cross and vertical polarisation, with $N = 32$ and a design

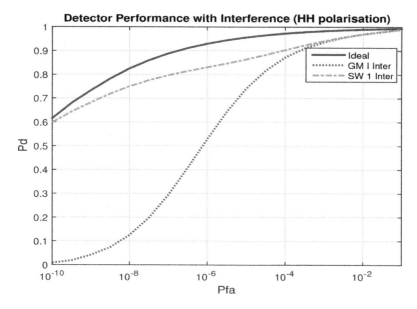

Figure 9.9: Interference in horizontal polarisation, where the CUT SCR is 30 dB and the interfering target has ICR of 20 dB.

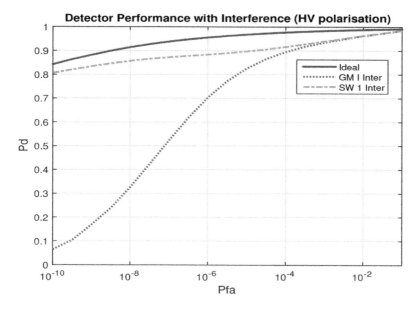

Figure 9.10: Switching CFAR performance in cross polarisation, under the same conditions as for Figure 9.9.

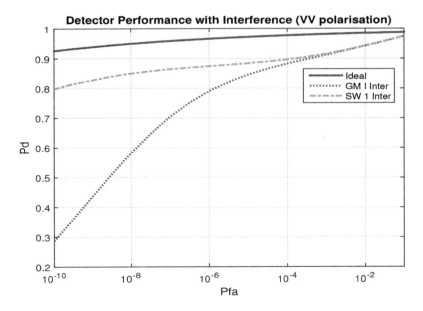

Figure 9.11: Switching detector performance in vertical polarisation with a 20 dB interfering target.

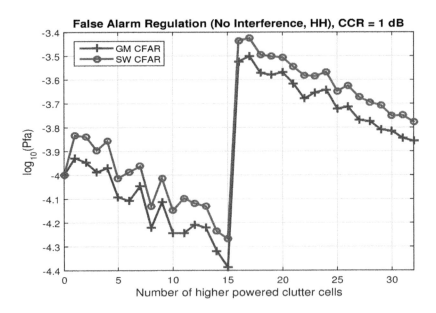

Figure 9.12: False alarm regulation in horizontal polarisation.

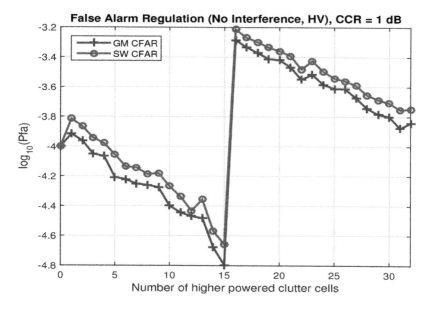

Figure 9.13: False alarm regulation in cross-polarisation.

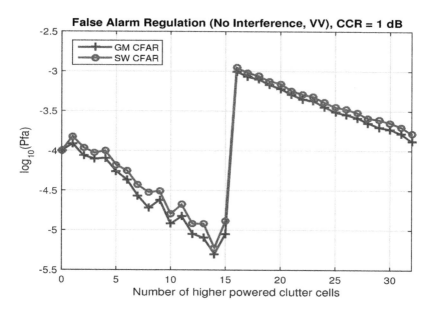

Figure 9.14: False alarm regulation of the switching detector in vertical polarisation.

Pfa of 10^{-4}. Here there is no substantial gain acquired with the switching-based approach. In fact, the switching CFAR can slightly increase the resultant Pfa during clutter transitions as shown. Such a phenomenon has been also reported in (Weinberg, 2014c), also for the Pareto Type I case.

In the presence of an interfering target in the CRP of sufficiently large ICR, the geometric mean detectors experience substantial detection losses, as has been documented in previous chapters. Hence it is worth investigating whether the switching CFAR manages the Pfa relatively well with interference. Toward this aim, Figure 9.15 applies a 20 dB interfering target to the CRP, where a 1 dB power level increase has been applied. This is in the case of horizontal polarisation, and the target is embedded within the lower powered clutter. Figure 9.15 shows that the GM-CFAR tends to reduce the number of false alarm more than the switching version. Once the CUT is affected with higher power level returns the GM-CFAR tends to have a smaller resultant Pfa than the switching detector.

Figure 9.16 repeats the same experiment except with a 30 dB interfering target. Here the GM-CFAR reduces the resultant Pfa more substantially, while the switching CFAR manages to regulate the Pfa reasonably well.

For a final example Figure 9.17 shows the resultant Pfa in vertical polarisation, with a 10 dB interfering target. With such a small level of interference there is very little difference between the resultant Pfa of both decision rules. Increasing the ICR results in similar performance as for the horizontal polarisation case.

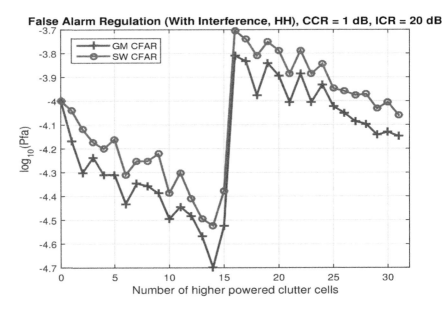

Figure 9.15: False alarm regulation with 20 dB interference.

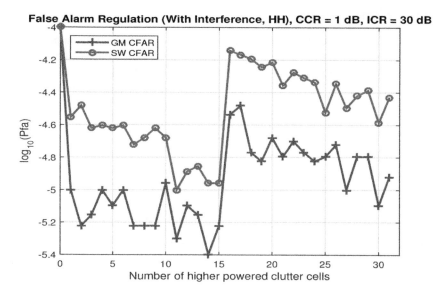

Figure 9.16: Second example of switching detector false alarm regulation in horizontal polarisation with 30 dB interference.

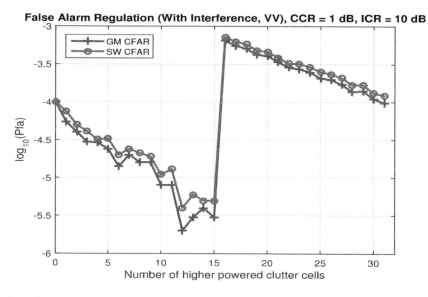

Figure 9.17: Switching CFAR false alarm regulation in vertical polarisation with 10 dB interference.

9.6   Conclusions

Switching based detectors were examined in this chapter, and the application of such a procedure for target detection in Pareto Type I distributed clutter was examined. It was shown that the original application in (Weinberg, 2014c) could be generalised to produce a generic switching detector. Furthermore it was demonstrated how the GM-CFAR (5.43) could be coupled with an alternative definition of the set S_0 to produce a switching CFAR for target detection in Pareto Type I clutter.

The performance of the latter was shown to match that of the GM-CFAR in ideal situations, while in the presence of interference the SW-CFAR had superior performance. During clutter transitions the switching CFAR provided better regulation of the Pfa in the presence of interference.

It is worth noting that the formulation of the SW-CFAR in Section 9.4 can be applied with the OS-CFAR (5.45), with the set (9.13). In such a situation the switching detector will either apply a given OS to the full CRP, or to elements of the set S_0. Given the success of an OS-based detector in the case of interference, a switched version of it may be of considerable utility. It is also worth observing that, in view of the fact that the maximum has the largest probability of detection for the class of OS detectors, a switching mechanism would enable the practical application of the maximum in the presence of interference. This is a possible line of investigation for future studies of sliding window switching CFAR detectors.

Chapter 10

Developments in Binary Integration

10.1 Introduction

Binary integration is a process which enables the results of several univariate detection schemes to be combined so that an overall detection decision can be realised with increased information. Integration is a common method used to enhance detection performance, and an example of what is known as classical non-coherent integrators is discussed in Appendix D. Binary integration was introduced in 1956 by Schwartz, who was examining what he referred to as a coincidence procedure for signal detection (Schwartz, 1956). It is interesting to note that the latter is a component of the Ph.D. thesis of Schwartz at Harvard University. Schwartz was interested in alternatives to the Neyman-Pearson optimal detector for targets in Gaussian clutter. His analysis showed that by integrating the results of a number of fixed threshold detectors, the performance of the optimal detector could also be replicated, but at an increased computational cost.

When binary integration is coupled with a radar detection process, the result can be a significant improvement in detection performance. Additionally, if the underlying decision process is CFAR with respect to a given clutter model parameter, then this feature is preserved in the binary integration process.

Subsequent to the original work of Schwartz there have been many advances and applications of binary integration relevant to the focus of this book. As an example, optimal choices for binary integration parameters was examined in (Worley, 1968). Further extensions of the method have been reported in (Weiner, 1991), (Shnidman, 2004), (Meng and Zhao, 2007), (Meng and Zhao, 2008),

(Meng, 2010) and (Detouche and Laroussi, 2011). Recent advances have focused on the application of it to enhance detection performance in a Weibull distributed clutter environment (Meng, 2013), although in this context the CFAR property is not achieved with respect to the Weibull shape parameter. In addition to this, (Weinberg and Kyprianou, 2015) examined the application of binary integration in a Pareto Type I clutter environment. The latter also included a mathematical analysis which establishes when binary integration will provide detection performance enhancement, as well as guidelines on when it actually reduces performance. This chapter will examine the work of (Weinberg and Kyprianou, 2015) in detail, and will apply binary integration with the new OS-CFAR (5.45), since in this case the CFAR property will also be achieved with the binary integration scheme.

10.2　Binary Integration

The general formulation of binary integration is mathematically simple, and is based upon the principle of counting the number of successes in a series of Bernoulli trials, where the latter is the result of independent detection decisions (Meng, 2013). Suppose that a series of M scans is available, where each consists of a series of measurements that are fed directly into a detection scheme as examined in previous chapters. Hence we suppose that each of these consists of N elements in the CRP, a CUT and guard cells on either side of the CUT. The process of binary integration is to declare a target present if at least S out of the M individual detection processes register a target present (Weiner, 1991). In order to analyse this process mathematically, we assume that each of the clutter returns in each range cell is independent. Additionally, it is assumed that the independence property holds between each of the M clutter range profiles, and the probability of detection remains uniform also, following the assumptions used in (Meng, 2013).

　　To illustrate the binary integration process in the context of sliding window decision rules, Figure 10.1 provides a schematic of such a process. It is assumed that a series of returns is passed to a sequence of decision rules, from which binary decisions on the presence of a target are made. These results are then passed to the integrator, after which a detection decision can be made. Each of the detection processes represents a sliding window decision rule as is illustrated in the subfigure as shown.

　　Suppose that p is the success probability in a series of M Bernoulli trials. Then the probability of at least S successes is given by

$$f(p) = \sum_{j=S}^{M} \binom{M}{j} p^j (1-p)^{M-j}, \tag{10.1}$$

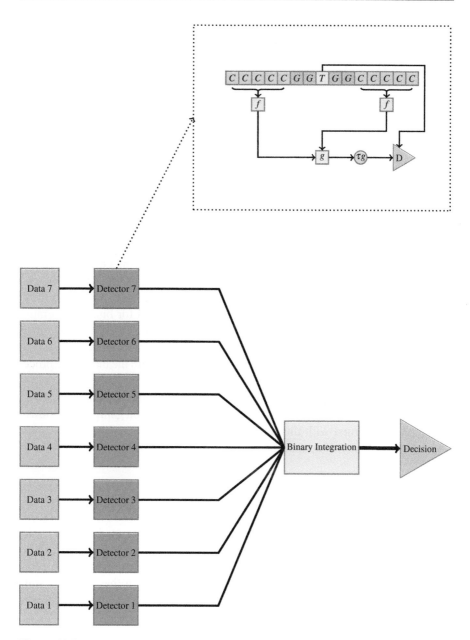

Figure 10.1: Structure of a binary integration detection process, showing inputs to a series of sliding window decision rules. These produce binary decisions which are passed to the integrator, and a detection decision is then made.

which is also identifiable as the cumulative binomial probability distribution function associated with (2.6). Under the assumption of independence, the false alarm probability for the binary integration scheme is given by $f(\text{Pfa})$, where Pfa is the univariate false alarm probability for the individual detection schemes, while if Pd denotes the detection probability associated with an individual univariate detection process, then $f(\text{Pd})$ is the detection probability of the binary integration scheme. Hence analysis in simulated clutter can be based upon (10.1) with the underlying Pfa or Pd of the individual detection processes. Observe that, in view of (10.1), if the underlying detection process has the CFAR property with respect to a given clutter parameter, then the binary integrated detection process will preserve this property. An additional advantage provided by binary integration is that the threshold multiplier is set via (10.1). If the univariate detection schemes under perform or saturate due to the magnitude of the threshold multiplier, the process of binary integration has the potential to rectify this.

Hence it is relatively straightforward to analyse binary integration's impact in a given clutter environment, with a particular detection process, with application of (10.1). To illustrate this, consider the order statistic CFAR (5.45) with Pfa set via inversion of (5.46). Then to determine the threshold for binary integration, one must determine the threshold through numerical inversion of Pfa $= f(p)$, where p is given by (5.46). The resultant probability of detection is given by $f(\text{Pd})$ where Pd is the detection probability associated with the univariate process (5.45).

It is important to investigate binary integration in a mathematical context, to determine whether such a process provides useful detection gains. The following section is concerned with this and demonstrates that binary integration gain is related to convexity of certain functions.

10.3 Mathematical Analysis of Binary Integration

A mathematical analysis of binary integration in general terms is now provided. Since the idea behind integration in a radar detection context is to improve performance via increasing information, one can ask the following question. When does the detection probability given by (10.1) improve on that from the single look/univariate detection process on which it is based? In the first instance, consider for a univariate detection probability p the requirement that

$$\sum_{j=S}^{M} \binom{M}{j} p^j (1-p)^{M-j} > p, \tag{10.2}$$

where M is considered fixed. We are interested in whether the inequality (10.2) can hold for any S, and whether there is an optimal choice for it. By simple analysis of binomial probabilities, the inequality (10.2) can be shown to be equivalent to

$$\sum_{j=0}^{S-1}\binom{M}{j}p^j(1-p)^{M-j}+p-1<0. \tag{10.3}$$

Hence define a function

$$g_S(p):=\sum_{j=0}^{S-1}\binom{M}{j}p^j(1-p)^{M-j}+p-1, \tag{10.4}$$

where $0 \le p \le 1$ and S is a fixed number in the set $\{1,2,\dots,M\}$. The binary integration process will improve on the underlying univariate detection process if there is an S for which the function (10.4) is negative. The following lemma encapsulates some of the main important properties of (10.4):

Lemma 10.1
The function $g_S(p)$ has the following properties:

1. *$g_S(0) = g_S(1) = 0$;*

2. *$g_1(p) = -(1-p)\left[1-(1-p)^{M-1}\right]$;*

3. *$g_M(p) = p\left(1-p^{M-1}\right)$;*

4. *$g_{S+1}(p) = g_S(p) + \binom{M}{S}p^S(1-p)^{M-S}$, for $2 \le S \le M-1$.*

The proof of the properties in Lemma 10.1 is relatively simple and hence is left as an exercise. Observe that property (1) in Lemma 10.1 implies that, with an application of the intermediate value theorem, for each S $g_S(p)$ will possess at least one stationary point. Property (2) implies that $g_1(p) < 0$ because $0 < p < 1$. Hence binary integration based upon $S = 1$ will always have a greater integrated detection probability.

By contrast, property (3) in Lemma 10.1 implies that $g_M(p)$ is always non-negative, and so binary integration based upon $S = M$ will always introduce a detection loss. Finally, note that the second term on the right hand side of property (4) in Lemma 10.1 is always nonnegative, therefore it follows that $g_S(p)$ is an increasing function of S, for each fixed p. Consequently one can conclude that the maximum binary integration gain will result from the $S = 1$ case.

To illustrate the behaviour of (10.4), Figures 10.2 and 10.3 plots it for the case where $M = 6$ and $M = 8$ respectively. The figures confirms the analysis presented above. What is interesting about these figures is that for each $2 \le S \le M-1$ there is a region, defined by $p > p_S$ where P_S is the root of $g_S(p)$ in $0 < p < 1$, for which binary integration will provide improvements in detection performance.

In a sliding window CFAR detection scheme one desires to maintain the Pfa at a fixed level, and thus it is necessary to determine an appropriate threshold

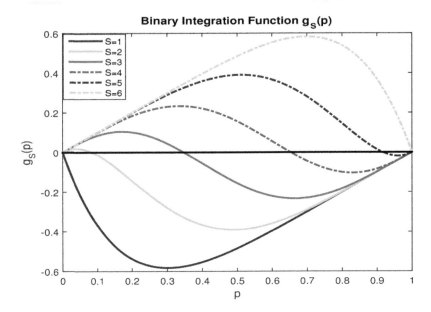

Figure 10.2: Binary integration gains occur in regions where g_S is negative. This function is shown for the case where $M = 6$, and for the six possible values for S.

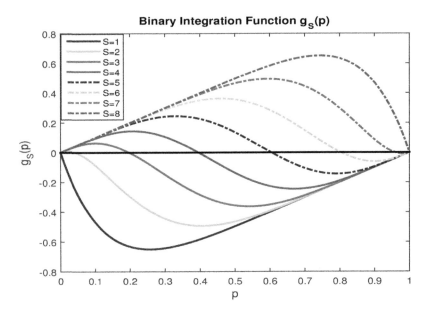

Figure 10.3: The function g_S when $M = 8$. Detection gains occur when g_S is negative.

multiplier to use in the underlying detection process based upon setting the Pfa through (10.1). As a result of this, the corresponding univariate detection probability, used in each binary integration and for the same Pfa, will be produced with a different threshold multiplier, and so one needs to account for this with the selection of an optimal S.

Consider the binary integration Pfa defined by $f(\text{Pfa})$ with (10.1). Due to the fact that in cases of interest, Pfa will be very small (such as 10^{-4} or smaller) one can apply a Poisson approximation to the binomial probability, as outlined in Chapter 2. Thus, applying a Poisson approximation to (10.1) one obtains

$$f(\text{Pfa}) \approx 1 - e^{-M\text{Pfa}} \sum_{k=0}^{S-1} \frac{(M\text{Pfa})^k}{k!} \approx 1 - e^{-M\text{Pfa}}, \qquad (10.5)$$

where the second approximation follows from the fact that $\text{Pfa}^k \ll \text{Pfa} \ll 1$ for all $k \geq 1$. Hence, based upon (10.5)

$$\text{Pfa} \approx -\frac{1}{M} \log\left(1 - f(\text{Pfa})\right), \qquad (10.6)$$

expressing the Pfa used in each integrated detection process in terms of the desired binary integration Pfa. The detection threshold multiplier, denoted τ_{BI}, used in each binary integrated process can then be obtained from

$$\frac{N!}{(N+1)(N-j)!} \frac{\Gamma(N-j+\tau_{BI}+1)}{\Gamma(N+\tau_{BI})} = -\frac{1}{M} \log\left(1 - f(\text{Pfa})\right), \qquad (10.7)$$

which has been obtained from the Pfa (1.9) and (10.6). In order to analyse (10.7), it is necessary to specialise the analysis to the underlying detection process to be used as the basis for binary integration. Hence for brevity the analysis will be restricted to the new OS-CFAR detector (5.45) in the following. Critical to this analysis is the following technical lemma:

Lemma 10.2
The probability of false alarm (5.46) of the OS detector (5.45) is a decreasing function of its threshold multiplier τ.

The proof of this lemma now follows, although it is very similar to the corresponding proof of the fact that (1.9) has the same property, which can be found in (Weinberg and Kyprianou, 2015). Suppppose that the order statistic index is j.

By the definition of the Gamma function, it can be shown that

$$\Gamma(\tau+N) = (\tau+N-1)(\tau+N-2)\ldots(\tau+N-(j-1))\Gamma(\tau+N-(j-1)), \qquad (10.8)$$

from which it follows that

$$\frac{\Gamma(\tau+N-j+1)}{\Gamma(\tau+N)} = \frac{1}{(\tau+N-1)(\tau+N-2)\ldots(\tau+N-(j-1))}. \qquad (10.9)$$

An application of this to (5.46), together with a logarithmic transformation, results in

$$-\log(\tau+N-1) \quad - \quad \log(\tau-N-2)-\cdots-\log(\tau+N-(j-1))$$

$$= \quad \log\left(\frac{(N+1)(N-j)!}{N!}\right)+\log(\text{Pfa}). \qquad (10.10)$$

Viewing the threshold multiplier as a function of the Pfa, one can differentiate (10.10) to obtain

$$\frac{d\tau}{d\text{Pfa}} = \frac{-\text{Pfa}^{-1}}{\left[\frac{1}{\tau+N-1}+\cdots+\frac{1}{\tau+N-(j-1)}\right]} < 0, \qquad (10.11)$$

which demonstrates that τ is a decreasing function of the Pfa, as required.

Returning to (10.7), based upon a Taylor series expansion,

$$\log(1-\theta) = -\sum_{n=1}^{\infty}\frac{\theta^n}{n} \approx -\theta, \qquad (10.12)$$

provided $0 < \theta << 1$. Applying this to the right hand side of (10.7), one concludes that

$$-\frac{1}{M}\log(1-f(\text{Pfa})) \approx \frac{f(\text{Pfa})}{M} < f(\text{Pfa}), \qquad (10.13)$$

because $M > 1$. From (10.13), the univariate detection processes used in binary integration are set with a Pfa smaller than the design Pfa, and hence it follows from the above considerations that the threshold multiplier τ_{BI} is greater than that used for univariate detection based upon (5.11). It is immediate also that the binary integration univariate probability of detection will be smaller than that of a single detection process (5.11). This can be realised by noting that if the underlying detection statistic is Z (for brevity), then if we let τ_{BIU} be the univariate binary integration threshold multiplier, and τ_U be that for the univariate detection process, since $\tau_{BIU} > \tau_U$, then $Z \geq \tau_{BIU}$ will imply $Z \geq \tau_U$. Hence $P(Z \geq \tau_{BIU}|H_1) \leq P(Z \geq \tau_U|H_1)$.

Consequently, for the case of detection with binary integration, the analysis in the preceding discussions needs to be reconsidered.

Hence towards acquiring an understanding of when binary integration will improve upon the performance of the univariate detection process on which it is based, consider the problem of finding an $S \in \{1,2,\ldots,M\}$ such that for fixed θ_1 and θ_2 (both of which take values in the unit interval and $\theta_1 < \theta_2$) and fixed M, the inequality

$$P(\text{Bin}(M,\theta_1) \geq S) > \theta_2 \qquad (10.14)$$

holds, where for brevity θ_1 is the univariate probability of detection set for binary integration, and θ_2 is the detection probability for the underlying process on which the integration is based, both of which have the same desired false alarm probability. Also for brevity the binomial random variable with parameters M and θ_1 is expressed in terms of its distributional representation. The specification in (10.14) is slightly more general than necessary, but facilitates the mathematical analysis. Due to the fact that the detection probabilities will depend on the clutter characteristics as well as the received target signature, and especially the target strength, it is difficult to determine a mathematically exact relationship between θ_1 and θ_2. However, one can show this general formulation can be used to determine whether detection preformance gains are possible with binary integration.

To begin, consider the case where $S = 1$. Then for binary integration to introduce gains, we require $\theta_1 < \theta_2$ and $1 - (1 - \theta_1)^M > \theta_2$. For convenience define $\phi_1 := 1 - \theta_1$ and $\phi_2 := 1 - \theta_2$. Thus, one requires $\phi_2 < \phi_1$ and $\phi_2 > \phi_1^M$. Observe that over the interval $[0, 1]$, the function $g(\phi) = \phi^M$ is convex for $M > 1$. Hence, these conditions can always be met, and so binary integration will always provide detection performance improvements when $S = 1$. This condition parallels the example considered previously, where $\theta_1 = \theta_2$. Note that if $M = 1$ the conditions above can never be met. To investigate this further, it is clear that when $S = M$,

$$P(\mathrm{Bin}(M, \theta_1) = M) = \theta_1^M < \theta_2^M < \theta_2 \qquad (10.15)$$

implying (10.14) can never hold, and thus binary integration will never yield improvements in detection performance. This result also parallels that for the case where $\theta_1 = \theta_2$.

For the scenario where $2 \le S \le M - 1$, the analysis becomes more involved. However, the key to it is related to the convexity property mentioned in the $S = 1$ case. One requires $\theta_1 < \theta_2$ and

$$\sum_{j=S}^{M} \binom{M}{j} \theta_1^j (1 - \theta_1)^{M-j} > \theta_2. \qquad (10.16)$$

Applying the definitions of ϕ_1 and ϕ_2 as used previously, then it is immediate that (10.16) is equivalent to

$$\sum_{j=0}^{S-1} \binom{M}{j} (1 - \phi_1)^j \phi_1^{M-j} < \phi_2, \qquad (10.17)$$

with the requirement that $\phi_1 > \phi_2$. Hence define a function h on the unit interval by

$$h(\theta) := \sum_{j=0}^{S-1} \binom{M}{j} (1 - \theta)^j \theta^{M-j}. \qquad (10.18)$$

Binary integration will provide an improvement in performance if the function h is convex in subsets of the unit interval. Hence an analysis of its derivatives is required. Firstly, observe that $h(0) = 0$ and $h(1) = 1$. By an application of differentiation, one can establish that

$$h'(\theta) = (M - M^2)\theta^{M-1} + (M^2 - M)\theta^{M-2} + \sum_{j=2}^{S-1} \binom{M}{j} \times$$

$$\left[-j(1-\theta)^{j-1}\theta^{M-j} + (M-j)(1-\theta)^j \theta^{M-j-1} \right]. \quad (10.19)$$

Note that the derivative (10.19) is zero at $\theta = 1$, implying a stationary point.

By a second application of differentiation to (10.19), one can arrive at the second derivative

$$h''(\theta) = \frac{1}{2}(M-1)^2 M(M-2)\theta^{M-2} + M(M-1)(M-2)\theta^{M-3}$$

$$+ M(M-1) \sum_{j=3}^{S-1} \binom{M}{j}(1-\theta)^{j-2}\theta^{M-j}$$

$$\sum_{j=2}^{S-1} \binom{M}{j}(1-\theta)^{j-2}\theta^{M-j-2} \times$$

$$(M-j)(M-j-1-2(M-j)\theta). \quad (10.20)$$

Since the function h is twice differentiable, it follows that it will be convex provided $h''(\theta) > 0$. Inspection of (10.20) shows that this will occur provided $(M-j)(M-j-1-2(M-j)\theta) > 0$. Since $j < M$ we require $M - j - 1 - 2(M - j)\theta > 0$. It therefore follows that

$$\theta < 1 - \frac{1}{M-j} < 1 - \frac{1}{M-1}, \quad (10.21)$$

for all $j > 1$, establishing an upper bound on the convexity limit point. Hence, the function h is convex provided the condition (10.21) holds.

This can now be applied to establish when binary integration can provide detection performance gains. Recalling that we require $\phi_2 < \phi_1$ and $\phi_2 > h(\phi_1)$, this will occur in the region of convexity of h, and so when $0 < \phi_1 < 1 - \frac{1}{M-1}$. Relating this back to the original θ_1, it requires $\theta_1 < \frac{1}{M-1}$.

Figures 10.4–10.7 illustrate the regions where binary integration can provide detection gains relative to an underlying detection process on which it is based. The regions plot the line $\phi_2 = \phi_1$ as well as the curve $\phi_2 = h(\phi_1)$, and shade the region where binary integration yields a greater detection probability. Figure 10.4 is for the case where $S = 1$, Figure 10.5 is for $S = 4$, Figure 10.6 corresponds to

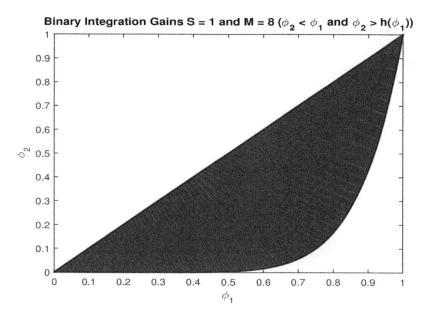

Figure 10.4: Binary integration for the case of $S = 1$ out of $M = 8$ always yields gains.

$S = 6$ while Figure 10.7 is for $S = 8$. In all cases $M = 8$, and it can be observed that the region of improvement decreases as S increases. The function h changes from a purely convex function for $S = 1$ to a purely concave function at $S = M = 8$. In the intermediate stages the improvement region is a function of the inflexion point of h on the unit interval.

The results in this section demonstrate that binary integration has the potential to improve detection performance. However it is difficult in general to determine the degree of improvement analytically. The next section is concerned with the practical selection of the parameter S in a binary integration scheme.

10.4   Binary Integration Parameter S

The question of relevance to now be considered is the selection of the parameter S, under the assumption that the maximum number of integrations M is fixed. The problem of optimal selection of S has been examined extensively in previous work; see for example (Worley, 1968), (Weiner, 1991) and (Frey, 1996). As discussed in the latter, for a specified univariate cumulative detection probability and false alarm rate, and a fixed number of maximum binary integration returns, there exists an optimal S which minimises the required SCR, and maximises the binary integration gain. This can be done visually or numerically by plotting the minimum SCR as a function of S, under the assumption of a certain signal model.

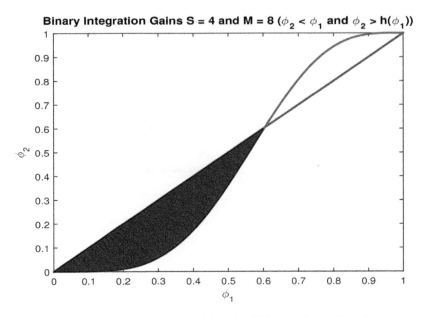

Figure 10.5: Gains for $S = 4$ out of $M = 8$, which occur in the shaded region.

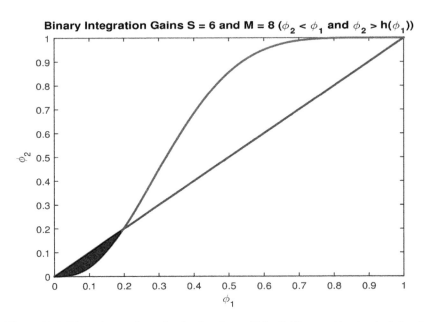

Figure 10.6: Binary integration gains for $S = 6$ out of $M = 8$, illustrated by the shaded region. This has significantly reduced from the previous two examples.

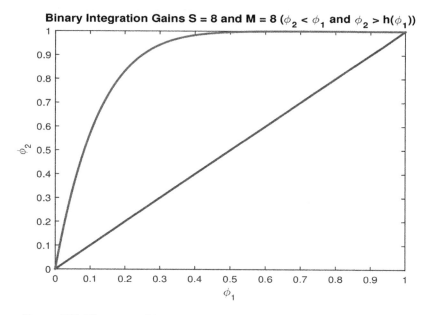

Figure 10.7: There are no binary integration gains for the case of $S = 8$ out of $M = 8$.

To illustrate this, for the three sets of Pareto distributional parameters considered throughout the book and with the two CFAR detectors (5.43) and (5.45), Figures 10.8–10.13 plot this minimum required SCR to produce the minimum univariate Pd, as a function of the binary integration parameter S. Figures 10.8, 10.10 and 10.12 are for the GM-CFAR (5.43) under HH- HV- and VV-polarisations respectively. The GM-CFAR is assumed to operate with a Pfa of 10^{-4}, with $N = 16$ and with $M = 6$, together with a Swerling I target model requiring a minimum Pd of 0.5. Based upon these figures, and for this particular detection scenario, one can conclude that $S = 3$ is optimal for HH-polarisation, while $S = 2$ is appropriate for the other two polarisations.

Figures 10.9, 10.11 and 10.13 are for the OS-CFAR (5.45) with $N = 32$, $k = N - 2$, and a Pfa of 10^{-4}. The binary integration maximum is $M = 8$ and a Swerling I target is again assumed, with a minimum required Pd of 0.5 as before. Based upon the series of figures the optimal choices for S are 3 (for HH-polarisation) and 2 for the other two polarisations. Such choices for S will be adopted in the analysis of detector performance to follow.

10.5   Performance in Homogeneous Clutter

Numerical analysis of binary integration is now undertaken, in the case of homogeneous Pareto Type I clutter. In this context the performance of both the GM-

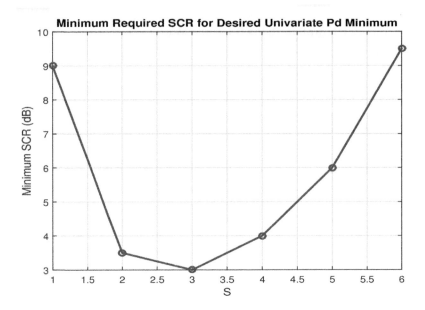

Figure 10.8: Optimal selection of S for the GM-CFAR in HH-polarisation. The figure shows that $S = 3$ will be the optimal choice.

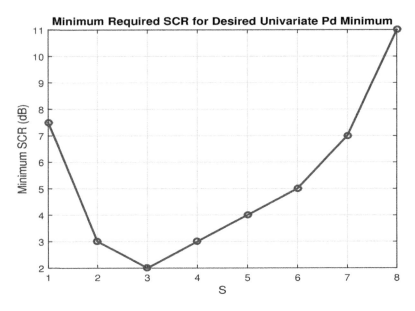

Figure 10.9: Optimal selection of S for the OS-CFAR in HH-polarisation. The optimal choice is again $S = 3$.

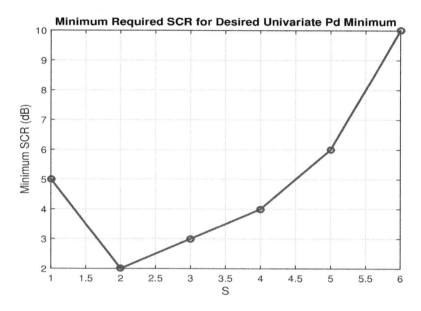

Figure 10.10: Optimal selection of S for the GM-CFAR in HV-polarisation. In this case $S = 2$ is optimal.

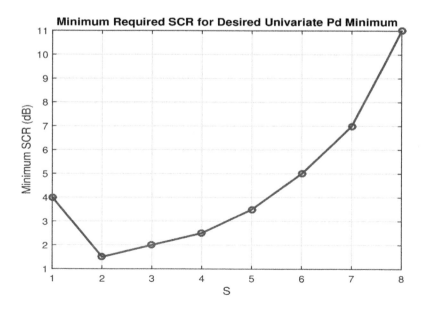

Figure 10.11: Optimal selection of S for the OS-CFAR in HV-polarisation, again indicating that the ideal choice is $S = 2$.

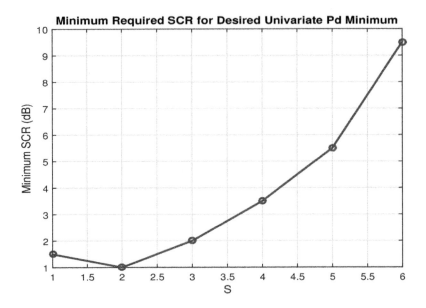

Figure 10.12: Optimal selection of S for the GM-CFAR in VV-polarisation. As in the case of cross polarisation the choice of $S = 2$ is ideal.

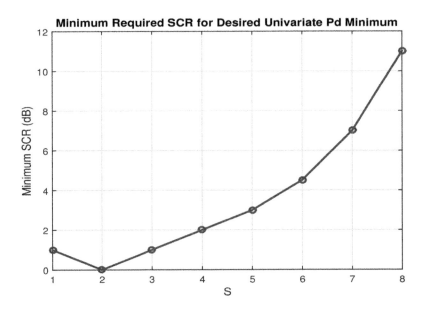

Figure 10.13: Optimal selection of S for the OS-CFAR in VV-polarisation. As previously, the optimal choice is $S = 2$.

and OS-CFARs will be examined where they are employed as the underlying univariate decision rules. For each scenario the Pfa is set to 10^{-4}, while for the OS-CFAR case, $N = 32$ and $k = N - 2$ with the maximum number of integrations set to $M = 8$. In this situation S is varied from 1 to 8 to illustrate performance. The GM-CFAR is based upon $N = 16$ with the same Pfa as for the OS-CFAR case, with $M = 6$. The reason for the reduction in N and S is one of computational complexity, since it has been found that binary integration coupled with the GM-CFAR requires an inordinate Monte Carlo sample size for larger N and M. A Swerling I target model will be assumed as in previous chapters, and Pareto clutter parameters have been sourced from Table 3.1.

Figure 10.14 shows the performance for horizontal polarisation with the OS-CFAR as basis detector. Here one observes that the maximum gain corresponds to $S = 3$, as expected.

Figure 10.15 is for the case where the basis detector is the GM-CFAR, operating in HH-polarisation, and as expected the optimal binary integration again occurs when $S = 3$.

Figure 10.16 examines the case where the univariate detection process is the OS-CFAR in cross-polarisation, showing the maximum binary integration gain is achieved when $S = 2$. The corresponding result for the GM-CFAR can be found in Figure 10.17, also showing that the maximum gain is at $S = 2$.

Figures 10.18 and 10.19 examine binary integration in vertical polarisation, with the OS- and GM-CFARs as basis detectors respectively. The figures confirm the maximum gain is achieved when $S = 2$.

Each of the optimal performing binary integration processes in Figures 10.14–10.19 can be compared with the performance of the univariate CFAR processs and shown to provide a substantial gain over the latter. The next section will examine interfering targets within the binary integration process.

10.6   Performance with Interference

Analysis of a binary integration scheme with interfering targets requires some careful considerations regarding the way in which such targets appear in the process. Consider again the binary integration schematic in Figure 10.1, where each of the univariate detection processes are processsing subsamples of incoming data. If there is a radar target present in the returns being passed to this cascaded detection scheme, then it will slowly permeate throughout each of the univariate detection processes. Simulation of such an occurrence is very complicated and so is usually not examined within the binary integration literature. For the purposes of computational ease, the way in which interfering targets will be studied is to assume that such targets arise from spillover from CUTs within the univariate detection processes. Thus one can consider the result of a single interfering target inserted into one of univariate detection processes' CRP. In addition to this,

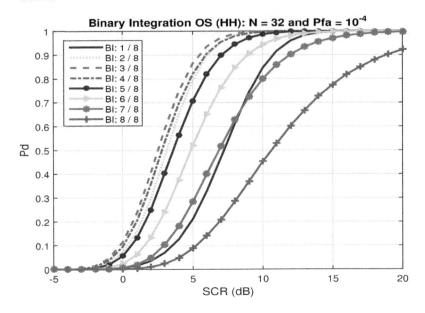

Figure 10.14: Binary integration performance, with an OS-CFAR, in horizontal polarisation. The figure confirms that the maximal binary integration gain is for $S = 3$.

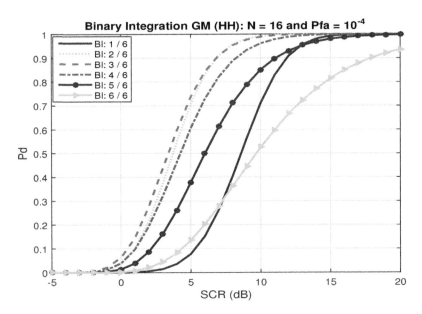

Figure 10.15: Binary integration with the GM-CFAR in horizontal polarisation. Clearly the maximum gain is achieved when $S = 3$.

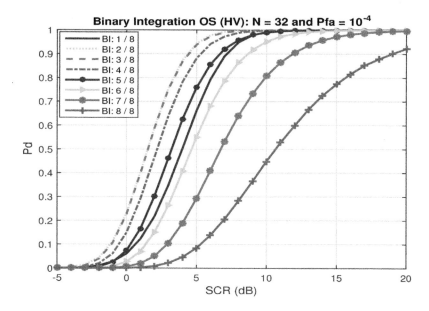

Figure 10.16: Binary integration performance in cross-polarisation with the OS-CFAR. The figure shows that maximal gains can be achieved with either $S = 2$ or 3.

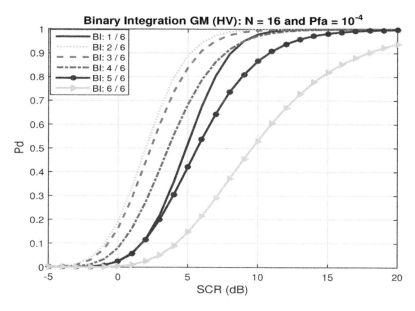

Figure 10.17: Example of binary integration in cross-polarisation with the GM-CFAR, from which it is clear that the optimal choice for S is 2.

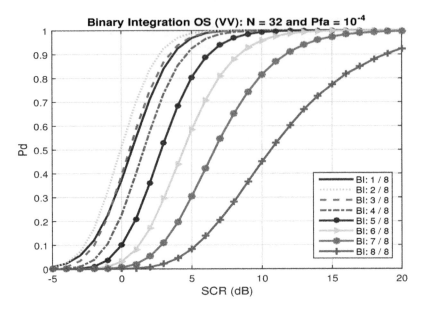

Figure 10.18: Binary integration in vertical polarisation based upon the OS-CFAR. This also shows that $S = 2$ is optimal.

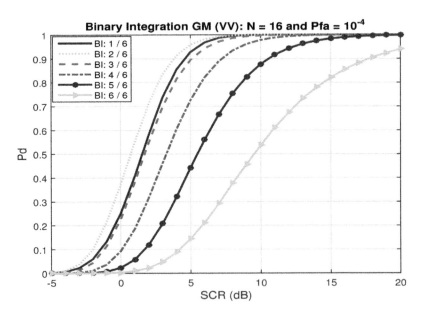

Figure 10.19: Binary integration in vertical polarisation, with the GM-CFAR. Here $S = 2$ has maximal probability of detection.

other univariate detectors can also be affected with interference, to examine the overall effect on the binary integration process.

Based upon this methodology, one can investigate the effects of interference using the following. Suppose that the binary integration process is based upon an S out of M criterion. For a single outcome of the binary integration process, let W count the number of the univariate detection processes that detect the presence of a target. If it is assumed that there are J processes that are subjected to interference, and $M - j$ that are not, then one can write

$$W = \sum_{j \notin J} W_j + \sum_{j \in J} W_j, \qquad (10.22)$$

where each W_j is a Bernoulli random variable, taking the values 0 or 1 depending on whether the corresponding univariate detection process has found a target present in the corresponding CUT. Adopting the assumption of independence, it follows that the first sum in (10.22) has a binomial distribution with parameters $M - J$ and p_U, where the latter is the univariate Pd for the binary integration scheme in the case of no interference. The second sum in (10.22) is a binomial-type distribution, except the associate Bernoulli random variables have Pd which corresponds to that for the univariate detection process with an assumed number of interfering targets present. Hence in order to simulate the effects of interference in the cascaded detection scheme, one can base the analysis upon (10.22). Throughout the following the Pfa is set to 10^{-4} and $N = 32$, with independent Swerling I fluctuation models used for all targets. Since the OS-CFAR is robust to interference, this basis detector will be used throughout for brevity. This univariate detection process has employed $k = N - 2$ throughout.

Figure 10.20 provides an example in the case of horizontal polarisation. In this situation all interfering targets have SCR of 10 dB, and the binary integration is based upon $S = 3$ out of $M = 8$. The figure shows the standard binary integration without interference (denoted Case 0), and then the result when one of the univariate detection processes has one interfering target in its CRP (Case 1). Case 2 illustrates the situation where two of the univariate detection processes have interference. One of these replicates the Case 1 example, while the second includes two interfering targets in a CRP. Hence in this case the cascaded scheme has a total of three interfering targets. The final example, denoted Case 3, combines Case 2 with another detection process which is subjected to three interfering targets. Thus in Case 3 there are a total of six interfering targets present. Examination of the figure shows that as the number of interfering targets increases, the performance decreases slightly. This same experiment is replicated in Figure 10.21, where the ICR of each interfering target has been increased to 50 dB. Due to the fact that the OS-CFAR is robust to interference, there is very little variation in the results.

It is also important to remark that the numerical methodology used to produce Figure 10.20, as well as the other interference examples to follow, differs considerably from that used in the analysis in homogeneous clutter. In the latter case one can exploit properties of the univariate decision rules to enhance the Monte Carlo estimation. When considering interference, this is not possible, and so there will be a Monte Carlo error incurred within the plots of this section, which explains slight deviation in the case of no interference. However, the examples still provide an outline of expected performance when the binary integration process is subjected to interference.

Figures 10.22 and 10.23 repeat the same numerical experiments illustrated in Figures 10.20 and 10.21 respectively, except in cross polarisation. The binary integration process has been based upon $S = 2$ out of $M = 8$. The results are very similar to the horizontal polarisation case.

Final examples of the performance of binary integration in the presence of interference is shown in Figures 10.24 and 10.25, which is for the case of vertical polarisation. In the case of Figure 10.24 the ICR of each interfering target is 10 dB, while in Figure 10.25 this has been increased to 50 dB. The binary integration scheme also uses $S = 2$ out of $M = 8$ as for cross-polarisation. The conclusions on performance are as for the previous two polarisation cases.

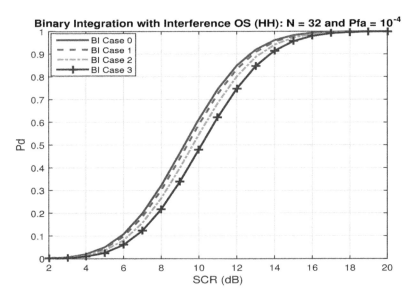

Figure 10.20: Interference in horizontal polarisation. Case 0 is standard binary integration, while Case I shows performance where one of the univariate detection processes has a single intefering target. Case 2 consists of two of the univariate detection processes having interfering targets (one and two respectively). Case 3 has three univariate detection processes with interference, consisting of one, two and three spurious targets respectively. Each of these have ICR of 10 dB.

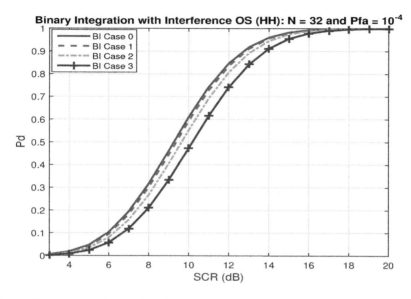

Figure 10.21: A second example of interference in binary integration with horizontal polarisation with the same test scenario as for Figure 10.20, with the exception that each ICR has been increased to 50 dB.

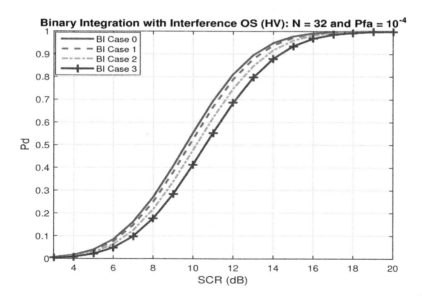

Figure 10.22: Binary integration: interference in cross-polarisation, where each interfering target has ICR of 10 dB.

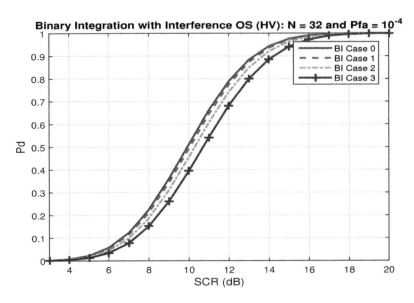

Figure 10.23: Second example of performance with binary integration in cross-polarisation. Here each interfering target has ICR of 50 dB.

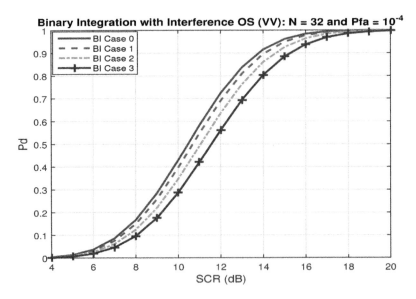

Figure 10.24: Binary integration detection performance in vertical polarisation, with ICR of 10 dB.

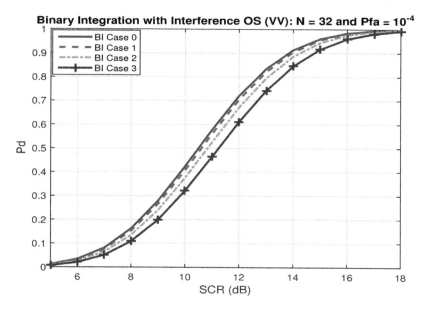

Figure 10.25: Second example of the effects of interference in vertical polarisation, with 50 dB interference.

10.7 Clutter Transitions

As for the analysis of interfering targets it can be a somewhat complicated undertaking to investigate clutter power shifts applied to the underlying univariate detection processes in a cascaded binary integration scheme. Again with reference to Figure 10.1, to study the resultant Pfa as the cascaded detection scheme is saturated progressively with higher powered clutter returns is problematic. The way in which this will be done is to investigate this by firstly assuming that one of the univariate detection processes is gradually filled with higher power clutter returns. Then it is assumed that a second detector in the cascaded scheme is also saturated with higher powered returns. As an upper bound, or indicator of the extreme case, the scenario where each of the univariate processes are affected with higher powered clutter is also included. These results, nonetheless, will provide an indicator of the effects of clutter power variations in a binary integration process. For simplicity only the case where the OS-CFAR is used as the basis univariate detection scheme will be considered. As in previous studies, the design Pfa is set to 10^{-4} and $N = 32$.

Figure 10.26 is for the case of horizontal polarisation, where the clutter power is increased by 1 dB, and the binary integration scheme is based upon $S = 3$ out of $M = 8$. Case 1 refers to the scenario where one of the univariate detection processes is subjected to clutter transitions, while the remainder are subjected to no variation in clutter power. Case 2 is for the situation where two of the univariate detectors experience clutter power level increases. Finally, Case 3 corresponds to all of the univariate detection processes being subjected to power level increases. The figure shows that when one or two of the univariate detection processes experience clutter power variation, the resultant Pfa is not an extreme deviation from the design Pfa. When all of the univariate processes experience clutter power variations, the resultant Pfa varies more considerably, as expected.

The same numerical experiment is repeated in Figure 10.27 with the exception that the clutter power level is increased by 5 dB. As can be observed, there is a sharper deviation from the design Pfa, especially in the scenario of Case 3.

Recall that throughout the numerical investigations included in this book, false alarm regulation is better managed generally in horizontal polarisation. In the cases of cross- and vertical polarisation there is usually a larger deviation from the design Pfa. The same phenomenon occurs with binary integration and so for these two polarisations only a 1 dB power level increase is examined. For these two polarisations the binary integration processes use $S = 2$ and $M = 8$. Figure 10.28 shows the results for cross-polarisation, while Figure 10.29 is for vertical polarisation. The same case scenario is employed as in the previous two examples. These two simulations demonstrate that false alarm regulation is achieved more effectively in horizontal polarisation, with a large deviation from the design Pfa in vertical polarisation.

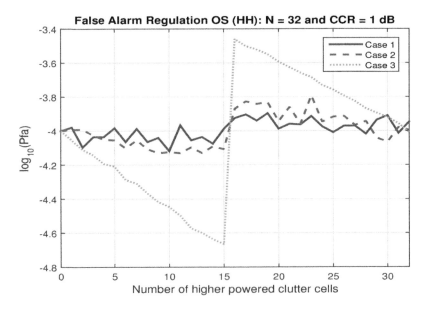

Figure 10.26: False alarm regulation in horizontal polarisation.

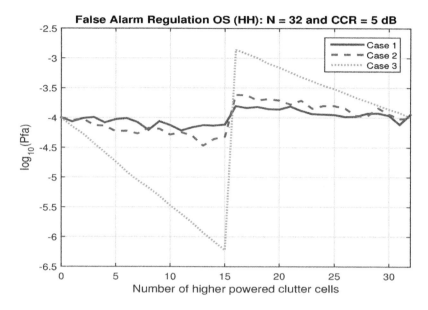

Figure 10.27: Second example of false alarm regulation in horizontal polarisation.

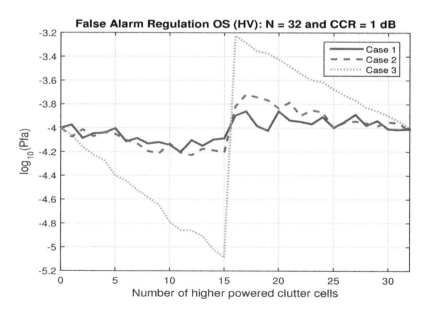

Figure 10.28: False alarm regulation in cross-polarisation.

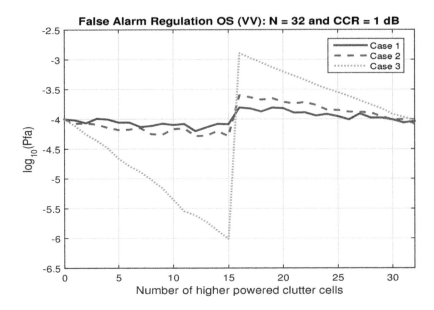

Figure 10.29: False alarm regulation in the vertically polarised case.

10.8   Conclusions

This chapter outlined recent advances in binary integration, with a view to enhancing the performance of sliding window CFAR detection processes in Pareto Type I clutter. It was shown, with an analysis of convex functions, that binary integration can enhance the performance of the univariate decision rule upon which it is based. However, in some cases it can actually result in a loss relative to the univariate detection process, if an inappropriate S was selected. Guidelines on selection of the optimal S were provided, and it was shown how this could be determined for a particular test scenario. This was demonstrated in the homogeneous clutter case, where plots of the estimated Pd for each of the considered binary integration processes confirmed the theoretical selection of an optimal S. Interference was studied and it was shown that binary integration with an OS-CFAR as basis detector could manage such spurious targets very well. Studies of false alarm regulation showed that the integration process managed clutter power level changes reasonably well until the entire cascaded detection scheme was affected with increased clutter power returns.

Chapter 11

Detection in Range Correlated Clutter

11.1 Introduction

A *sine qua non* underlying the development of the decision rules derived in this book thus far has been the existence of independent and identically distributed clutter statistics for the clutter range profile. Real radar clutter returns will often be correlated spatially and hence it is important to account for this in the design of decision rules. This will result in an inevitable complication in the derivation of sliding window CFAR detectors. Nonetheless it is possible to produce detectors to account for correlated clutter. If one makes the assumption that the CUT and CRP are independent then it is possible to use results from (David and Nagaraja, 2003) to construct an OS-based detector in the case of correlated clutter. This was analysed in (Weinberg, 2017c) and the results are described here. To illustrate the results from a practical perspective a particular multivariate Pareto distribution, attributed to (Mardia, 1962), will be used as the underlying model for the CRP statistics. This model has marginal Pareto Type I distributions such that there is a constant correlation between any two such marginal distributions. This common correlation is exactly the reciprocal of the Pareto shape parameter, providing a multivariate model whose correlation decreases with increasing Pareto shape parameters. This is consistent with the fact that X-band maritime surveillance radar clutter tends to be spikier in the horizontally polarised case (small shape parameters) while less spiky in the vertically polarised case (larger Pareto shape parameters); see for example (Weinberg, 2011a).

The analysis begins with derivation of a general form of an OS-based detector for correlated clutter.

11.2 Decision Rule in Correlated Clutter

Due to the existence of expressions for the kth OS for a series of correlated random variables, in terms of joint distribution functions of the underlying univariate random variable, it is convenient to consider the development of an OS-based detector. This is also further motivated by the fact that an OS-based decision rule is robust to strong interference. The key requirement is the determination of an appropriate threshold multiplier for application in an OS-detector. Here the focus will be on the fundamental decision rule of the form

$$Z_0 \underset{H_0}{\overset{H_1}{\gtrless}} \tau Z_{(k)} \tag{11.1}$$

which was considered in the case of independent clutter in Chapter 4. Here the CUT statistic is Z_0 and $Z_{(k)}$ is the kth OS for the CRP $\{Z_1, Z_2, \ldots, Z_N\}$ as before. It is worth remarking that one can apply a similar analysis to the decision rule (5.11), however, it has been found that the detector (11.1), when applied to the case of correlated Pareto distributed clutter, preserves the CFAR property with respect to the Pareto scale parameter but not its shape. The decision rule (5.11), as may be recalled, is dependent on the scale parameter but CFAR with respect to the shape parameter, when operating in independently distributed clutter. When applied to the case of correlated clutter, this detector loses the CFAR property completely. In addition to this, its threshold multiplier tends to be rather large resulting in detector saturation issues. Hence this decision rule is not considered in the following developments.

The CFAR decision rule (5.45) could also be used in the context of this chapter. However, the fact that this is dependent on the minimum of the correlated CRP, adds an extra dimension of complexity to the problem. This development is thus deferred to subsequent studies of detector development in correlated clutter.

The key to applying (11.1) to the situation of interest is to express its Pfa in terms of the distribution function of the kth order statistic of correlated returns, allowing the determination of an appropriate τ for (11.1). Towards this objective, adopting the assumption that the CUT and CRP statistics are independent, by conditioning on the CUT statistic under the null hypothesis, the Pfa of (11.1) is

$$\text{Pfa} = P(Z_0 > \tau Z_{(k)} | H_0)$$

$$= \int_\beta^\infty \frac{\alpha \beta^\alpha}{t^{\alpha+1}} F_{Z_{(k)}}\left(\frac{t}{\tau}\right) dt, \tag{11.2}$$

where the Pareto Type I density has been applied.

The assumption that the CUT and CRP statistics are independent has a practical justification relative to the manner in which sliding window detection schemes are implemented. Specifically, the guard cells serve a dual purpose of reducing the correlation between the CUT and CRP statistics, as well as limiting the effects of a range spread target.

Noting that the OS must exceed the Pareto minimum β, the lower integrand in (11.2) can be changed to $\beta\tau$ since the OS distribution function is zero to the left of this point. Hence by applying this, together with a change of variables given by $z = \frac{t}{\beta\tau}$, the Pfa reduces to

$$\text{Pfa} = \frac{\alpha\beta^{\alpha}}{(\beta\tau)^{\alpha}} \int_{1}^{\infty} z^{-\alpha-1} F_{Z_{(k)}}(\beta z) dz. \tag{11.3}$$

Finally, changing variables via $\omega = z^{-\alpha}$, the Pfa is reduced to

$$\text{Pfa} = \tau^{-\alpha} \int_{0}^{1} F_{Z_{(k)}}\left(\beta\omega^{-\frac{1}{\alpha}}\right) d\omega. \tag{11.4}$$

Suppose that $F_{Z_{i_1}, Z_{i_2}, \ldots, Z_{i_j}}(z_{i_1}, z_{i_2}, \ldots, z_{i_j}) = P(Z_{i_1} \leq z_{i_1}, Z_{i_2} \leq z_{i_2}, \ldots, Z_{i_j} \leq z_{i_j})$ is the multivariate distribution of the jointly distributed random variables $Z_{i_1}, Z_{i_2}, \ldots Z_{i_j}$ for indices i and j. Then the kth OS for a series of N random variables $Z_1, Z_2, \ldots Z_N$ is given by

$$F_{Z_{(k)}}(t) = \sum_{j=k}^{N} (-1)^{j-k} \binom{j-1}{k-1} H_j(t), \tag{11.5}$$

where the function $H_j(t)$ is defined by

$$H_j(t) = \sum_{1 \leq i_{j+1} < i_{j+2} < \ldots < i_N \leq N} F_{Z_{i_1}, Z_{i_2}, \ldots, Z_{i_j}}(t, t, \ldots, t),$$

$$\tag{11.6}$$

where it is important to note that the sum is over the indices which form the complement of the set of indices over which the distribution function is evaluated. This result can be found in (David and Nagaraja, 2003). Additionally, it is informative to note that the distribution function in (11.5) is equivalent to a maximum order statistic, over the underlying sample, evaluated at t. Hence an application of (11.5) to (11.3) results in

$$\text{Pfa} = \tau^{-\alpha} \sum_{j=k}^{N} (-1)^{j-k} \binom{j-1}{k-1} \int_{0}^{1} H_j\left(\beta z^{-\frac{1}{\alpha}}\right) dz. \tag{11.7}$$

Clearly one requires a specific choice for the multivariate Pareto distribution to evaluate (11.7) in practice. However it is clear from (11.7) that it is relatively simple to solve for the threshold multiplier τ. In particular, this is given by

$$\tau = \left(\mathrm{Pfa}^{-1} \sum_{j=k}^{N} (-1)^{j-k} \binom{j-1}{k-1} \int_0^1 H_j \left(\beta z^{-\frac{1}{\alpha}} \right) dz \right)^{1/\alpha} . \qquad (11.8)$$

Whether the CFAR property is preserved with respect to the underlying multivariate Pareto model is a question which can only be answered with respect to a particular assumed model. Hence the next section considers Mardia's multivariate Pareto distribution.

11.3   Mardia's Multivariate Pareto Model

Although (11.8) provides a convenient way in which to determine the threshold multiplier τ for (11.1) it is necessary to specify a suitable joint distribution function for the statistics in the CRP. Here a simple multivariate Pareto model is introduced, due to Mardia in 1962, which has a very convenient form for application to (11.8). In this section this model is introduced, together with interesting properties of it, useful for the analysis to follow. This multivariate distribution has been shown to fit the compound Gaussian model framework in (Weinberg, 2017c), where the interested reader can find a more general multivariate Pareto model with non-constant correlation.

The first variation of multivariate distributions to be considered is the bivariate Pareto Type I distribution. Suppose that Z_1 and Z_2 are two random variables with joint density

$$f_{(Z_1,Z_2)}(z_1,z_2) = \alpha(\alpha+1)(\beta_1\beta_2)^{\alpha+1}(\beta_2 z_1 + \beta_1 z_2 - \beta_1\beta_2)^{-(\alpha+2)}, \qquad (11.9)$$

where each $z_j > \beta_j$ and α, β_j are all positive parameters. Then the random variables are said to have a bivariate Pareto Type I distribution. By integrating (11.9) over appropriate limits, it is possible to demonstrate that the corresponding distribution function is given by

$$F_{(Z_1,Z_2)}(z_1,z_2) = 1 - \sum_{j=1}^{2} \left(\frac{z_j}{\beta_j} \right)^{-\alpha} + \left(\sum_{j=1}^{2} \frac{z_j}{\beta_j} - 1 \right)^{-\alpha}, \qquad (11.10)$$

for $z_j \geq \beta_j$. Based upon (11.10), it is clear that the marginal distributions $Z_j \overset{d}{=} \mathrm{Pareto}(\alpha, \beta_j)$. Hence α functions as a common shape parameter, while the β_j allow for different marginal scale parameters.

It is relatively straightforward to prove that the mean of the product of the two marginal distributions is

$$E(Z_1 Z_2) = \beta_1 \beta_2 \left(\frac{\alpha^2 - \alpha - 1}{(\alpha - 1)(\alpha - 2)} \right), \tag{11.11}$$

from which the covariance can be derived, which is

$$Cov(Z_1, Z_2) = \frac{\beta_1 \beta_2}{(\alpha - 1)^2 (\alpha - 2)} \tag{11.12}$$

and therefore the correlation between the two marginal random variables is

$$Corr(Z_1, Z_2) = \frac{1}{\alpha}. \tag{11.13}$$

Using the fact that

$$f_{Z_1 | Z_2}(z_1 | z_2) = \frac{f_{(Z_1, Z_2)}(z_1, z_2)}{f_{Z_2}(z_2)} \tag{11.14}$$

it can be demonstrated that the density of $Z_1 | Z_2$ is

$$f_{Z_1 | Z_2}(z_1 | z_2) = \frac{\pi_1(z_2)}{(z_1 + \pi_2(z_2))^{\alpha+2}}, \tag{11.15}$$

which proves that the conditional distribution is also Pareto distributed, but of Type II, where

$$\pi_1(z_2) \quad = \quad (\alpha + 1) \left(\frac{\beta_1}{\beta_2} z_2 \right)^{\alpha+1}$$

and

$$\pi_2(z_2) \quad = \quad \frac{\beta_1}{\beta_2} z_2 - \beta_1. \tag{11.16}$$

Hence the conditional distribution can be written

$$Z_1 | Z_2 = W - \pi_2(z_2) \tag{11.17}$$

where $W \overset{d}{=} \text{Pareto}\left(\alpha + 1, \frac{\beta_1}{\beta_2} z_2 \right)$.

Next the survival function is examined, which turns out to have a rather simple expression for the bivariate Pareto model. This is given by the expression

$$P(Z_1 > z_1, Z_2 > z_2) = \int_{z_1}^{\infty} \int_{z_2}^{\infty} f_{(Z_1, Z_2)}(t_1, t_2) dt_2 dt_1 \tag{11.18}$$

where the density in (11.18) is (11.9), where it is assumed that $z_j > \beta_j$. By evaluating this double integral, it follows that

$$P(Z_1 > z_1, Z_2 > z_2) = \left(\frac{z_1}{\beta_1} + \frac{z_2}{\beta_2} - 1 \right)^{-\alpha}, \tag{11.19}$$

which provides an elegant expression for the survival function. This function implies that the minimum of the two marginal Pareto distributions is also Pareto distributed. To see this, note that the probability that the minimum of Z_1 and Z_2 exceeds t is exactly equal to the survival function (11.19) with $z_1 = z_2 = t$. Hence it is relatively easy to demonstrate that

$$P(Z_{(1)} \leq t) = 1 - \left(\frac{\theta}{t - \theta} \right)^{\alpha}, \tag{11.20}$$

where $\theta = \frac{\beta_1 \beta_2}{\beta_1 + \beta_2}$. In (11.20) it is clear that $t > \theta$ since $t > \beta_j$ for each j. Consequently it follows that the minimum $Z_{(1)} \stackrel{d}{=} \theta + \text{Pareto}(\alpha, \theta)$ as required.

The extension of the bivariate model to multivariate is now examined, which was also investigated in (Mardia, 1962). The multivariate Pareto distribution, for an $m \in \mathbf{N}$, is specified through the joint probability density function

$$f_{Z_1, Z_2, \ldots, Z_m} (z_1, z_2, \ldots, z_m) =$$

$$\frac{\alpha(\alpha + 1) \cdots (\alpha + m - 1)}{\left(\prod_{j=1}^{m} \beta_j \right) \left\{ \left(\sum_{j=1}^{m} \beta_j^{-1} z_j \right) - m + 1 \right\}^{\alpha + m}}, \tag{11.21}$$

where $\alpha > 0$ is a shape parameter, each $\beta_i > 0$ are scale parameters with $z_i > \beta_i$ for all i. This model reduces to the bivariate version for the case of two variables, and so the results derived in the analysis of the bivariate model apply to this multivariate distribution. For convenience the case where $\beta_j = \beta$ for all j will be used subsequently.

By successive integrations of (11.21), the cumulative distribution function of Mardia's multivariate Pareto model can be shown to be

$$F_{Z_1, Z_2, \ldots, Z_m}(z_1, z_2, \ldots, z_m) = 1 + \sum_{k=1}^{m} (-1)^k$$

$$\sum_{1 \leq i_1 < i_2 < \ldots < i_k \leq N} \left(\frac{\beta}{z_{i_1} + z_{i_2} + \cdots + z_{i_k} - (k-1)\beta} \right)^{\alpha} \tag{11.22}$$

where $\beta_i = \beta$ in (11.21). In the case where each of the $Z_{i_j} = t$, which will be required in the analysis to follow, the second sum in (11.22) reduces to counting the number of sequences of subsets of $\{1,2,\ldots,N\}$ order k such that $1 \leq i_1 < i_2 < \ldots < i_k \leq N$. Hence this can be written terms of combinatorial coefficients, resulting in

$$F_{Z_1,Z_2,\ldots,Z_m}(t,t,\ldots,t) = \sum_{k=0}^{m} \binom{m}{k}(-1)^k \left(\frac{\beta}{kt - (k-1)\beta}\right)^{\alpha}, \qquad (11.23)$$

which can also be identified as the distribution function of the maximum of m such random variables. The multivariate survival function can also be shown to have the form

$$P(Z_1 > z_1, Z_2 > z_2, \ldots, Z_m > z_m) = \left(\sum_{j=1}^{m} \frac{z_j}{\beta} - m + 1\right)^{-\alpha}, \qquad (11.24)$$

from which it can be deduced that the minimum is also Pareto Type II distributed. Specifically, it can be shown that

$$Z_{(1)} \overset{d}{=} \left(\frac{m-1}{m}\right)\beta + \text{Pareto}\left(\alpha, \frac{\beta}{m}\right). \qquad (11.25)$$

11.4 Order Statistic Decision Rule Thresholds

It is now possible, under the assumption that clutter evolves according to the multivariate Mardia model, to determine detection threshold multipliers for application in (11.1). It is necessary to specify an order statistic index k and then apply (11.23) to (11.8). It is convenient to define an integral function

$$I(m,\alpha) := \int_0^1 \left(mz^{-\frac{1}{\alpha}} - m + 1\right)^{-\alpha} dz, \qquad (11.26)$$

which can be evaluated numerically and is well-behaved. It is clear that the application of (11.23) to (11.8) will result in the second summation in the latter reducing to counting the number of sequences of indices in the set $\{1,2,\ldots,N\}$ satisfying the inequality $1 \leq i_{m+1} < i_{m+2} < \ldots < i_N \leq N$ for a given m.

Here three selections for the OS statistic parameter are considered, namely $k \in \{N, N-1, N-2\}$. The maximum is fundamental to the construction of the lower-ordered OS cases, while the selection of the latter can be used to manage a given number of interfering targets.

To determine the maximum OS detector, the distribution of the maximum OS is given by (11.23) with $m = N$. Consequently applying (11.23) to (11.7), together with (11.26), it follows that

$$\text{Pfa} = \tau^{-\alpha} \sum_{k=0}^{N} \binom{N}{k}(-1)^k I(k,\alpha). \qquad (11.27)$$

Therefore it follows by inversion of (11.27) that the relevant τ is given by

$$\tau = \left(\text{Pfa} \sum_{k=0}^{N} \binom{N}{k} (-1)^k I(k, \alpha) \right)^{\frac{1}{\alpha}}. \tag{11.28}$$

Based upon (11.27) one can conclude that the detection process (11.1) will be CFAR with respect to the multivariate scale parameter β, but requires *a priori* knowledge of the multivariate shape parameter.

Next OS detector choices that can manage strong interference are examined. In order to determine the distribution function corresponding to an OS with $k = N - 1$, it follows from (11.5) that

$$
\begin{aligned}
F_{Z_{(N-1)}}(t) &= \sum_{j=N-1}^{N} (-1)^{j-(N-1)} \binom{j-1}{N-2} H_j(t) \\
&= H_{N-1}(t) - (N-1)H_N(t), \tag{11.29}
\end{aligned}
$$

and

$$H_{N-1}(t) = N\text{P}(Z_{i_1} \leq t, Z_{i_2} \leq t, \dots, Z_{i_{N-1}} \leq t), \tag{11.30}$$

since the sum in (11.5) is over only one index. Since

$$H_N(t) = \text{P}(Z_{i_1} \leq t, Z_{i_2} \leq t, \dots, Z_{i_N} \leq t), \tag{11.31}$$

the distribution function (11.29) is equivalent to

$$
\begin{aligned}
F_{Z_{(N-1)}}(t) &= N\text{P}(Z_{i_1} \leq t, Z_{i_2} \leq t, \dots, Z_{i_{N-1}} \leq t) \\
&\quad - (N-1)\text{P}(Z_{i_1} \leq t, Z_{i_2} \leq t, \dots, Z_{i_N} \leq t). \tag{11.32}
\end{aligned}
$$

Finally an application of (11.32) to (11.4) results in

$$
\begin{aligned}
\text{Pfa} &= \tau^{-\alpha} \left(\sum_{k=0}^{N-1} (-1)^{k+1}(k-1)\binom{N}{k} I(k, \alpha) \right. \\
&\quad \left. - (N-1)(-1)^N I(N, \alpha) \right). \tag{11.33}
\end{aligned}
$$

which again can be inverted to determine τ for (11.1). As for the case of the maximum, it is clear that the detector (11.1), in the situation where $k = N - 1$, will be CFAR only with respect to β.

For the choice of $k = N - 2$ similar calculations can be used to demonstrate that the appropriate distribution function is

$$F_{Z_{(N-2)}}(t)$$

$$= \frac{N(N-1)}{2} P(Z_{i_1} \leq t, Z_{i_2} \leq t, \ldots, Z_{i_{N-2}} \leq t)$$

$$- N(N-2) P(Z_{i_1} \leq t, Z_{i_2} \leq t, \ldots, Z_{i_{N-1}} \leq t)$$

$$\frac{(N-1)(N-2)}{2} P(Z_{i_1} \leq t, Z_{i_2} \leq t, \ldots, Z_{i_N} \leq t).$$

$$(11.34)$$

Based upon this, with an application of (11.23), it can be demonstrated that the corresponding Pfa is given by

$$\text{Pfa} = \tau^{-\alpha} \times$$

$$\left\{ \sum_{k=0}^{N-2} (-1)^k \frac{1}{2}(k-1)(k-2) \binom{N}{k} I(k, \alpha) \right.$$

$$+ (1)^{N-1} \frac{1}{2} N(N-2)(N-3) I(N-1, \alpha)$$

$$+ (-1)^N \frac{1}{2}(N-1)(N-2) I(N, \alpha) \Bigg\}.$$

$$(11.35)$$

As in previous choices for the OS index, it is relatively simple to invert (11.35) to produce the appropriate τ for application in (11.1). Additionally, the detector (11.1) with $k = N - 2$, will also only achieve the CFAR property with respect to β.

11.5 Performance Analysis

Analysis of the performance of the decision rule (11.1) with the choices of $k \in \{N-2, N-1, N\}$ and utilising (11.28), (11.33) and (11.35) respectively for setting of the appropriate τ is now undertaken. Throughout the following the design Pfa is set to 10^{-4}, while the length of the clutter range profile is examined for two particular cases. The first is $N = 32$, which has been investigated in previous chapters. However, most of the results will be focused on the case of $N = 16$.

The major reason for this is to minimise the numerical complexity in simulation. However, as will be observed, it appears increasing N results in worse performance. An additional assumption adopted throughout is *a priori* knowledge of the shape parameter α, which has been sourced from Table 3.1 as in previous chapters.

A complication in this chapter is the generation of correlated Pareto samples, distributed according to Mardia's multivariate model, for the CRP. However, there is a well-known technique which can be used in this case, since the multivariate distribution function is known and conditional distributions, associated with its marginal random variables, are also Pareto distributed as demonstrated earlier. In general terms, suppose that $F_{(Y_1, Y_2, \ldots, Y_N)}(y_1, y_2, \ldots, y_N)$ is a multivariate distribution function. Then to generate a realisation from this distribution the following algorithm may be employed:

1. Generate y_1 from the marginal distribution Y_1;

2. Generate y_k from the conditional distribution $Y_k | \{Y_1 = y_1, \ldots, Y_{k-1} = y_{k-1}\}$ for each $k \in \{2, \ldots, N\}$;

3. Then the sequence $\{y_1, y_2, \ldots, y_N\}$ will be a realisation of this multivariate random variable.

By an application of Mardia's multivariate Pareto density with $\beta_j = \beta$, it is not difficult to show that the marginal variable Z_k, conditioned on $\{Z_1 = z_1, Z_2 = z_2, \ldots, Z_{k-1} = z_{k-1}\}$, can be generated via

$$-\lambda + (\beta + \lambda) R^{-\frac{1}{\alpha + k - 1}}, \tag{11.36}$$

where R is uniformly distributed on the unit interval and $\lambda = \sum_{i=1}^{k-1} z_i + \beta(1 - i)$ for $k \geq 2$. Hence (11.36) is used in the analysis to follow.

11.5.1 *Performance in Homogeneous Clutter*

As in previous chapters the numerical analysis begins with performance in homogeneous clutter, which is simulated via (11.36). The CUT target model is Swerling I as assumed previously, which is embedded in clutter which is independent of that in the CRP. Figures 11.1–11.3 demonstrate detection performance in the case of horizontal, cross and vertical polarisation for the corresponding Pareto clutter parameter estimates used throughout the book. For these three figures the choice of $N = 32$ is utilised. It is clear that as the polarisation is shifted from horizontal to vertical, the difference between the three decision rules is reduced. In all cases the maximum has the best performance, while the other two cases perform somewhat similarly. It is interesting to observe that there is a reduction in performance overall, in contrast to the case of independently distributed clutter.

Figure 11.4 provides an example of the performance of the three decision rules when N is reduced to 16, in horizontally polarised clutter. It is interesting to observe that the performance has improved considerably, with the three decision rules performing very similarly. When the polarisation is changed to cross- and then vertical, the three detectors have an even smaller detection loss relative to the ideal decision rule (figures omitted for brevity).

An explanation for the improvements in performance as the polarisation is changed from horizontal to vertical is that the correlation decreases, so that in the vertically polarised case, the clutter is approximately uncorrelated.

To complement these results a series of ROC curves are now examined for the detectors introduced in this chapter. The first is Figure 11.5, which shows an ROC for the case where $N = 32$, with horizontal polarisation, and when the independent Swerling I target model in the CUT has SCR of 15 dB. Figures 11.6 and 11.7 correspond to the case where $N = 16$, with horizontal polarisation, and when the CUT target has SCR of 15 and 30 dB respectively. These ROC curves complement the results in the plots of the Pd as a function of the SCR. The cases of cross- and vertical polarisation are omitted for brevity.

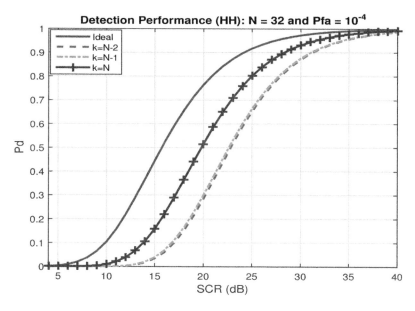

Figure 11.1: Performance of the OS-detector in correlated Pareto distributed clutter, corresponding to horizontal polarisation. The figure plots the three cases of OS-detector together with the ideal decision rule.

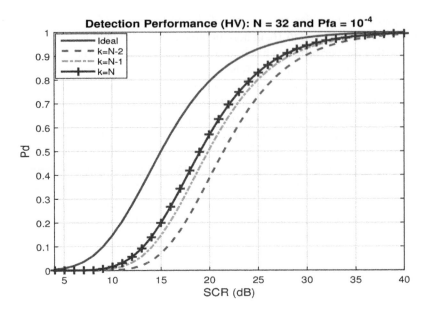

Figure 11.2: Repeat of the simulation in Figure 11.1 except in cross-polarisation.

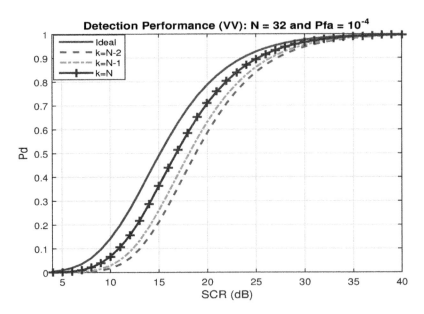

Figure 11.3: Performance example of the OS-detector in vertically polarised correlated clutter, with the same decision rules as for Figures 11.1 and 11.2.

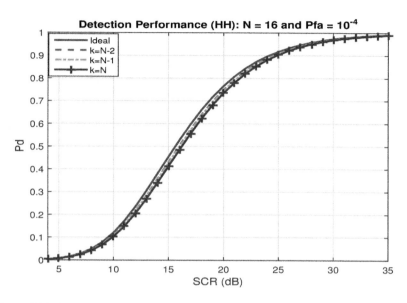

Figure 11.4: Second example of performance of the OS-detectors in correlated horizontally polarised clutter. In this case the CRP size has been reduced to 16.

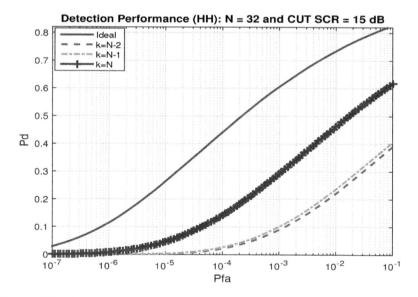

Figure 11.5: ROC perspective on detection performance, for the case of horizontal polarisation and when the CUT target has SCR of 15 dB. In this simulation the CRP has length 32.

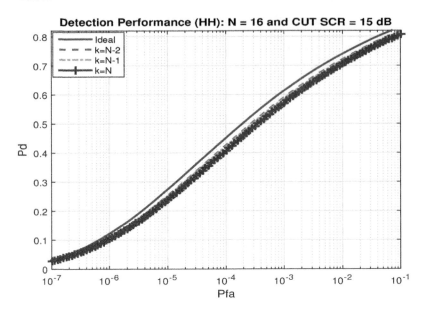

Figure 11.6: Second ROC for detection in correlated clutter, for horizontal polarisation and when the CUT target has SCR of 15 dB. Here the CRP length has been reduced to 16.

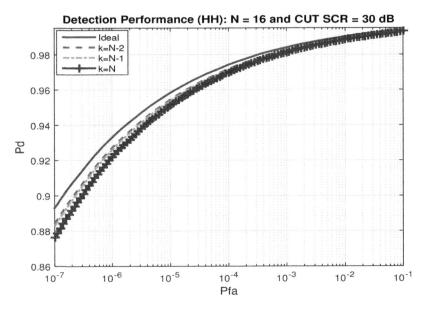

Figure 11.7: Third ROC for detection performance, also in horizontal polarisation but when the CUT SCR is 30 dB, and when the CRP length is 16.

11.5.2 Performance with Interference in the CRP

Several examples are now considered to investigate performance when the CRP is subjected to interfering targets. Embedding such targets within correlated clutter is a difficult exercise, since it would be necessary to ensure that the CRP is generated via (11.36), such that an interfering target is generated in the complex domain, and combined with an element of the CRP but when it corresponds to a realisation of a compound Gaussian model with inverse gamma texture. As a simpler alternative, the approach adopted in this chapter is to generate the CRP via Mardia's model, for the number of cells for which there is no interference, and then generate the remainder similarly to the generation of the CUT. Hence one is assuming that the interference occurs in independent Pareto distributed clutter. Although this is a simplification, it permits a computationally achievable solution, and still provides one with an understanding of the performance of the decision rules in correlated clutter with interference present.

Each of the three detectors are subjected to one and then two such interfering targets, both of which are modelled by a Swerling I fluctuation model. These interfering targets have ICR of 10 and 20 dB respectively. Throughout the Pfa is set to 10^{-4}.

The first example of performance in the presence of interference can be found in Figure 11.8, which is for the case where $N = 16$ and horizontal polarisation. It is clear that the maximum-based detector suffers in the presence of interference, while the other decision rules tend to manage two interfering targets well. It is interesting to note that the OS detector, with $k = N - 1$ manages the interference very well despite the fact that it is being subjected to the 10 dB interfering target.

Figure 11.9 repeats the same scenario in Figure 11.8 with the exception that $N = 32$. This figure highlights some interesting results, especially concerning the maximum-based decision rule. One can see that the latter detector only experiences a small detection loss, relative to the same decision rule without interference, when subjected to a 10 dB interfering target. However, when the maximum is subjected to the two interfering targets, it performs very poorly. This phenomenon has been observed in the case of GM-based detectors, which one may recall could manage single interfering targets with ICR no more than roughly 10 dB.

In addition to this, Figure 11.9 shows that the decision rule with $k = N - 2$ experiences very little detection loss, while that with $k = N - 1$ experiences a loss when subjected to the pair of interfering targets.

Next the case of cross-polarisation is examined, with Figure 11.10 replicating the scenario of Figure 11.8. The conclusions are much the same, except the overall detection performance has improved due to the polarisation. Figure 11.11 shows the performance in cross-polarisation and when N is increased to 32. The conclusions are analogous to those as for Figure 11.9 with an overall increase in performance as for the $N = 16$ case.

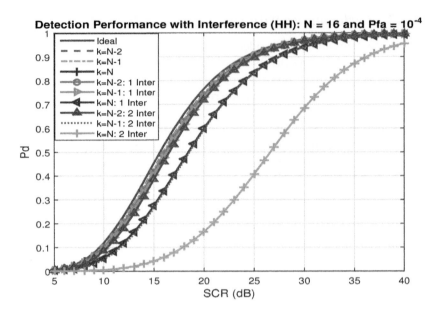

Figure 11.8: Interfering targets in correlated clutter, with horizontal polarisation. The first interfering target has ICR of 10 dB, while the second has ICR of 20 dB.

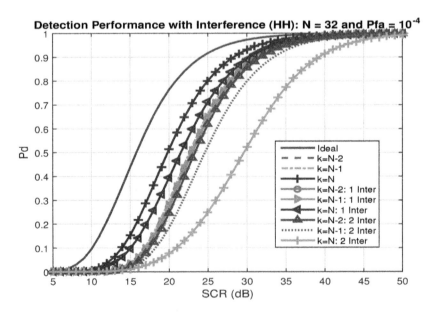

Figure 11.9: Second example of interference in correlated clutter (HH polarisation). The two interfering targets have ICR of 10 and 20 dB respectively.

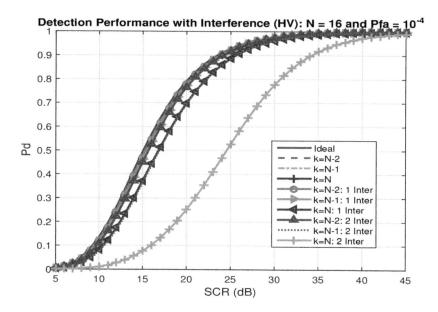

Figure 11.10: Interfering targets in cross-polarised correlated clutter. As for previous examples, the interference levels are 10 and 20 dB respectively.

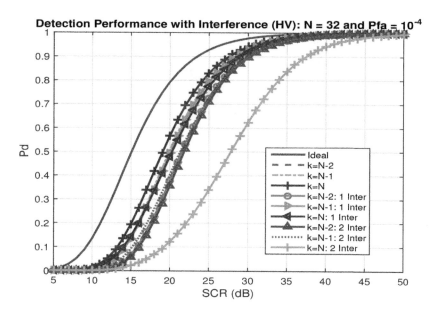

Figure 11.11: Second example of interfering targets in cross-polarisation, with the same ICR as previously employed.

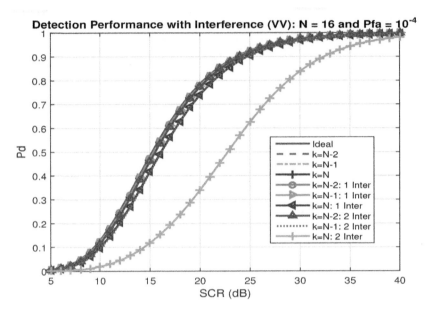

Figure 11.12: Interfering targets in vertically polarised correlated clutter, with ICR of 10 and 20 dB respectively.

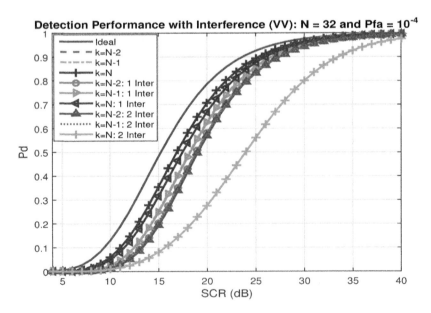

Figure 11.13: Second example of interference in vertical polarisation, with the same levels of interference as in the previous figures.

Finally, the case of vertical polarisation is considered, with a replication of the experimental factors used as for the previous two polarisations considered. Figure 11.12 is for $N = 16$, while Figure 11.13 corresponds to $N = 32$. The conclusion reached is that the performance has improved due to the polarisation, while the overall detector characteristics are much the same.

11.5.3 False Alarm Considerations

The performance of the three detectors, relative to the maintenance of the design Pfa, is now examined. Throughout the following $N = 16$, since if N is chosen to be larger, it is difficult to acquire reasonable Monte Carlo estimates without an inordinate simulation run size. In the first case it is useful to examine the detectors to investigate whether they are maintaining the desired Pfa in clutter which is correlated but contains no irregularities. Figure 11.14 plots the estimated Pfa, for each α and when $\beta = 0.1$. For each such α and β each of the detectors is run on clutter generated by Mardia's multivariate Pareto model, while the CUT is generated by an independent Pareto observation. This is done 10^6 times so that a Monte Carlo estimate of the resultant Pfa can be produced. Here we observe that the Pfa is maintained well by the three decision rules.

Figure 11.15 plots the appropriate threshold multiplier, as a function of α, for the three detectors, showing that the thresholds decrease as a function of α. Additionally, it is suggested by Figure 11.14 that the thresholds tend to decrease as a function of the OS index k.

To provide a slightly different perspective on the behaviour of the Pfa, Figure 11.16 plots the estimated Pfa as a function of the CRP length N, for the case of horizontal polarisation. Figure 11.17 is a similar plot except for the vertically polarised case. These figures confirm that the decision rules manage to maintain the design Pfa in ideal scenarios.

As a second stage analysis of the management of the design Pfa, clutter transitions are examined. In order to accomplish this it is necessary to generate two components of the CRP. The first consists of lower powered clutter returns, or clutter for which the decision rule has been designed for operation. The second series of returns are higher powered, representing a transition from one clutter environment to another. In order to make this computationally simpler, these are generated via two separate and independent Mardia models, and combined to form one CRP for analysis. The usual practice, with respect to the CUT, is adopted throughout.

Figure 11.18 shows the estimated Pfa for the case of horizontal polarisation and in the case where the CCR is 0.5 dB. Here we observe that the maximum tends to always reduce the resultant Pfa, while the other two decision rules regulate it until the CUT is affected by higher powered clutter. Once this has occurred the resultant Pfa is increased to around $10^{-3.55}$.

Next the same scenario is examined but in the situation where the CCR is now 1 dB. It can be observed that the three detectors tend to reduce gradually the number of false alarms until the CUT is affected. Once this has occurred, the decision rules have a rough Pfa of $10^{-3.3}$ for the detectors with $k = N - 1$ and $k = N - 2$, while the maximum has a Pfa slightly smaller than this.

A final example, for the case of horizontal polarisation, can be found in Figure 11.20, for which the CCR has been set to 2 dB. The increased CCR results in a larger increase in the resultant Pfa once the CUT is affected with higher powered clutter, which is roughly $10^{-2.8}$ for the non-maximum decision rules.

To conclude the analysis an example of clutter transitions in vertically polarised clutter is also provided. Figure 11.21 corresponds to a 1 dB power increase in vertical polarisation, also with $N = 16$. It is observed that each of the decision rules tend to reduce the number of false alarms until the CUT is affected. However, there is a larger overall increase in the resultant Pfa once the CUT is affected. In particular, the resultant Pfa is larger than $10^{-2.5}$ for each of the detectors, which is significantly larger than the same power level increase in horizontal polarisation.

Results for cross-polarisation are similar to those for the cases of horizontal and vertical polarisation and hence are omitted for brevity.

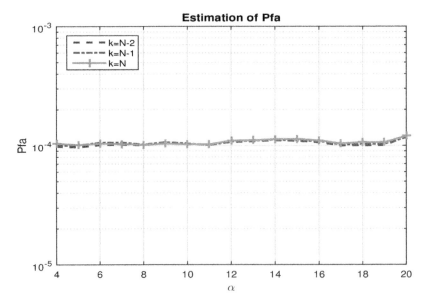

Figure 11.14: Resultant Pfa for the three OS-detectors as a function of the Pareto shape parameter, with scale parameter set to $\beta = 0.1$.

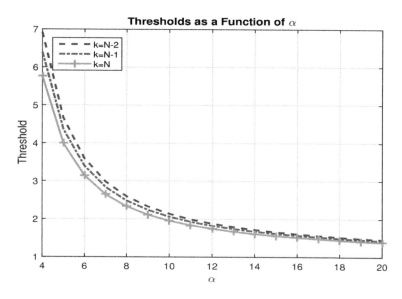

Figure 11.15: Thresholds for the three OS-detectors as a function of α with $\beta = 0.1$, with $N = 16$.

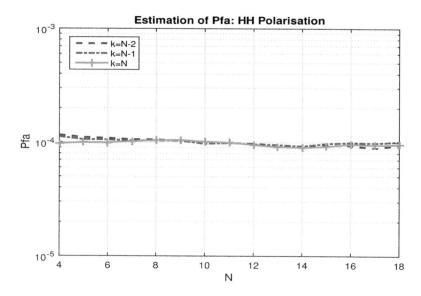

Figure 11.16: Estimation of the resultant Pfa for the three OS-detectors, as a function of the CRP length N, for the case of horizontal polarisation.

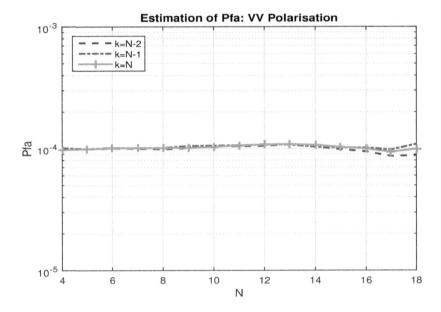

Figure 11.17: Estimation of the resultant Pfa in vertical polarisation.

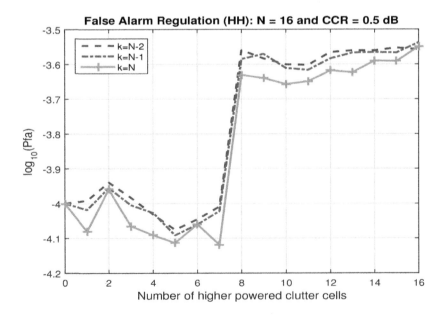

Figure 11.18: False alarm regulation in horizontal polarisation.

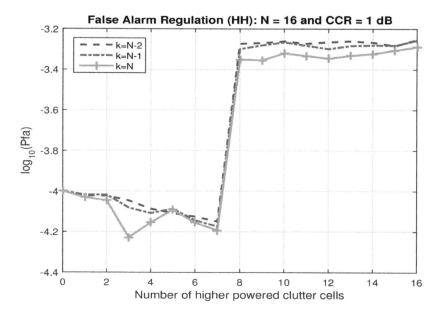

Figure 11.19: A second example of clutter transitions in horizontal polarisation.

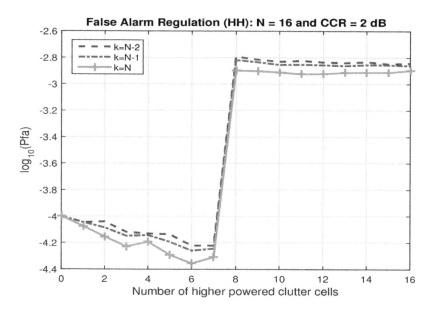

Figure 11.20: Clutter transitions with a 2 dB power level increase, in horizontal polarisation.

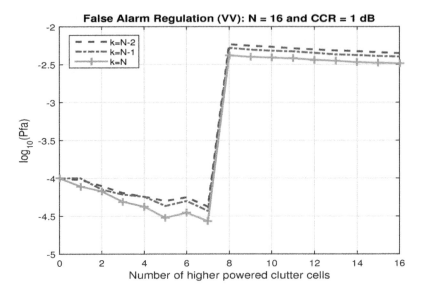

Figure 11.21: An example of false alarm regulation in vertical polarisation.

11.6 Analysis of the Minimum-Based Detector

In view of the fact that (11.25) demonstrates that the minimum of a series of cutter statistics, generated from Mardia's multivariate model, is also Pareto distributed, it is thus of interest to investigate the minimum OS detector. This section is thus concerned with an analysis of this decision rule, based upon (11.25).

In view of (11.25) the distribution function of the minimum can be seen to be

$$F_{Z_{(1)}}(t) = 1 - \left(\frac{\beta}{Nt - \beta(N-1)} \right)^{\alpha}, \qquad (11.37)$$

for a series of N correlated clutter statistics, distributed according to Mardia's multivariate Pareto model. Hence an application of (11.37) to (11.4) with $k = 1$ results in

$$\tau = \mathrm{Pfa}^{-\frac{1}{\alpha}} \left(1 - I(N, \alpha) \right)^{\frac{1}{\alpha}}, \qquad (11.38)$$

where (11.26) has been utilised. Thus (11.38) provides an expression for the threshold multiplier for the minimum based detector.

Some simulations are now considered to examine the performance of this minimum-based detector. Figure 11.22 shows the three OS-based detectors, for the case where $N = 16$ with Pfa $= 10^{-4}$, and for the case of horizontal polarisation with no interference. Figure 11.23 corresponds to the vertical polarisation case. These two figures demonstrate that the minimum matches the performance of an ideal detector in this scenario. In fact, each of the detectors have almost ideal performance, in the absence of interference, when $N = 16$.

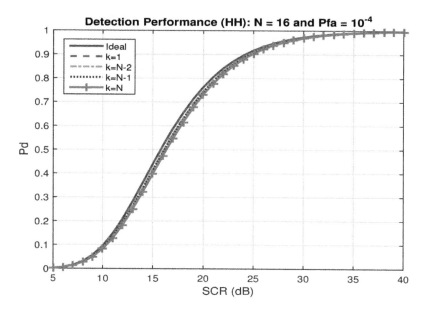

Figure 11.22: Comparison of the three OS-detectors in horizontal polarisation, with the inclusion of the minimum-based detector.

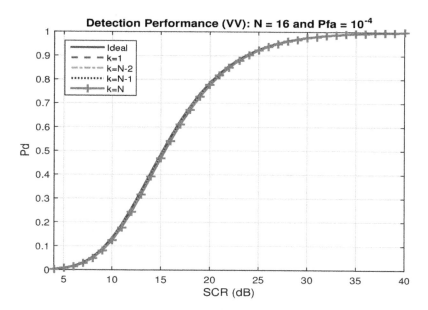

Figure 11.23: Detector performance in vertical polarisation, showing the three OS-based decision rules together with the minimum detector.

Next an example to examine the effects of strong interference is considered, with Figure 11.24 for the case where there are up to 2 interfering targets with ICR of 40 dB, in horizontal polarisation. Here the design Pfa is 10^{-4} and $N = 16$. Figure 11.25 provides clarification that the minimum is very robust to strong interference.

To complete the analysis, clutter transitions are investigated, beginning with a 0.5 dB power level increase in horizontal polarisation. The estimated Pfa is plotted in Figure 11.26, with the minimum compared with the other three OS detectors. It is clear that the minimum regulates the Pfa reasonably well until the CUT is affected. Thereafter, the minimum decision rule has the worst performance, with the largest increase in the design Pfa.

Figure 11.27 repeats the same scenario except with a 1 dB power level increase. As observed previously the minimum has the worst performance once the CUT is affected. The increase in the resultant Pfa is also larger due to the increased CCR.

As a final example, Figure 11.28 repeats the scenario in Figure 11.27 except in vertical polarisation. The results indicate that the minimum can result in a huge increase in the number of false alarms once the CUT is affected.

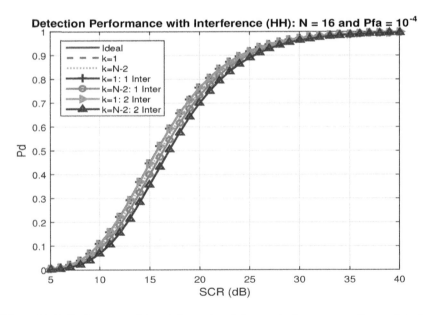

Figure 11.24: Performance of the minimum-based detector in the presence of strong interference. In this simulation the minimum is compared with the OS-based detector with $k = N - 2$, since the latter is robust to two interfering targets. In this example both of the detectors are subjected to one and then two interfering targets, both with ICR of 40 dB.

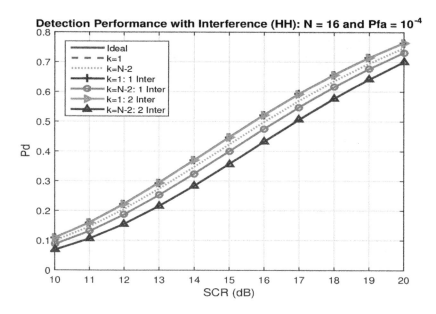

Figure 11.25: Enlargement of the plot in Figure 11.24.

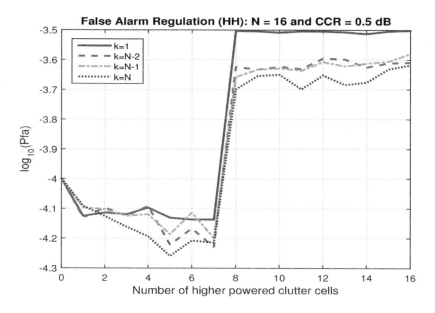

Figure 11.26: Comparison of the minimum with the three other OS detectors during clutter transitions, in horizontal polarisation.

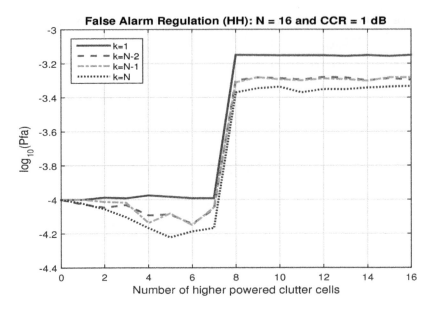

Figure 11.27: Second example of the false alarm regulation with the minimum detector in horizontal polarisation.

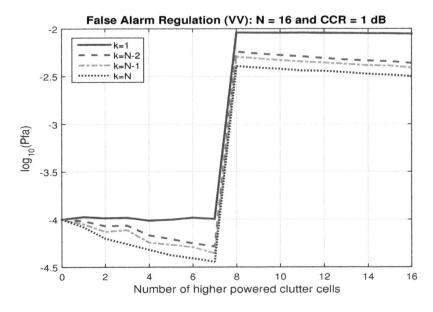

Figure 11.28: Examination of the false alarm regulation of the minimum and three OS-based detectors in vertical polarisation.

11.7 Achieving CFAR in Correlated Pareto Distributed Clutter

The developments in the preceding sections have shown that the CFAR property can only be achieved for the Pareto scale parameter, for the case of correlated Pareto distributed clutter modelled by the Mardia multivariate model. The choice of the decision rule (11.1) was motivated by the fact that the Mardia model introduced Pareto shape parameter dependence via its correlation being exactly the reciprocal of this parameter. However, it is possible to acquire the full CFAR property in the case of a correlated CRP. This section outlines this, and indicates suitable avenues for investigation towards this objective.

Recall the OS-CFAR (5.45), and suppose that for each j, $X_j \overset{d}{=} \text{Exp}(1)$ is the dual of each Z_j, where Z_0 is the CUT statistic and $Z_1, Z_2, \ldots, Z_N$ is the CRP. Then the Pfa of (5.45) can be written

$$\text{Pfa} = \mathbf{P}(X_0 > X_{(1)} + \tau\left(X_{(k)} - X_{(1)}\right)), \qquad (11.39)$$

where $2 \leq k \leq N$ is the OS index. Observe that no assumption of independence has been used in the derivation of (11.39). However, it is convenient to assume that the CUT and CRP statistics are independent as justified previously. Additionally it is being assumed that the individual statistics in the CRP possess a common Pareto Type I univariate distribution, but are correlated. This is equivalent to assuming that the exponential random variables in the duality relationship form a sequence of correlated exponential random variables with unity mean.

Suppose that $f_{X_{(1)}}(t)$ is the density of the minimum of the correlated dual CRP $X_1, X_2, \ldots, X_N$. Then by conditioning on this minimum the Pfa (11.39) becomes

$$\text{Pfa} = \int_0^\infty f_{X_{(1)}}(t)\mathbf{P}(X_0 > t + \tau(X_{(k)} - t)|\{X_{(1)} = t\})dt. \qquad (11.40)$$

Next define a random variable $\Xi_k = X_{(k)}|\{X_{(1)} = t\}$, then (11.40) becomes

$$\text{Pfa} = \int_0^\infty f_{X_{(1)}}(t)\mathbf{P}\left(\Xi_k < \frac{X_0 - t + t\tau}{\tau}\right)dt. \qquad (11.41)$$

Finally, since it is assumed that X_0 is independent of the CRP, (11.41) reduces to

$$\text{Pfa} = \int_0^\infty \int_0^\infty e^{-w} f_{X_{(1)}}(t) F_{\Xi_k}\left(\frac{w - t + \tau t}{\tau}\right)dt\,dw, \qquad (11.42)$$

where F_{Ξ_k} is the distribution function of Ξ_k.

Therefore if one specifies the joint distribution of the exponential duals, then the density of the minimum and the distribution function of Ξ_k can be applied to (11.42), which allows the numerical determination of τ for a given Pfa and OS index k. Additionally, this will result in the acquisition of the full CFAR property with the decision rule (5.45).

These observations suggest that suitable avenues of investigation should focus on the generation of correlated sequences of exponential random variables with unity mean, and then understanding the effect on the resultant sequence of Pareto distributed random variables. It should then be possible to appeal to the memoryless nonlinear transformation to gain an understanding of the relationship between the respective correlation functions. Provided the correlation does not depend on the Pareto shape and scale parameters it should be possible to produce a model of correlated Pareto clutter which achieves the CFAR property with (5.45). A suitable correlation function is one that is the reciprocal of the CRP length; assuming a stationary sequence, one could assume that $E(X_j X_1) = \frac{1}{j}$ for $j \geq 1$ for instance. Some useful references, on the construction of multivariate exponential distributions, include (Mallik, 2003), (Marshall and Olkin, 1967), (Raftery, 1984) and (Basu and Sun, 1997). This is a planned future direction for research at DST Group.

11.8   Conclusions

This chapter made a sharp deviation from the central theme of independent and identically distributed clutter returns in the CRP. A formulation of a general OS-based decision rule was provided, which was then specialised to the case of clutter modelled by Mardia's multivariate Pareto model. Performance analysis showed that acceptable detection performance could be achieved.

An assumption made throughout was *a priori* knowledge of the Pareto shape parameter. The decision rules examined are CFAR with respect to the Pareto scale parameter. Additionally, the case of constant correlation was only examined, due to the application of Mardia's multivariate Pareto model. This investigation highlighted a number of interesting avenues for future research. It would be interesting to develop a multivariate Pareto model whose correlation matched those observed in real data. As an example, sequential measurements of clutter may experience spatial correlations which decrease exponentially, and so an autocorrelation which reflects this could be investigated. It is expected that this could be achieved through a memoryless nonlinear transformation approach, beginning with a Gaussian process which is then transformed to a multivariate Pareto model. This could then be coupled with the approach utilised in this chapter to investigate detection performance. Alternatively one can examine the relationship between the Pareto and exponential random variables, in terms of generation of correlated sequences, and apply the OS-CFAR (5.45) as outlined in the previous section.

FURTHER CONCEPTS

Chapter 12

Invariance and the CFAR Property

12.1 Introduction

The concept of an invariant family of distributions and invariant statistics can be used to explain or demonstrate the CFAR property of the decision rules that have been examined throughout this book. The concept of an invariant statistic has been investigated in mathematical statistics since the 1960s as documented in (Ferguson, 1967) and (Lehmann, 1986). Invariance was identified as providing an explanation for equivalence between hypothesis tests, when the underlying test statistic was subjected to a transformation via a group action. In the context of signal processing, this was identified as a technique which could be used to construct CFAR detection schemes (Scharf, 1973). A good example of the construction of sliding window CFAR detectors is provided in (Guan and He, 2000), who demonstrates that based upon a class of scale-invariant clutter models a general specification of a CFAR decision rule can be established.

A substantial development to ensure that the CFAR property is acquired at the detector design stage can be found in (Conte et al., 2003), who examine multidimensional signal detection problems. Invariance in a radar detection context was examined in (De Maio and Iommelli, 2008), with the concept of generalised CFAR and existence of a uniformly most powerful invariant detector introduced in (De Maio, 2013). An interesting study in the context of signal detection can be found in (Gabriel and Kay, 2005), who show that the performance of the generalised likelihood ratio test detector tends to that of a uniformly most powerful invariant test as the Pfa limits to zero, or as the SCR tends to infinity. Another substantial development of the concept of invariance in radar detection can be found in (Ghobadzadeh et al., 2014), who demonstrated the existence of an optimal CFAR detector and examined its relationship to what is referred to as a maximal invariant statistic with respect to a minimal invariant group of transformations.

This chapter presents an overview of invariance, specialised to the central theme of this book, which can provide further insight into the CFAR property of some of the sliding window detection processes examined thus far. Since this is specialised to Pareto Type I models the derivation of the invariance property can be stated more simply. In order to facilitate the discussion of invariance it is useful to review the basic principles of group theory.

12.2   Group Theory Basics

It is necessary to introduce briefly some of the definitions and results from the theory of groups. A classic text to which the reader can refer is (Fraleigh, 1967), on which the following discussion is based. Suppose that G is a set, then a binary operation, denoted $\circledast$, is a mapping which assigns elements of the product space of G with itself to the set G. That is $\circledast : G \times G \longrightarrow G$. A group, denoted $< G, \circledast >$ consists of a set together with a binary operation on the set which satisfies the following three requirements:

1. if $g_1, g_2, g_3 \in G$ then $(g_1 \circledast g_2) \circledast g_3 = g_1 \circledast (g_2 \circledast g_3)$;

2. there exists an $e \in G$ such that for all $g \in G$, $g \circledast e = e \circledast g = g$;

3. for every $g \in G$ there exists an element $\tilde{g} \in G$ such that $g \circledast \tilde{g} = \tilde{g} \circledast g = e$.

Property 1 is known as associativity, Property 2 is the existence of an identity element, while Property 3 is the requirement that each element of G possesses an inverse under the group action. By the definition of the binary operation, the closure property is satified immediately and hence does not need to be listed within the above. Groups are an abstraction of addition and multiplication of the real numbers.

If the group operation is commutative, meaning that $g_1 \circledast g_2 = g_2 \circledast g_1$ for all $g_1, g_2 \in G$ then the group is called Abelian. Many properties observed in the real numbers can be formulated and proven in the context of groups, such as the uniqueness of the identity, inverses, and solutions to linear equations (Fraleigh, 1967).

If $H \subset G$ is such that it is closed under the group operation $\circledast$ and satisfies each of the three requirements above, then it is called a subgroup of the group based upon set G.

Suppose that $< G_1, \circledast_1 >$ and $< G_2, \circledast_2 >$ are two groups. Then a mapping $f : G_1 \longrightarrow G_2$ is called a homomorphism if for any $g_1, g_2 \in G_1$, $f(g_1 \circledast_1 g_2) =$

$f(g_1) \circledast_2 f(g_2)$. In the case where the mapping f is a bijection, the two groups are called isomorphic.

As an example of a group that will be useful in the analysis to follow, consider the set of real-valued functions $G = \{g : g(t) = \lambda_1 t^{\lambda_2}, t \geq 0, \lambda_1, \lambda_2 \in \mathbf{R}\}$. Hence this set consists of scale and power transformations. Suppose the binary operation is functional composition, so that for $g_1, g_2 \in G$, $g_1 \circledast g_2 = g_1 \circ g_2$. Then $< G, \circ >$ is a group, which is relatively simple to demonstrate. This can also be extended to the situation where multivariate sequences are transformed via the same operation. In particular, $G = \{g : g(t_j) = \lambda_1 t_j^{\lambda_2}, j \in \{1, 2, \ldots, N\}, t_j \geq 0, \lambda_1, \lambda_2 \in \mathbf{R}\}$ with functional composition as the group operation is also a group.

12.3   The Invariance Property

The following account of invariance follows closely the corresponding development in (Ghobadzadeh et al., 2014), which has been based upon that in (Lehmann, 1986). Here the main concepts are overviewed, and the key results from (Ghobadzadeh et al., 2014) are reviewed.

Define a family of probability density functions by $\mathcal{P} = \{f(\mathbf{t}; \boldsymbol{\theta}) : \mathbf{t} \in \mathcal{T}, \boldsymbol{\theta} \in \Theta \subset \mathbf{R}^M\}$, where $\mathbf{t}$ is the N-dimensional observation vector in sample space $\mathcal{T}$, $\boldsymbol{\theta}$ is the distributional parameter vector and Θ is the M-dimensional parameter space. Then $\mathcal{P}$ is called an invariant family of distributions (IFD) under a group of transformations $< G, \circledast >$ if for every $g \in G$ there exists a $\bar{g} \in \bar{G}$ such that $\bar{g} \in \Theta$ and

$$\int_A f(g(\mathbf{t}); \boldsymbol{\theta}) dg(\mathbf{t}) = \int_A f(\mathbf{t}; \bar{g}(\boldsymbol{\theta})) d\mathbf{t}, \tag{12.1}$$

where A is an arbitrary subset of the nonnegative real line and g is a bijection on the nonnegative real line, and the integration is understood to be with respect to Lebesgue measure on $\mathbf{R}^N$. The set of functions $\bar{g} : \Theta \to \Theta$, denoted $\bar{G}$ forms what is known as the induced group. This group is homomorphic to the original group (Ghobadzadeh et al., 2014). Suppose that $T : \mathbf{R}^+ \times \mathbf{R}^+ \times \cdots \times \mathbf{R}^+ \longrightarrow \mathbf{R}^+$ is a test statistic, acting on an N-dimensional observation vector. Then the test is called invariant under a group of transformations $< G, \circledast >$ if for all observational vectors $\mathbf{x}$ and all $g \in G$, $T(g(\mathbf{x})) = T(\mathbf{x})$. An invariant test T is called maximally invariant if for any pair of observational vectors $\mathbf{x}_1$ and $\mathbf{x}_2$ if $T(\mathbf{x}_1) = T(\mathbf{x}_2)$ then there exists a $g \in G$ such that $\mathbf{x}_2 = g(\mathbf{x}_1)$.

For an IFD as defined above, with respect to the induced group $\bar{G}$, the family is referred to as being minimally invariant if for all $\boldsymbol{\theta}, \boldsymbol{\theta}' \in \Theta$ there exists a $\bar{g} \in \bar{G}$ such that $\bar{g}(\boldsymbol{\theta}) = \boldsymbol{\theta}'$. Based upon this definition, (Ghobadzadeh et al., 2014) defines a minimal invariant group G_m as a subgroup of the group G, for the IFD, such that if $h : G_m \to \bar{G}$ is a injective (one-to-one) homomorphism, where $h(g_m) = \bar{g}_m$ for all $g_m \in G_m$. The key results in (Ghobadzadeh et al., 2014) can now be stated. The first is that unknown parameters from a class of distributions can be eliminated from a maximal invariant statistic under the minimally invariant group, while preserving the maximum information about the observed signal. Additionally, it is shown that any invariant test with respect to a minimally invariant group is CFAR, and that the converse also holds under some conditions. Based upon this development it is then shown how enhanced tests can be derived systematically, giving a unified approach to the design of CFAR detectors in a very general context.

Further to this, the concept of a uniformly most powerful CFAR is introduced in (Ghobadzadeh et al., 2014), which is defined as the optimal CFAR test among all such decision rules for a specific problem space. It is then shown that the uniformly most powerful CFAR test for a minimally invariant hypothesis testing problem is given by the likelihood ratio of the maximal invariant under the minimally invariant group.

From the perspective of producing CFAR detectors for operation in Pareto Type I distributed clutter, these results suggest that a suitable focus is to begin by identifying statistics which are invariant with respect to a group of suitable transformations. This will be the focus in the following section.

12.4   Some Invariant Statistics

Throughout the following the group consisting of the set $G = \{g : g(z_j) = \alpha_1 z_j^{\alpha_2}, j = 1, 2, \ldots, N, \alpha_1, \alpha_2 > 0\}$, together with functional composition as the group action, will be used, and it will thus be assumed that each measurement of clutter is modelled via $Z_j = \Phi_1 W_j^{\Phi_2}$ as in Chapter 5. The latter model of clutter is under the assumption that each of the statistics W_j have a known distribution with no unknown parameters.

To understand the role invariant statistics can play in the construction of sliding window CFAR detection processes, suppose that we have a test statistic $T = T(Z_0, Z_1, \ldots, Z_N)$, where Z_0 is the CUT statistic and $Z_1, \ldots, Z_N$ are the statistics in the clutter range profile. Assume that under H_0 the CUT is $Z_0 = \Phi_1 W_0^{\Phi_2}$. Then there exists a $g \in G$ such that $g(Z_0, Z_1, \ldots, Z_N) = (W_0, W_1, \ldots, W_N)$ under H_0. Specifically, $g(Z_0, Z_1, \ldots, Z_N) = \left(\left(\frac{Z_0}{\Phi_1} \right)^{\frac{1}{\Phi_2}}, \ldots, \left(\frac{Z_N}{\Phi_1} \right)^{\frac{1}{\Phi_2}} \right)$ will achieve this, and is in G. Hence if the statistic T is invariant with respect to this group of transformations, then under H_0, $T(g(Z_0, Z_1, \ldots, Z_N)) = T(W_0, W_1, \ldots, W_N)$, and the latter is independent of Φ_1 and Φ_2. Since T is invariant, $T(Z_0, Z_1, \ldots, Z_N) = T(g(Z_0, Z_1, \ldots, Z_N))$, implying the detector

$$T(Z_0, Z_1, \ldots, Z_N) \underset{H_0}{\overset{H_1}{\gtrless}} \tau \qquad (12.2)$$

will achieve the CFAR property with respect to Φ_1 and Φ_2. Hence, in the search for sliding window CFAR detectors under the generic clutter model assumed throughout, one can investigate invariant decision rules.

Towards this objective, consider the test statistic

$$T(Z_0, Z_1, \ldots, Z_N) = \frac{\log\left(\frac{Z_0}{Z_{(1)}}\right)}{f\left(\log\left(\frac{Z_1}{Z_{(1)}}\right), \ldots, \log\left(\frac{Z_N}{Z_{(1)}}\right)\right)} \underset{H_0}{\overset{H_1}{\gtrless}} \tau, \tag{12.3}$$

which has been based upon (5.38) in Chapter 5, where f satisfies the scale invariance property as before. Observe that since each $Z_j > Z_{(1)}$, the logarithmic term is always nonnegative, and since f is a nonnegative valued function, the term on the bottom of (12.3) is nonnegative. This is important to observe due to the fact that this implies the detector (12.3) is equivalent to (5.38).

Now suppose that $g \in G$ such that $g(z_j) = \alpha_1 z_j^{\alpha_2}$. Then since $g(Z_j) = (\alpha_1 \Phi_1^{\alpha_2}) W_j^{\alpha_2 \Phi_2}$ it follows that the minimum of the series of $g(Z_j)$ will be $(\alpha_1 \Phi_1^{\alpha_2}) W_{(1)}^{\alpha_2 \Phi_2}$, and consequently it is not difficult to show that the statistic T in (12.3) is invariant with respect to G, implying it will achieve the CFAR property as expected. This result implies that the detector (12.3) will achieve the CFAR property when operating in Pareto Type I distributed clutter, or if it is operating in Weibull distributed clutter.

In the search for other invariant statistics for clutter of the form considered here, an examination of (12.3), and careful considerations as to why it is an invariant statistic, results in the consideration of the alternative statistic

$$T(Z_0, Z_1, \ldots, Z_N) = \frac{\log\left(\frac{Z_0}{h(Z_1, Z_2, \ldots, Z_N)}\right)}{f\left(\log\left(\frac{Z_1}{h(Z_1, Z_2, \ldots, Z_N)}\right), \ldots, \log\left(\frac{Z_N}{h(Z_1, Z_2, \ldots, Z_N)}\right)\right)}, \tag{12.4}$$

where h is a secondary function of the clutter range profile. Then the following is relatively straightforward to prove:

Lemma 12.1
Suppose that the function $h : \mathbf{R}^+ \times \mathbf{R}^+ \times \cdots \times \mathbf{R}^+ \to \mathbf{R}^+$ satisfies the condition that $h(g(Z_1, Z_2, \ldots, Z_N)) = g(h(Z_1, Z_2, \ldots, Z_N))$ for all $g \in G$. Then the statistic T defined by (12.4) is invariant.

Lemma 12.1 encompases the case where h is selected to be a minimum as in (12.3), but also implies the same result will hold if an arbitrary order statistic is selected to replace the minimum. What is interesting is that there are also other choices for h which will satisfy the requirements of Lemma 12.1. Observe that, in view of the group of transformations considered for the Pareto Type I model, it can be shown that functions h that satisfy Lemma 12.1 must be such that

$$h(Z_1, Z_2, \ldots, Z_N) = \left[\frac{1}{\lambda_1} h(\lambda_1 Z_1^{\lambda_2}, \ldots, \lambda_1 Z_N^{\lambda_2})\right]^{\frac{1}{\lambda_2}}, \tag{12.5}$$

for all $\lambda_1, \lambda_2 > 0$. Other admissible choices for h that satisfy the condition (12.5) include $h(Z_1, Z_2, \ldots, Z_N) = Z_j$ for some $1 \leq j \leq N$ and $h(Z_1, Z_2, \ldots, Z_N) = (Z_1 Z_2 \ldots Z_N)^{\frac{1}{N}}$. The latter is clearly a geometric mean.

To explore this further, let $\mathcal{H}$ be the class of admissible functions, defined through (12.5). Then it is clear that this is equivalent to

$$\mathcal{H} = \{h(z_1, \ldots, z_N) : h(\lambda_1 z_1^{\lambda_2}, \ldots, \lambda_1 z_N^{\lambda_2}) = \lambda_1 h(z_1, \ldots, z_N)^{\lambda_2}, \lambda_1, \lambda_2 > 0 \text{ and } z_j \geq 0\}. \tag{12.6}$$

The trivial function $h(z_1, \ldots, z_N) = 0$ is an element of $\mathcal{H}$, but it is interesting to note that linear scale transformations do not fall within this class. To see this, observe that if h is selected to be the minimum and $\kappa > 0$ then with reference to Lemma 12.1 $g(h(Z_1, Z_2, \ldots, Z_N)) = \kappa^{\lambda_2} \lambda_1 Z_{(1)}^{\lambda_2}$ and $h(g(Z_1, Z_2, \ldots, Z_N)) = \kappa \lambda_1 Z_{(1)}^{\lambda_2}$, where g is the group action applied previously. Since these two do not coincide, it follows that the detector with $h(Z_1, Z_2, \ldots, Z_N) = \kappa Z_{(1)}$ is not invariant.

It is also interesting to observe that if $h \in \mathcal{H}$ then powers of this are not elements of $\mathcal{H}$. Consider for example $k(Z_1, Z_2, \ldots, Z_N) = h(Z_1, Z_2, \ldots, Z_N)^{\frac{1}{p}}$ for some $p > 0$. Then for any $\lambda_1, \lambda_2 > 0$, $k(\lambda_1 Z_1^{\lambda_2}, \lambda_1 Z_2^{\lambda_2}, \ldots, \lambda_1 Z_N^{\lambda_2}) = \lambda_1^{\frac{1}{p}} h(Z_1, Z_2, \ldots, Z_N)^{\frac{\lambda_2}{p}}$ while $\lambda_1 k(Z_1, Z_2, \ldots, Z_N)^{\lambda_2} = \lambda_1 h(Z_1, Z_2, \ldots, Z_N)^{\frac{\lambda_2}{p}}$, showing that $k \notin \mathcal{H}$.

By an application of a similar argument it can be demonstrated that the class is not closed with respect to addition of functions. However, it is closed with respect to geometric means, as the following result demonstrates:

Lemma 12.2
Suppose that $\{h_j, j \in \{1, 2, \ldots m\}\}$ is a sequence of functions in $\mathcal{H}$, for some $m \geq 1$. Then the function

$$h(z_1, \ldots, z_N) = \prod_{j=1}^{m} h_j(z_1, \ldots, z_N)^{\frac{1}{m}} \tag{12.7}$$

is an element of $\mathcal{H}$.

To demonstrate the validity of Lemma 12.2 observe that for any $\lambda_1, \lambda_2 > 0$,

$$h(\lambda_1 z_1^{\lambda_2}, \ldots \lambda_1 z_N^{\lambda_2}) = \prod_{j=1}^{m} h_j(\lambda_1 z_1^{\lambda_2}, \ldots, \lambda_1 z_N^{\lambda_2})^{\frac{1}{m}}$$

$$= \lambda_1 \left(\prod_{j=1}^{m} h_j^{\frac{1}{m}}(z_1, \ldots, z_N) \right)^{\lambda_2}$$

$$= \lambda_1 h(z_1, \ldots, z_N)^{\lambda_2}, \tag{12.8}$$

where the fact that each $h_j \in \mathcal{H}$ has been used. Based upon (12.8) it follows that $h \in \mathcal{H}$, as required.

The geometric mean is a specific example of a function representable in the form given by (12.7); however Lemma 12.2 indicates that the class of functions $\mathcal{H}$ consists of many possible elements. It is speculated that the minimum function is the smallest non-trivial member of $\mathcal{H}$. The discussion above shows that a constant times the minimum is not an admissible function in $\mathcal{H}$, nor are powers of it, providing support for this claim. The proof of this result appears to be rather challenging and is left to the reader to pursue.

Observe that if it is assumed f is nonnegative the decision rule (12.4) can be written in the equivalent form

$$Z_0 \underset{H_0}{\overset{H_1}{\gtrless}} h(Z_1, Z_2, \ldots, Z_N) e^{\tau f \left(\log \left(\frac{Z_1}{h(Z_1, Z_2, \ldots, Z_N)} \right), \ldots, \log \left(\frac{Z_N}{h(Z_1, Z_2, \ldots, Z_N)} \right) \right)} \tag{12.9}$$

Hence it is clear that in order to maximise the detection performance of (12.9) the best choice for h will be that which reduces the right hand side of (12.9) to its least possible value. If h is selected to be a constant, then the original transformed decision rules are recovered, with h selected to be the distributional minimum value, which in the Pareto case is β. As such, especially in horizontal polarisation, this will be quite small resulting in better detection performance than a decision rule where h is selected to be a function of the CRP, such as the minimum. In order to produce a full CFAR detector, it is clear that the ideal choice for h will be the sample minimum. This analysis suggests that the decision rule (12.3) may in fact be the best possible choice. Further work is required to clarify this claim.

In order to investigate this further, the next section derives a series of new CFAR detectors based upon (12.9) with some suitable choices for h.

12.5 Examples of Invariant CFAR Detectors

This section examines some choices for h that can be applied to the decision rule (12.9), and in some cases the Pfa is derived. The cases where this is performed are those for which the analysis is relatively simple. To begin, consider the selection of $h(z_1, z_2, \ldots, z_N) = (z_1 z_2 \ldots z_N)^{\frac{1}{N}}$, which is the geometric mean. The test statistic, with such a choice for h, reduces to

$$Z_0 \underset{H_0}{\overset{H_1}{\gtrless}} (Z_1 Z_2 \cdots Z_N)^{\frac{1}{N}} e^{\tau f(W_1, W_2, \ldots, W_N)}, \tag{12.10}$$

where for each $1 \le j \le N$

$$W_j = \log \left(\frac{Z_j}{(Z_1 Z_2 \ldots Z_N)^{\frac{1}{N}}} \right) = \left(1 - \frac{1}{N} \right) \log Z_j + \frac{1}{N} \sum_{k \ne j} \log Z_k. \tag{12.11}$$

Suppose that under H_0 the random variables $X_0, X_1, \ldots, X_N$ are the duals of the corresponding Pareto Type I distributed random variables. Here this means that an appeal to the Pareto-exponential duality property (3.21) has produced the latter sequence of exponentially distributed statistics. Hence $Z_j = \beta e^{\alpha^{-1} X_j}$, where each X_j has an exponential distribution with unity mean. Then under H_0

$$h(Z_1, Z_2, \ldots, Z_N) \equiv \beta e^{\frac{1}{\alpha N} \sum_{j=1}^{N} X_j} \tag{12.12}$$

and for each $1 \leq j \leq N$,

$$\log\left(\frac{Z_j}{h(Z_1, Z_2, \ldots, Z_N)}\right) \equiv \alpha^{-1}\left[X_j - \frac{1}{N}\sum_{i=1}^{N} X_i\right]. \tag{12.13}$$

Hence an application of (12.12) and (12.13) to the Pfa expression for (12.10) results in

$$\mathrm{Pfa} = \mathrm{P}\left(X_0 > \frac{1}{N}\sum_{i=1}^{N} X_i + \tau f\left(X_1 - \frac{1}{N}\sum_{i=1}^{N} X_i, \ldots, X_N - \frac{1}{N}\sum_{i=1}^{N} X_i\right)\right), \tag{12.14}$$

where the scale invariance property of f has been applied. As expected, this is independent of both the Pareto distributional parameters. Observe that if f is selected to be a sum then $f\left(X_1 - \frac{1}{N}\sum_{i=1}^{N} X_i, \ldots, X_N - \frac{1}{N}\sum_{i=1}^{N} X_i\right) = 0$, implying the corresponding detector cannot have its Pfa set via τ. Hence this choice is redundant. Alternatives to this is to select f to be a partial sum such as a trimmed mean, or to apply a smallest-of or greatest-of procedure based upon partial sums. Here these are not considered, but the option of selecting f as the kth order statistic is instead investigated.

With f selected to be the kth OS, the Pfa (12.14) reduces to

$$\mathrm{Pfa} = \mathrm{P}\left(X_0 > \frac{1 - \tau}{N}\sum_{i=1}^{N} X_i + \tau X_{(k)}\right). \tag{12.15}$$

As a simple choice the case where $k = 1$ is investigated, corresponding to f being a minimum. Based upon this it is not difficult to evaluate (12.15) to produce a closed form expression for τ. Recall that since it has been shown in Chapter 5 that each $X_j | \{X_{(1)} = t\} \overset{d}{=} t + Y_j$, where Y_j is exponentially distributed with unity mean, and consequently the sum $\sum_{i=1}^{N} X_i | \{X_{(1)} = t\} \overset{d}{=} Nt + \gamma(N, 1)$ and so

$$\begin{aligned}
\mathrm{Pfa} &= \int_0^\infty N e^{-Nt} \mathrm{P}\left(X_0 > \frac{1 - \tau}{N}\sum_{i=1}^{N} X_i + \tau t \,\middle|\, X_{(1)} = t\right) dt \\
&= \int_0^\infty \int_0^\infty N e^{-Nt} f_{\gamma(N,1)}(w) \mathrm{P}\left(X_0 > t + \left(\frac{1 - \tau}{N}\right) w\right) dt\, dw \\
&= \int_0^\infty N e^{-(N+1)t}\, dt \int_0^\infty e^{\left(\frac{\tau - 1}{N}\right) w} f_{\gamma(N,1)}(w)\, dw, \tag{12.16}
\end{aligned}$$

where $f_{\gamma(N,1)}(w)$ is the density of a gamma distributed random variable, denoted $\gamma(N,1)$, with parameters N and 1. If it is assumed that $\tau < 1$ then the second integral exists and coincides with the moment generating function of the associated gamma distribution. Based upon this, the Pfa reduces to

$$\text{Pfa} = \frac{N}{N+1}\left(1 - \frac{\tau-1}{N}\right)^{-N} \tag{12.17}$$

and by inversion of (12.17) the threshold is

$$\tau = 1 + N\left[1 - \left(\frac{N+1}{N}\text{Pfa}\right)^{-\frac{1}{N}}\right]. \tag{12.18}$$

The corresponding decision rule is given by

$$Z_0 \underset{H_0}{\overset{H_1}{\gtrless}} (Z_1 Z_2 \ldots Z_N)^{\frac{1-\tau}{N}} Z_{(1)}^{\tau}. \tag{12.19}$$

It is important to note that the detector (12.19) is exactly the same process as the GM-CFAR (5.43) derived previously. This can be realised by defining $v = \frac{1-\tau}{N}$ in (12.19), and also applying the same transformation to (12.17) and (12.18). This derivation provides a slightly different validation of the Pfa expression for the decision rule (5.43).

Next the case where h is an arbitrary order statistic is examined, so suppose that $h(z_1, z_2, \ldots, z_N) = z_{(k)}$ for some $1 \le k \le N$. Then the decision rule (12.9) becomes

$$Z_0 \underset{H_0}{\overset{H_1}{\gtrless}} Z_{(k)} e^{\tau f\left(\log\left(\frac{Z_1}{Z_{(k)}}\right), \ldots, \log\left(\frac{Z_N}{Z_{(k)}}\right)\right)}. \tag{12.20}$$

Observe that with the choice of $f(z_1, \ldots, z_N) = z_{(i)}$ for some $i \ne k$ the detector (12.20) reduces to

$$Z_0 \underset{H_0}{\overset{H_1}{\gtrless}} Z_{(k)}^{1-\tau} Z_{(i)}^{\tau}, \tag{12.21}$$

which is readily identified as the dual order statistic CFAR examined in Chapter 7.

Here it is possible to examine the case where f is a sum. Based upon this choice it is relatively straightforward to show that

$$f\left(\log\left(\frac{Z_1}{Z_{(k)}}\right), \ldots, \log\left(\frac{Z_N}{Z_{(k)}}\right)\right) = \sum_{i=1}^{N} \log(Z_i) - N\log(Z_{(k)}) \tag{12.22}$$

which when applied to (12.20) reduces to the decision rule

$$Z_0 \underset{H_0}{\overset{H_1}{\gtrless}} Z_{(k)}^{1-N\tau} \prod_{j=1}^{N} Z_j^{\tau} \tag{12.23}$$

which provides a generalisation of the GM-CFAR (5.43). Derivation of the corresponding Pfa of (12.23) involves extensive conditional probability and is thus omitted and left to the interested researcher.

The case where $h(z_1, z_2, \ldots, z_N) = z_k$ is now investigated, for some fixed $1 \leq k \leq N$. Based upon such a choice, the detector (12.9) becomes

$$Z_0 \underset{H_0}{\overset{H_1}{\gtrless}} Z_k e^{\tau f \left(\log \left(\frac{z_1}{z_k} \right), \ldots, \log \left(\frac{z_1}{z_k} \right) \right)}. \tag{12.24}$$

It is clear that the selection of f as an order statistic will not produce a useful detector, due to the fact that there is the possibility that this order statistic will coincide with Z_k, resulting in the elimination of τ from the corresponding expression for the Pfa. However, the case where f is a sum produces a viable detector. In this case it is simple to show that the corresponding decision rule is

$$Z_0 \underset{H_0}{\overset{H_1}{\gtrless}} Z_k \left(\prod_{j \neq k} \frac{Z_j}{Z_k} \right)^{\tau}. \tag{12.25}$$

Derivation of the Pfa of (12.25) proceeds as in previous analyses. It is immediate that the Pfa is equivalent to

$$\text{Pfa} = \text{P} \left(X_0 > [1 - \tau(N-1)]X_k + \tau \sum_{j \neq k} X_j \right). \tag{12.26}$$

By conditioning on X_k and the sum of exponentially distributed variables in (12.26), the Pfa reduces to

$$\text{Pfa} = \int_0^\infty e^{-t(2 - \tau(N-1))} dt \int_0^\infty e^{-\tau w} f_{\gamma(N-1,1)}(w) dw, \tag{12.27}$$

where $f_{\gamma(N-1,1)}(w)$ is the density of the associated gamma random variable. With the proviso that $\tau < \frac{1}{N-1}$ and $\tau > -1$ this Pfa can be reduced to the form

$$\text{Pfa} = (2 - \tau(N-1))^{-1}(1 + \tau)^{-(N-1)}. \tag{12.28}$$

Although providing one with a CFAR detector, the issue with (12.25) is the appropriate selection of the statistic Z_k. One can show that a randomised selection of this statistic results in loss of the CFAR property, since if h is selected to be a random element of the CRP, $h \notin \mathcal{H}$.

As a final example of an invariant detector, one can consider the selection of h to be a geometric mean of order statistics, in view of Lemma 12.7. Suppose that $I \subset \{1, 2, \ldots, N\}$ and select h to be $h(z_1, z_2, \ldots, z_N) = \prod_{i \in I} z_{(i)}^{\frac{1}{|I|}}$, where $|I|$ is the cardinality of the set I. Then by Lemma 12.2 it follows that $h \in \mathcal{H}$. Based upon this choice, and with the selection of f as a sum, it is immediate that the corresponding decision rule is

$$Z_0 \underset{H_0}{\overset{H_1}{\gtrless}} \left(\prod_{i \in I} Z_{(i)} \right)^{\frac{1-N\tau}{|I|}} \left(\prod_{j=1}^{N} Z_j \right)^{\tau}. \tag{12.29}$$

Clearly (12.29) provides an extension of the GM-CFAR (5.43); however there is increased complexity in the derivation of its corresponding Pfa. As a simple example, the situation where $I = \{1,2\}$ is examined, with a view to investigate whether there is any merit in the extension of I beyond a single element set. The corresponding decision rule is

$$Z_0 \underset{H_0}{\overset{H_1}{\gtrless}} \left(Z_{(1)} Z_{(2)} \right)^{\frac{1-N\tau}{2}} \left(\prod_{j=1}^{N} Z_j \right)^{\tau}. \tag{12.30}$$

In order to derive the Pfa of (12.30) one applies the Pareto-exponential duality property (3.21), so that $Z_j = \beta e^{\alpha^{-1} X_j}$ for all $0 \leq j \leq N$ as well as the fact that $Z_{(k)} = \beta e^{\alpha^{-1} X_{(k)}}$ for $k = 1$ and 2, so that

$$\text{Pfa} = \mathbf{P}\left(X_0 > \left(\frac{1-N\tau}{2} \right) [X_{(1)} + X_{(2)}] + \tau \sum_{j=1}^{N} X_j \right). \tag{12.31}$$

It is clear that in order to evaluate (12.31) it will be necessary to utilise the joint distribution of $X_{(1)}$ and $X_{(2)}$ as well as the conditional distribution of sum of the X_j given the joint order statistics. Firstly, note that in view of (2.98) in Chapter 2 this joint density is

$$f_{(X_{(1)},X_{(2)})}(x,y) = N(N-1)e^{-x} e^{-y(N-1)}, \tag{12.32}$$

on the region where $y \geq x$. In order to derive the distribution of the sum conditioned on the pair of order statistics, note that

$$\mathbf{P}(X_1 \leq z | X_{(1)} = x, X_{(2)} = y) \;=\; \mathbf{P}(X_1 \leq z | X_1 \geq x, X_2, X_3, \ldots X_N \geq y)$$

$$=\; \mathbf{P}(X_1 \leq z | X_1 \geq x)$$

$$=\; 1 - e^{-(z-x)}, \tag{12.33}$$

provided $z \geq x$ where independence has been used. Similarly for the case where $k \geq 2$,

$$
\begin{aligned}
P(X_k \leq z | X_{(1)} = x, X_{(2)} = y) &= P(X_k \leq z | X_1 \geq x, X_2, X_3, \ldots X_N \geq y) \\
&= P(X_k \leq z | X_k \geq y) \\
&= 1 - e^{-(z-y)},
\end{aligned}
\tag{12.34}
$$

again where independence has been employed, and provided $z \geq y$. Based upon these results it follows that for each k $X_k | \{X_{(1)} = x, X_{(2)} = y\} = xI[k = 1] + yI[k \neq 1] + Y_k$ where each Y_k is exponentially distributed with mean unity, and $I[A]$ is the indicator function on the set A. Hence the latter is unity if the condition on the set A is met, and is zero otherwise. In adddition to this, the Y_k form an independent sequence. By applying these results it follows immediately that

$$
\sum_{j=1}^{N} X_j | \{X_{(1)} = x, X_{(2)} = y\} \stackrel{d}{=} x + (N-1)y + W,
\tag{12.35}
$$

where $W \stackrel{d}{=} \gamma(N, 1)$, provided $y > x$. One can now apply (12.32) and (12.35) to evaluate (12.31). Hence by conditioning on the joint distribution of the order statistics,

$$
\begin{aligned}
\text{Pfa} &= \int_{y=0}^{\infty} \int_{x=0}^{y} P\left(X_0 > \left(\frac{1-N\tau}{2}\right)[X_{(1)} + X_{(2)}] + \tau \sum_{j=1}^{N} X_j \middle| X_{(1)} = x, X_{(2)} = y\right) \times \\
&\quad f_{(X_{(1)}, X_{(2)})}(x, y) dx dy \\
&= \int_{y=0}^{\infty} \int_{x=0}^{y} \int_{z=0}^{\infty} f_{\gamma(N,1)}(z) P\left(X_0 > \left(\frac{1-N\tau}{2}\right)(x+y) + \tau(x + (N-1)y + z) \times \right. \\
&\quad \left. f_{(X_{(1)}, X_{(2)})}(x, y)\right) dz dx dy \\
&= M_{\gamma(N,1)}(-\tau) N(N-1) \int_{y=0}^{\infty} e^{-y\left[\frac{1-N\tau}{2} + (N-1)\tau + N - 1\right]} \times \\
&\quad \int_{x=0}^{y} e^{-x\left[\frac{1-N\tau}{2} + \tau + 1\right]} dx dy \\
&= \frac{N(N-1)(1 + \tau^{-N})}{(N+1)\left(\frac{1-N\tau}{2} + (\tau+1)(N-1)\right)},
\end{aligned}
\tag{12.36}
$$

where $M_{\gamma(N,1)}$ is the moment generating function of the gamma distribution, and simplification has been performed. Thus one can solve numerically for τ via (12.36), for application to (12.30).

The next section examines the decision rule (12.30) and compares it with detectors based upon (12.3) and some other test statistics.

12.6 Performance of Invariant Detectors

The same test procedure is applied to examine the relative merits of the new decision rule (12.30). Hence the CUT is modelled as a Swerling I fluctuation and the clutter is simulated with the Pareto clutter parameters applied previously. For the case of interference, the additional target is again modelled by a Swerling I fluctuation with a fixed ICR and independent of the target in the CUT.

12.6.1 Homogeneous Clutter

Several examples are provided for the case of homogeneous clutter returns. A series of detectors are included to permit a benchmark comparison with the invariant detector. These are the original GM detector (5.10), the OS detector (5.11), the GM-CFAR (5.43), the OS-CFAR (5.45) and the OS detector (4.8). The ideal detector is again used as an upper bound on performance. In all cases the Pfa is set to 10^{-4}, while the order statistic index is selected to be $k = N - 2$ in all such detectors. The nomenclature employed within the following figures is to specify the detector in terms of which parameters it assumes known *a priori*. Hence, for example, the detector (4.8) is specified as *OS α known*, while (5.45) is referred to as *OS CFAR*. Clutter parameters are again sourced from Table 3.1, and the following discussion is based upon performance of the detectors operating in clutter simulated with parameters matched to those obtained in the three polarisation cases. Hence for convenience performance will be discussed relative to the respective polarisation. The new invariant detector (12.30) is denoted GM-DOS CFAR, to indicate that it is based upon two order statistics as well as a geometric mean.

In the case of horizontal polarisation, a series of simulations are provided, which can be found in Figures 12.1–12.5. Figure 12.1 is for the case where $N = 32$, and it can be observed that the OS detector (4.8) matches the ideal decision rule, while the remaining detectors have clustered into two bands. The OS-based detectors (5.11) and (5.45) tend to match and have the worst performance, although the difference between the ideal decision rule and these two is roughly 0.1 or a 10 dB loss. To clarify this Figure 12.2 provides an enlargement. As can be observed, the new detector (12.30) does not provide a substantial gain on the GM-CFAR (5.43), but is slightly better than the OS-CFAR (5.45).

Next the CRP length is reduced to 16, while maintaining all other parameters. Figure 12.3 shows the detection performance, while Figure 12.4 provides a magnification of this plot. Similar performance is observed as for the scenario where

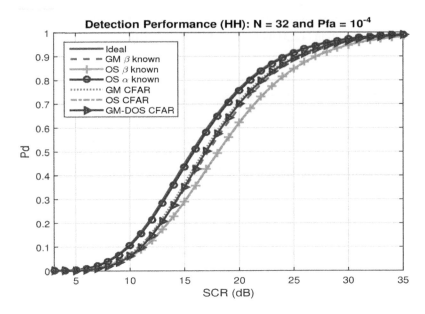

Figure 12.1: Invariant detector performance in homogeneous horizontally polarised clutter. The new decision rule is compared with a selection of detectors previously analysed throughout the book. The latter are labelled by the assumptions made relative to clutter parameters. The ideal decision rule is also included as an upper bound on performance.

$N = 32$. The GM-CFAR is slightly better than the invariant detector (12.30), while the latter is slightly better than the OS-CFAR (5.45).

As a final example of performance in horizontally polarised clutter Figure 12.5 shows the consequence of a further reduction in the CRP length; in this case $N = 8$. Here it is observed that the OS-CFAR has the worst performance, while the OS-detector (5.11) has marginally better performance than its CFAR analogue. The GM detector (5.10) is the third worst performing decision rule, despite having *a priori* knowledge of β. The proposed detector (12.30) and the GM-CFAR (5.43) are matching in terms of detection performance. Hence it is clear, when examining the relative merits in the application of a second order statistic as in (12.30), that there is not a significant detection gain relative to the GM-CFAR (5.43).

Similar results are observed in the case of cross polarisation, as is illustrated in Figure 12.6 for $N = 16$. Figure 12.7 provides an enlargement of Figure 12.6 to clarify this. The major difference, when compared with the horizontally polarised case, is that the decision rules experience slightly less detection loss.

To complete the analysis in homogeneous clutter Figure 12.8 repeats the same scenario in vertical polarisation, while Figure 12.9 provides an enlargement. The detection loss is reduced due to the fact that the clutter is less spiky, but the same conclusions are reached as in the other two cases.

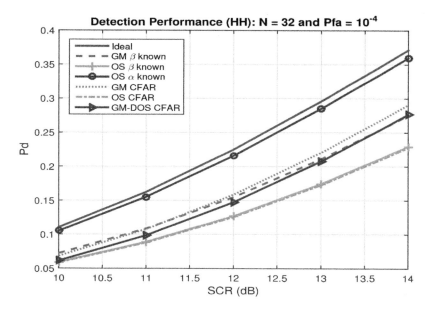

Figure 12.2: Enlargement of the results of Figure 12.1.

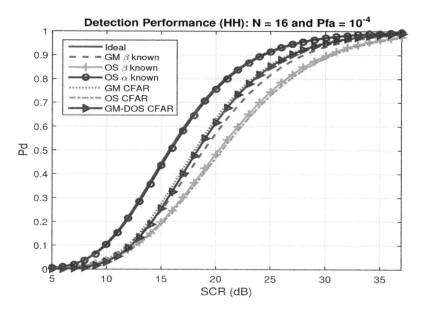

Figure 12.3: Second example of invariant detector performance in horizontally polarised clutter. In this experiment the CRP length has been halved.

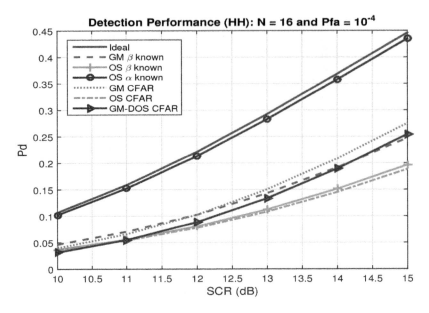

Figure 12.4: Enlargement of Figure 12.3.

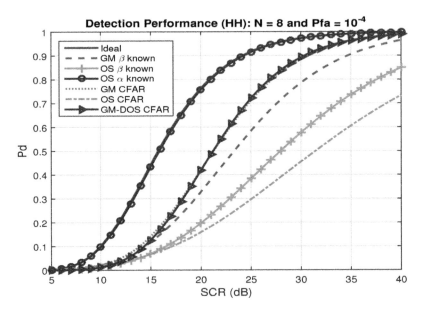

Figure 12.5: Third example of invariant detector performance in horizontally polarised clutter, showing the effects of a further reduction in the CRP length.

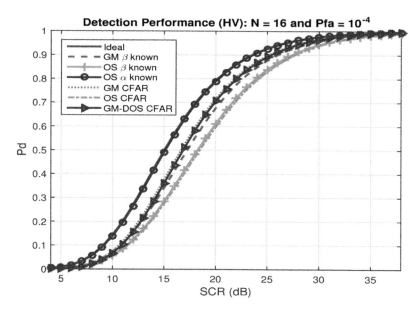

Figure 12.6: An example of detector performance in cross-polarisation.

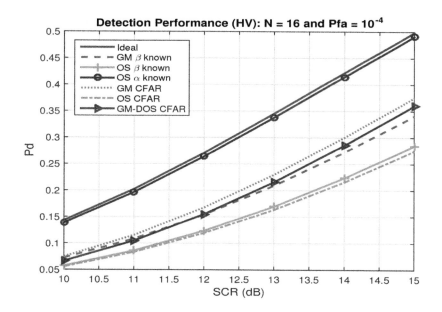

Figure 12.7: Enlargement of Figure 12.6.

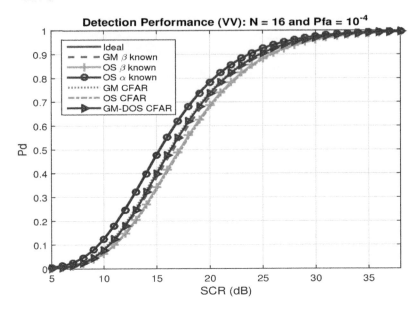

Figure 12.8: An example of detector performance in vertical polarisation.

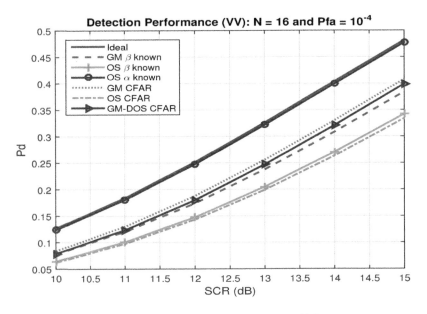

Figure 12.9: Magnification of Figure 12.8.

Thus the conclusion reached is that in the case of homogeneous clutter the detector (12.30) does not provide a justification for its application in practice. The next subsection examines whether it provides any gains in the case of interference.

12.6.2 Interference

Next the effects of an independent interfering target in the CRP is examined. Given the fact that the invariant detector (12.30) is based upon a geometric mean it is expected that detection losses will be experienced when the CRP has strong interference. Beginning with the case of horizontal polarisation Figure 12.10 shows performance when a 10 dB Swerling I target fluctuation is inserted into the CRP. The ideal decision rule is not subjected to this, but provides an indicator of the cost on performance as shown. The detector (4.8) has the greatest immunity to the effects of the interference, but requires *a priori* knowledge of α. Next the detectors (5.10) and (5.11) rank as the second and third best performing decision rules. A 10 dB interfering target is not sufficiently strong to affect the performance of a geometric mean detector significantly. The new invariant detector (12.30) is the best performing CFAR. The OS-CFAR (5.45) has the worst performance. It is expected that the latter is due to the fact that a small interfering target may be selected to be the minimum in a large number of simulations, which results in the clutter measurement based upon the interfering target, which can have a detrimental effect on performance.

Figure 12.11 shows the effect of increasing the ICR to 20 dB in horizontal polarisation and with $N = 16$. Due to the strong level of interference the geometric mean-based detectors are experiencing significant detection degradation. This includes the new invariant detector (12.30). By contrast the OS-based detectors are performing well. It is interesting to note that the performance of each of the three OS detectors reduces as the *a priori* assumptions are altered. The decision rule (4.8) assumes that α is known, and has the best performance. The detector (5.11) requires *a priori* knowledge of β while (5.45) is a CFAR decision rule.

Figure 12.12 examines the consequence of increasing the level of interference to 40 dB, in horizontal polarisation. The results are similar to those as in the previous example except the detection losses of the geometric mean based detectors are more severe.

The results are similar in the other two polarisations. Figure 12.13 shows this for the case of cross-polarisation and when $N = 16$ and ICR = 20 dB, while Figure 12.14 is for the vertically polarised case.

Hence in the management of interference in the clutter range profile the OS-CFAR (5.45) provides the most robust solution.

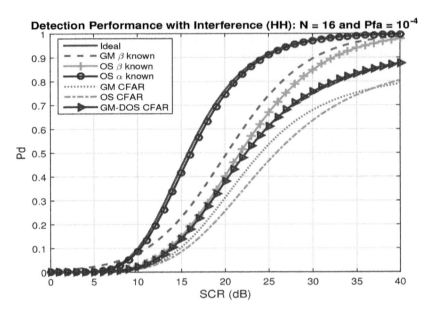

Figure 12.10: Performance analysis of the decision rules examined in this chapter in the case of horizontal polarisation, with small ICR of 10 dB.

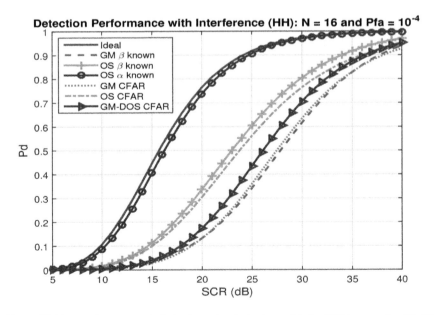

Figure 12.11: Interference, again in horizontal polarisation, with the ICR increased to 20 dB.

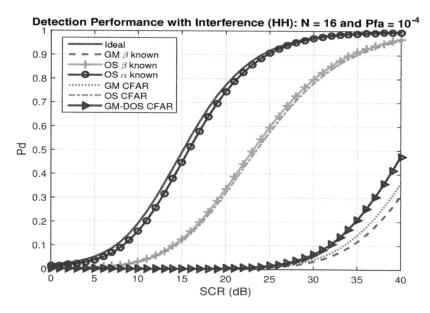

Figure 12.12: Detector performance in the presence of a 40 dB interfering target, again in horizontal polarisation.

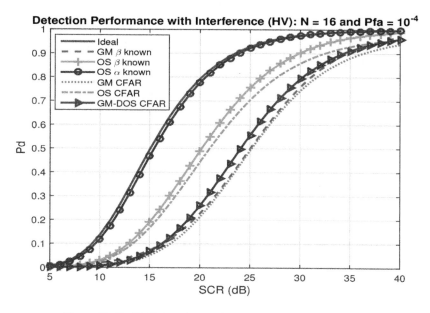

Figure 12.13: 20 dB interference in the case of cross-polarisation.

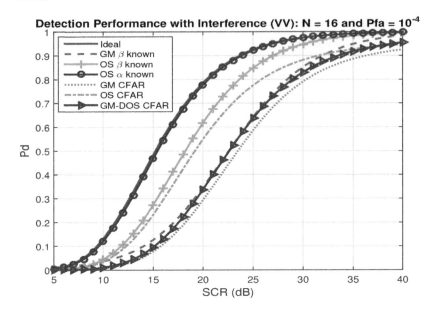

Figure 12.14: Interference in vertical polarisation, where the ICR = 20 dB.

12.6.3 False Alarm Regulation

To complete the analysis of the invariant detector (12.30) its performance during clutter transitions is examined. Figure 12.15 shows the estimated Pfa for the case of a 1 dB clutter power level increase, when the design Pfa is 10^{-4} and $N = 16$, and for horizontal polarisation. It is interesting to observe that the new decision rule does not manage the Pfa very well, compared with the other CFAR detectors. The OS detector (4.8) and (12.30) tend to have the worst performance once the CUT is affected. The GM-CFAR also has a tendency to increase the number of false alarms as previously noted.

Next Figure 12.16 repeats the scenario in Figure 12.15 except with the CRP length increased to 32. Although the performance of the new invariant detector has improved, the original CFAR detectors (5.43) and (5.45) tend to regulate the Pfa better.

Figure 12.17 shows the estimated Pfa again in horizontal polarisation with $N = 16$ but will a CCR of 0.5 dB. Comparing Figures 12.17 and 12.15 it is interesting to note that less of a clutter power level increase does not result in improvements in the new decision rule.

In the case of cross-polarisation the results are similar, as can be observed from Figure 12.18, which is for $N = 16$ with a CCR of 1 dB. The CFAR detector (5.45) tends to reduce the Pfa prior to the saturation of the CUT, after which it tends to regulate the Pfa reasonably well.

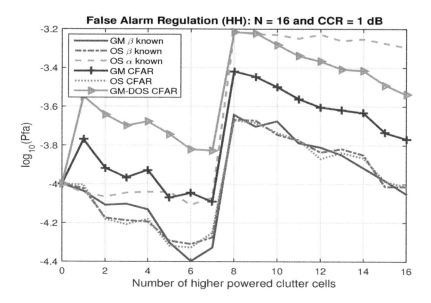

Figure 12.15: False alarm regulation in horizontal polarisation.

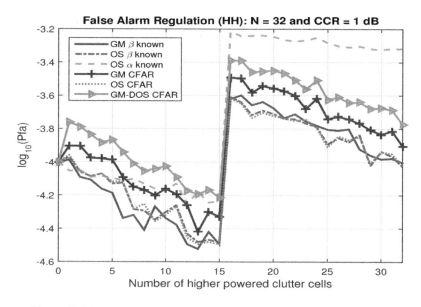

Figure 12.16: Second example of false alarm regulation in horizontal polarisation.

Figure 12.19 shows the effect of increasing N to 32. Here one observes that the OS detector (4.8) regulates the Pfa very well until the CUT is affected. The

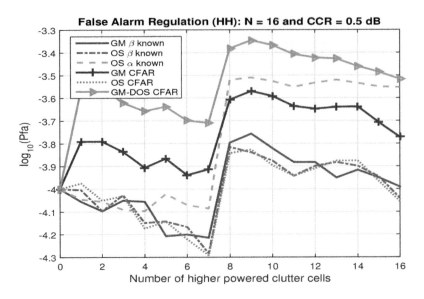

Figure 12.17: Third example of false alarm regulation in horizontally polarised clutter.

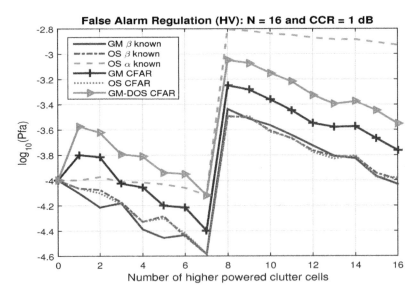

Figure 12.18: False alarm regulation in cross-polarisation.

other decision rules are experiencing less variable performance. The CFAR detectors (5.43) and (5.45) still provide a robust solution to the problem of false alarm regulation.

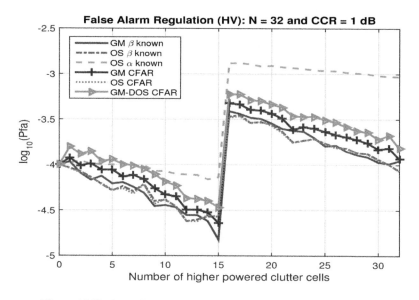

Figure 12.19: Second example of false alarm regulation in cross polarisation.

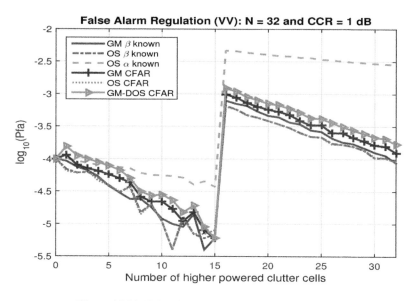

Figure 12.20: False alarm regulation in vertical polarisation.

In vertical polarisation the results are less variable, as can be observed from Figure 12.20, which is for $N = 32$ with a 1 dB power level increase. The conclusions are similar to those reached in the cross-polarisation case.

12.7 Conclusions

The concept of an invariant statistic and an invariant family of distributions has been introduced, and it has been shown that the transformed detectors derived in Chapter 5, with a complete sufficient statistic used as a replacement for the Pareto scale parameter, are invariant statistics. It is somewhat difficult to demonstrate the maximal invariance of such decision rules, however, the fact that they are invariant detectors explains why they achieve the CFAR property, and also suggests that they are in the class of the best CFAR processes available for the Pareto Type I clutter model.

An analysis of a generic decision rule resulted in further insights into admissible invariant detection statistics. This allowed the specification of a new CFAR detection processes, given by (12.30). Performance analysis showed that (12.30) did not provide significant performance gains on the detector (5.43). The best performing CFAR detectors are (5.43) and (5.45), although the latter detector is the best CFAR solution to the management of interference and false alarm regulation.

Chapter 13

Convergence and Approximation of the Pareto Model

13.1 Introduction

There has been much interest in the asymptotic behaviour of radar clutter models, since if the model applied to real data can be approximated by a simpler limiting distribution, then it may be justifiable to apply decision rules constructed for this limiting distribution. As an example, it was established in (Watts, 1985) that the K-distribution limits to that of an exponential random variable when the K-distribution's shape parameter exceeds roughly 20. Given the difficulty in achieving the CFAR property with respect to the K-distribution's shape parameter, this limiting result implies that for situations where a K-distribution's fitted shape parameter is large, one may assume that the data is roughly exponentially distributed. Hence the original sliding window CFAR detection processes may be applied to data in such cases, as an approximation. Consequently this provides the radar engineer with a practical means of achieving CFAR, when it is otherwise not possible.

The same line of reasoning can be applied to the context of this book. Hence one can consider the possibility of applying detectors, designed to operate in exponentially distributed clutter, when the Pareto model is approximately exponentially distributed. Towards this objective, several recent studies have attempted to quantify the validity of the Pareto-exponential approximation. As a result of this two principle methods have been explored for this purpose. The first is based upon the Stein-Chen Method from mathematical probability (Barbour et al., 1992). This method involves identifying an equation characterising the limiting distribution, and then bounds on it are used to measure the distributional differences, which then provide rules of thumb on the validity of the approximation. The case of the exponential approximation of the Pareto model via the Stein-Chen Method has been reported in (Weinberg, 2012), while the scenario of the Rayleigh approximation of the K-distribution has been examined in (Weinberg, 2016d). The second approach, which has only recently been investigated in the open literature, is the application of the Kullback-Leibler divergence to measuring the distributional approximations. This technique, for the two distributional approximations described above, has been reported in (Weinberg, 2016e) and (Weinberg and Glenny, 2016) respectively. The conclusion reached from these analyses is that the Kullback-Leibler divergence approach is not only simpler, but provides a better understanding of the validity of the approximation. This chapter is based upon the research reported in (Weinberg, 2012) and (Weinberg, 2016e). However, recent advances in Stein approximation have yielded better bounds for product-based distributions. In particular, (Gaunt, 2016) has investigated the Rayleigh-K distributional approximation. The techniques employed in (Gaunt, 2016) could also be applied to the Pareto-exponential approximation, since the Pareto model also follows a product-type structure. This is left to the interested reader to pursue.

It is important to emphasize that it is often quite difficult to bound distributional differences directly, explaining why alternative methods are neccessary to achieve suitable bounds on distributional differences.

13.2   Problem Specification

The fundamental problem to be considered is the exponential distributional approximation of the Pareto clutter model. However, it is necessary to compare distributions with the same support. Hence this requires analysis of a Pareto Type II model. Thus the focus is to compare a random variable X with distribution function

$$F_X(t) = 1 - \left(\frac{\beta}{t + \beta} \right)^{\alpha}, \tag{13.1}$$

for $t \geq 0$ to a random variable Y with distribution function

$$F_Y(t) = P(Y \leq t) = 1 - e^{-\lambda t} \tag{13.2}$$

also for $t \geq 0$. The objective is to develop an understanding of when Y is a suitable approximation for X in terms of distributions. Hence it is desired to understand when $P(X \in A) \approx P(Y \in A)$, where A is an arbitrary subset of the nonnegative integers. It would be useful if one could establish approximate rules of thumb on the validity of this approximation, in terms of the Pareto clutter parameters. Additionally, it would be of relevance to establish whether an optimal choice for λ in (13.2) exists.

Throughout it will be assumed that $\alpha > 3$, which implies that the relevant moments of the Pareto distributional model exist. The exponential distribution arises as a limit of the Pareto as the latter's shape parameter increases. This can be seen through a reparameterisation of $\beta = \frac{\alpha(1+o_\alpha(1))}{\lambda}$ where $o_\alpha(1) \to 0$ as $\alpha \to \infty$. Applying this to the complementary distribution function of X yields

$$
\begin{aligned}
\lim_{\alpha \to \infty} P(X > t) &= \lim_{\alpha \to \infty} \left(1 + \frac{t}{\beta}\right)^{-\alpha} \\
&= \lim_{\alpha \to \infty} \left(1 + \frac{\lambda t}{\alpha(1 + o_\alpha(1))}\right)^{-\alpha} \\
&= e^{-\lambda t} = P(Y > t),
\end{aligned}
\tag{13.3}
$$

from which it can be concluded that the distribution function of X limits to that of Y as the Pareto shape parameter increases without bound. The limit (13.3) can be quantified by establishing bounds on the distributional differences. Towards this aim, Stein's Method (Barbour et al., 1992), (Barbour and Chen, 2005) can be used to measure the rate of convergence of (13.1) to a limiting distribution of the form (13.2). This method starts with a differential equation characterising the exponential distribution, and bounds on this are then used to measure the rate of convergence. In particular, it is shown in (Weinberg, 2012) that the two distributions above satisfy the inequality

$$
-(1 - e^{-\lambda})\frac{1}{\alpha - 1} \le F_X(t) - F_Y(t) \le \frac{3}{\alpha},
\tag{13.4}
$$

which shows that the rate of convergence is controlled by the Pareto shape parameter. It is clear that as α increases, the bounds in (13.4) decrease to zero, implying the exponential approximation to the Pareto model is valid for large shape parameters.

The problem with the Stein approach is that the bounds do not suggest a suitable way in which, for a given Pareto model, an appropriate approximating exponential distribution can be specified. This shortcoming can be addressed with an application of the Kullback-Leibler divergence as an alternative to analysing distributional approximations directly.

13.3   Information Theory

Information Theory is concerned with the study of entropy as a measure of uncertainty, and was introduced into the engineering community by Shannon (Shannon, 1948a), (Shannon, 1948b) and has had a profound effect on the understanding and optimisation of data networks (Arndt, 2004). In particular, the Kullback-Leibler divergence, introduced in (Kullback and Leibler, 1951), has found application in signal processing analysis and statistical model fitting (Weinberg, 2016e). Only a brief introduction to the measurement of entropy is included, since the focus will be on the Kullback-Leibler divergence.

Suppose that X is a discrete random variable, taking values in a set $S = \{s_j, j \in J\}$ where $J \subset \mathbf{N}$ is an indexing set. Assume that for each $j \in J$ the point probabilities are given by $\mathbf{P}(X = s_j) = p_j$, where it is necessary to impose the conditions that the latter are nonnegative and their sum over the index set is unity. Then the entropy of X is defined to be

$$\mathcal{H}(X) = -\sum_{j \in J} p_j \log(p_j). \tag{13.5}$$

The definition (13.5) has a natural extension to continuous random variables. Suppose that X is such a random variable with density f_X. Then the entropy of X is defined to be the integral

$$\mathcal{H}(X) = -\int_0^\infty f_X(t) \log(f_X(t)) dt. \tag{13.6}$$

The two definitions of entropy given above can be interpreted in terms of expectations of the underlying random variables. To illustrate this, the integral in (13.6) is the expectation of $\log(f_X(X))$, viewing the latter as a random variable, and so the entropy is the negative of this expectation. Focusing on the continuous case, suppose that one has two random variables X and Y with densities f_X and f_Y respectively. Assume that both these random variables have support the nonnegative real line. Then the cross entropy of X and Y is given by

$$\mathcal{H}(X,Y) = -\int_0^\infty f_X(t) \log(f_Y(t)) dt. \tag{13.7}$$

The Kullback-Leibler divergence is a measure of the information lost when one distribution is approximated by another. Hence, for two random variables X and Y with densities f_X and f_Y respectively, the information lost when Y is used to approximate X is defined to be

$$D_{KL}(X||Y) := \int_0^\infty f_X(t) \log\left(\frac{f_X(t)}{f_Y(t)}\right) dt. \tag{13.8}$$

It is not difficult to show that (13.8) is the difference between the cross entropy of X and Y, and the entropy of X. Hence, based upon (13.6) and (13.7) one can show that

$$D_{KL}(X||Y) = \mathcal{H}(X,Y) - \mathcal{H}(X). \tag{13.9}$$

Since (13.8) measures the information lost in the approximation of X by Y, it can be used to assess the convergence of these distributions. It can be shown that $D_{KL}(X||Y) \geq 0$, a result known as Gibb's Inequality, which follows from Jensen's Inequality (Arndt, 2004). It is clear that if the two random variables X and Y coincide then $D_{KL}(X||Y) = 0$. The converse of this can also be demonstrated to be true. However, it is clear from (13.8) that the Kullback-Leibler divergence is not symmetric, nor satisfies a triangle inequality. Consequently it is not a metric in the usual analysis sense but is a pseudo-metric. Its merit in assessing convergence in distribution follows from the Pinsker-Csiszár Inequality (Pinsker, 1964), (Csiszár, 1967), (Kullback, 1967). Suppose that for the two random variables X and Y their distribution functions are $F_X(t)$ and $F_Y(t)$ respectively, with support the nonnegative real line. Then this inequality states that

$$\|F_X - F_Y\|_\infty = \sup_{t \geq 0} |F_X(t) - F_Y(t)| \leq \sqrt{2D_{KL}(X||Y)}, \tag{13.10}$$

where the norm on the left hand side of (13.10) is the supremum norm over the domain of the distribution functions. Clearly if the Kullback-Leibler divergence is close to zero, the supremum norm inherits this and thus implies the random variables X and Y are close in distribution.

Observe that (13.10) also implies that if a sequence of random variables X_n is such that $\lim_{n \to \infty} D_{KL}(X_n||Y) = 0$, for some random variable Y, then the limiting distribution of X_n and Y coincide, which can be justified with an application of Lebesgue's Dominated Convergence Theorem.

These results justify using the Kullback-Leibler divergence to measure distributional approximations. It is worth noting that although the triangle inequality is not achievable with this divergence, it is possible to construct a measure which is symmetric. This can be produced by defining the distance

$$\tilde{D}_{KL}(X||Y) = D_{KL}(X||Y) + D_{KL}(Y||X), \tag{13.11}$$

which has been utilised in (Seghouane, 2006). However, as will be shown in the next section, it is sufficient to apply (13.8) to the problem under investigation.

13.4 Kullback-Leibler Divergence

In the analysis of the Rayleigh-K distributional approximation it was found that (13.8) is very difficult to calculate analytically. An expression for it is derived in (Weinberg, 2016e), which shows that the underlying issue is with the determination of the entropy of the K-distribution. However, in the case of the Pareto distribution considered in this chapter, it is possible to derive a closed form expression for the entropy of the Pareto model, and consequently for (13.8), which is the focus in this section.

Towards this objective, note that with an application of (13.1) and (13.2),

$$\frac{f_X(t)}{f_Y(t)} = \frac{\alpha\beta^\alpha}{\lambda}e^{\lambda t}(t+\beta)^{-\alpha-1}, \tag{13.12}$$

from which it follows by applying logarithms to (13.12) and substituting the result into (13.8) that

$$D_{KL}(X||Y) = \log\left(\frac{\alpha\beta^\alpha}{\lambda}\right) + \lambda E(X) - (\alpha+1)E(\log(X+\beta)), \tag{13.13}$$

where the fact that the density of X integrates to unity has been applied. The mean of X can be shown to be

$$E(X) = \frac{\beta}{\alpha-1}, \tag{13.14}$$

with the proviso that $\alpha > 1$, while the mean of $\log(X+\beta)$ is given by

$$E(\log(X+\beta)) = \int_0^\infty \log(t+\beta)\frac{\alpha\beta^\alpha}{(t+\beta)^{\alpha+1}}dt. \tag{13.15}$$

By applying a transformation $u = \log(t+\beta)$, followed by integration by parts, it can be shown that (13.15) reduces to

$$E(\log(X+\beta)) = \log(\beta) + \frac{1}{\alpha}. \tag{13.16}$$

An application of (13.14) and (13.16) to (13.13) demonstrates that the Kullback-Leibler divergence reduces to

$$D_{KL}(X||Y) = \log\left(\frac{\alpha}{\lambda\beta}\right) + \frac{\lambda\beta}{\alpha-1} - \left(\frac{\alpha+1}{\alpha}\right). \tag{13.17}$$

Figures 13.1–13.4 plot the Kullback-Leibler divergence (13.17) as a function of λ, for a series of Pareto shape and scale parameters. Each figure shows curves for a specified β, with $\alpha \in \{5,10,15,20,25,30\}$. Figure 13.1 is for the case where $\beta = 0.1$, Figure 13.2 is for $\beta = 0.5$, Figure 13.3 corresponds to $\beta = 0.95$ while the situation where $\beta = 10$ features in Figure 13.4. These figures show a common structure to the Kullback-Leibler divergence. In particular, for each α and β there exists a λ which minimises (13.17). It is also interesting to observe the effect β has on the Kullback-Leibler divergence. For an upper bound of approximately 10^{-3} on the Kullback-Leibler divergence, it is clear from Figure 13.1 that for $\alpha \approx 30$, one must select $\lambda \approx 300$. For the case of $\beta = 0.5$, Figure 13.2 suggests that $\alpha \approx 30$ and $\lambda \approx 50$. For the case of $\beta = 0.95$, Figure 13.3 implies that $\alpha \approx 30$ and $\lambda \approx 30$. Finally, as shown in Figure 13.4, for $\beta = 10$ we require $\alpha \approx 30$ and $\lambda \approx 3$.

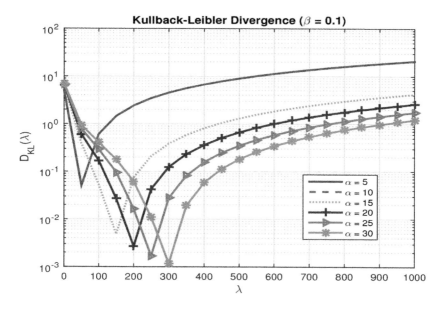

Figure 13.1: Kullback-Leibler divergence for the Pareto-exponential approximation, in the case where $\beta = 0.1$.

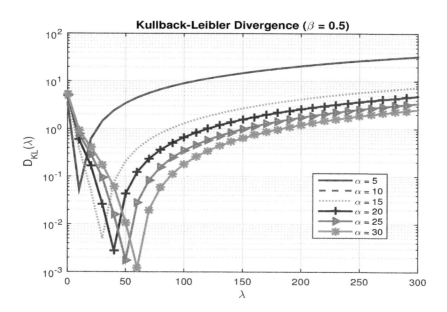

Figure 13.2: Kullback-Leibler divergence for $\beta = 0.5$.

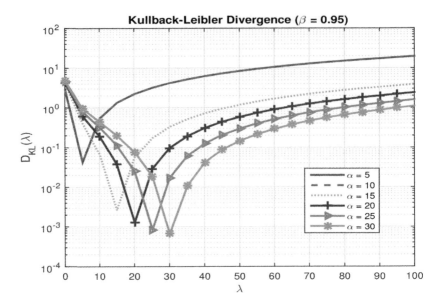

Figure 13.3: Kullback-Leibler divergence where $\beta = 0.95$.

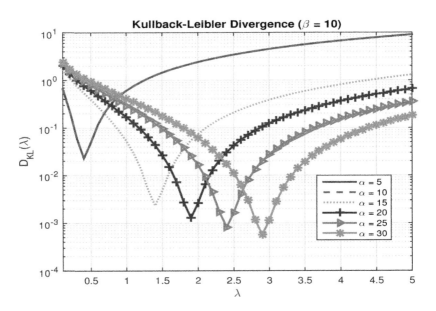

Figure 13.4: An example of the Kullback-Leibler divergence for the scenario of large β. In this example $\beta = 10$.

To understand this mathematically, differentiating (13.17) with respect to λ yields

$$\frac{\partial D_{KL}(X\|Y)}{\partial \lambda} = -\frac{1}{\lambda} + \frac{\beta}{\alpha - 1}, \tag{13.18}$$

which is zero when $\lambda = \frac{\alpha-1}{\beta}$, where it is necessary to assume $\alpha > 1$. Applying a second differentiation to (13.18) shows that this is a point where a minimum occurs. This explains the phenomenon observed in these plots. In order to investigate these results further, Figures 13.5–13.8 plot a series of Pareto Type II distributions, together with the optimal exponential approximation. Here optimal is used in the sense that the Kullback-Leibler divergence is minimised with an appropriate selection of exponential distribution shape parameter. Figure 13.5 is for the case where the Pareto scale parameter is $\beta = 0.1$, with shape parameter varying from 5, 15 to 30. It can be observed that as the Pareto shape parameter increases, the optimal exponential distribution is a better fit. This is consistent with the results illustrated in Figure 13.1. Figure 13.6 is for the case where $\beta = 0.5$, Figure 13.7 corresponds to $\beta = 0.95$ and Figure 13.8 corresponds to $\beta = 10$. Observe in all figures that for $\alpha = 5$, the approximation is poor, while for $\alpha = 15$ the approximation has improved significantly. When $\alpha = 30$ it is very difficult to see a difference between the two distributions.

Returning to the analysis of the minimum achievable divergence, by applying $\lambda = \frac{\alpha-1}{\beta}$ to (13.17), it can be shown that the minimum divergence is

$$D_{KL}(X\|Y)_{\min} = \log\left(1 + \frac{1}{\alpha - 1}\right) - \frac{1}{\alpha}. \tag{13.19}$$

Since for any $x > 0$ we have the bound $\log(1+x) \le x$, an application of this to (13.19) yields the upper bound

$$D_{KL}(X\|Y)_{\min} \le \frac{1}{\alpha(\alpha - 1)}. \tag{13.20}$$

An application of (13.20) to (13.10) results in

$$\|F_X - F_Y\|_\infty \le \sqrt{\frac{2}{\alpha(\alpha - 1)}}. \tag{13.21}$$

One can compare the upper bound provided by (13.21) to that obtained via Stein's Method, given by the upper bound $\frac{3}{\alpha}$ in (13.4). With an application of some simple analysis one can show that the upper bound based upon (13.21) improves on that from (13.4) whenever $\alpha^2 - \frac{9}{7}\alpha > 0$. This occurs when $\alpha > \frac{9}{7}$, and since in most cases of interest $\alpha > 2$, as shown in (Weinberg, 2011a), it follows that the upper bound attained by the Kullback-Leibler divergence is smaller than that obtained with Stein's Method.

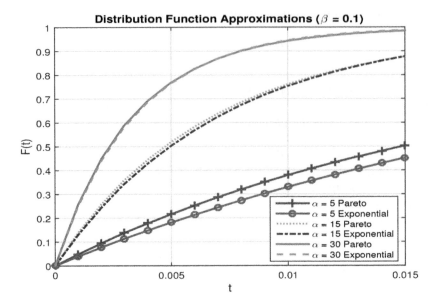

Figure 13.5: Comparison of distribution functions where the exponential distribution has shape parameter selected so that it is the optimal fit for each Pareto case. Here $\beta = 0.1$.

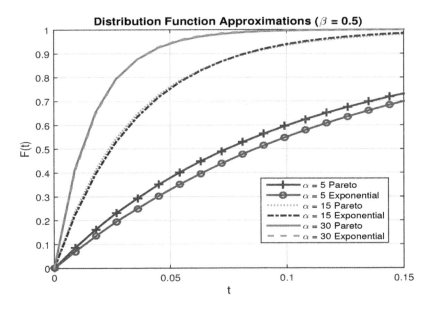

Figure 13.6: Second comparison of distribution functions, under the same conditions as for Figure 13.5, except with $\beta = 0.5$.

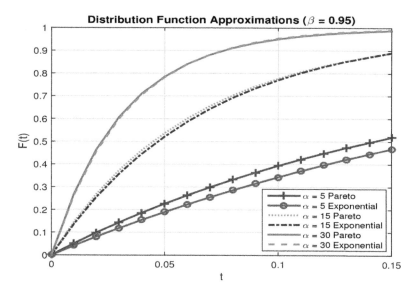

Figure 13.7: Further comparison of distribution functions, with $\beta = 0.95$.

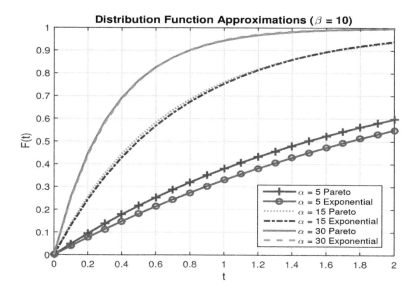

Figure 13.8: Final comparison of distribution functions, where $\beta = 10$.

To illustrate the differences between the upper bounds, Figure 13.9 plots the two upper bounds as a function of α. It can be observed that the upper bound (13.21) is better than that based upon (13.4).

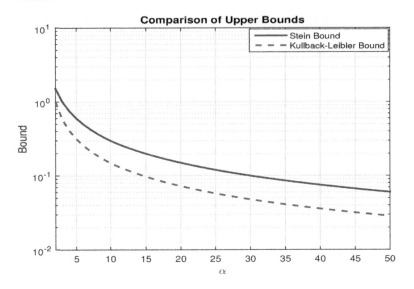

Figure 13.9: Upper bounds, based upon Stein's Method and the Kullback-Leibler divergence approaches.

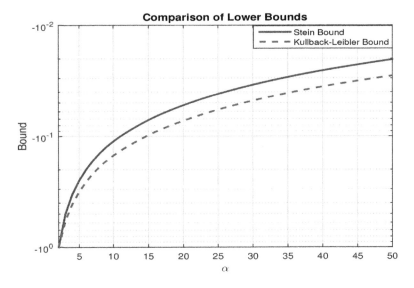

Figure 13.10: Comparison of the Stein and Kullback-Leibler divergence lower bounds.

Using a similar analysis it can be shown that the Stein lower bound, namely $-\frac{1}{\alpha-1}$, tends to be closer to zero than that obtained by the Kullback-Leibler divergence, as illustrated in Figure 13.10.

13.5   Conclusions

The Kullback-Leibler divergence was used to assess the discrepancy between the Pareto Type II and exponential distributions, in order to better understand the validity of the exponential approximation of the Pareto model. It was shown that for any given Pareto model an optimal exponential approximation exists. This approximation was shown to improve as the Pareto shape parameter increased, for any fixed Pareto scale parameter. This means that in cases where the Pareto shape parameter exceeds 30, it is acceptable to apply sliding window CFAR detection schemes based upon an exponential clutter model assumption. In terms of X-band maritime surveillance radar, this will occur typically in the case of vertical polarisation.

13.5 Conclusions

Appendices

A   Neyman-Pearson Lemma

This appendix discusses the Neyman-Pearson Lemma to illustrate the issues with the construction of optimal non-coherent detectors, and to justify the development of the sliding window approach outlined in this book. The following is the main result which has been taken from (Beaumont, 1980):

Lemma A.1 *Suppose that $Z_1, Z_2, \ldots, Z_N$ are statistics with joint density function $f(z_1, z_2, \ldots, z_N | \theta_1, \theta_2, \ldots, \theta_k)$ where the θ_j are k unknown distributional parameters. Suppose we want to test the hypothesis that $H_0 : \theta_j = \theta_j^0$ against the alternative $H_1 : \theta_j = \theta_j^1$, where the parameters θ_j^0 and θ_j^1 are known. Suppose U is the set of N-tuples $(z_1, z_2, \ldots, z_N)$ such that*

1. $\dfrac{f(z_1, z_2, \ldots, z_N | H_1)}{f(z_1, z_2, \ldots, z_N | H_0)} > \tau > 0;$

2. $P((Z_1, Z_2, \ldots, Z_N) \in U | H_0) = \kappa.$

Then U is the optimal critical region of size κ, meaning that no other test with the same size has greater power when H_1 is true.

From a signal processing perspective, this result implies that a likelihood ratio test will yield the optimal decision rule for a given probability of false alarm. Hence suppose one has a return $\mathbf{z} = (z_1, z_2, \ldots, z_N)$, consisting of intensity measurements, and we want to test whether this multilook return is just clutter, or whether it contains a signal embedded within the clutter. Then Lemma A.1 implies that the optimal test is a likelihood ratio of the returns under the respective hypotheses, compared to a threshold determined from the desired probability of false alarm. In general terms it is somewhat difficult to derive the density of a multivariate intensity model, especially if one accounts for correlations between the univariate marginal distributions. Mardia's multivariate Pareto model analysed in Chapter 11 provides one such example. Multivariate Rayleigh and exponential distributions are analysed in (Mallik, 2003). These can be used to derive distributional densities under H_0; the complexity arises when one requires the density under H_1. As a simple example, to illustrate the complexity in constructing such a detector, suppose that the return consists of independent and identically distributed Pareto Type I random variables, with shape and scale parameters α and β respectively. Then the joint density of the return under H_0 is

$$f(z_1, z_2, \ldots, z_N | H_0) = \frac{\alpha^N \beta^{\alpha N}}{\left(\prod_{j=1}^N z_j \right)^{\alpha+1}}. \tag{A.1}$$

The return under a general target model is analytically intractable and requires a specific target model, or an approximation for it. For simplicty, we take the approach in (Weinberg, 2013d), who derives the distribution for a single Gaussian target embedded within compound Gaussian clutter with inverse gamma texture. Hence we assume the signal is a bivariate Gaussian process with zero mean and covariance matrix $1/(2\lambda)I_2$, for some $\lambda > 0$ determining the signal strength, while the compound Gaussian clutter has speckle a bivariate Gaussian distribution, also with zero mean and covariance matrix $(1/2\mu)I_2$, for some $\mu > 0$. Throughout I_2 is the 2×2 identity matrix. Then based upon equation (14) of (Weinberg, 2013d), one can show that if we define

$$I(t|\alpha,\beta,\lambda,\mu) = \int_0^\infty u^{\alpha-1} e^{-\beta u} e^{-\frac{t}{\frac{1}{\mu}+\frac{1}{\mu u}}} \frac{1}{\frac{1}{\lambda}+\frac{1}{\mu u}} du \qquad (A.2)$$

then the density under H_1 is

$$f(z_1, z_2, \ldots, z_N | H_1) = \frac{\beta^{\alpha N}}{(\Gamma(\alpha))^N} \prod_{j=1}^N z_j I(z_j | \alpha, \beta, \mu, \lambda). \qquad (A.3)$$

Hence, by taking the ratio of (A.3) with (A.1), the likelihood ratio test reduces to a test of the form

$$(\alpha+2) \sum_{j=1}^N \log z_j + \sum_{j=1}^N \log I(z_j | \alpha, \beta, \lambda, \mu) \underset{H_0}{\overset{H_1}{\gtrless}} \rho, \qquad (A.4)$$

where ρ is the detection threshold. Inspection of (A.4) reveals that in order to determine ρ, it is necessary to assume *a priori* knowledge of the target parameter λ. Additionally, it is clear that *a priori* knowledge of all clutter parameters will be required, as well as the speckle parameter μ. One strategy that can be adopted is to apply a series of approximations to (A.4), which will result in a suboptimal decision rule. Examination of the integral (A.2) indicates this may be a complicated exercise. Recalling that the detector (A.4) was constructed under the assumption of a simple Gaussian target model in independent returns, it is clear that there will be much complexity in the extension of this approach to other target models. This would especially be the case in the scenario of correlated clutter returns. Hence, in terms of non-coherent detection, the sliding window approach is a useful suboptimal decision strategy, especially if it can be designed to achieve the CFAR property.

B   CA- and OS-CFAR in Exponentially Distributed Clutter

This appendix includes a derivation of the probability of false alarm for both the sliding window CA- and OS-CFARs for target detection in independent exponentially distributed clutter. In adddition to this, closed form expressions will be derived for the probability of detection under the assumption of an independent Gaussian target model, or equivalently a target model following a Swerling I fluctuation law. The approach is similar to that in (Levanon, 1988), to which the reader is referred for further details.

Suppose that the clutter is exponentially distributed with parameter $\lambda > 0$, and that the target model is Gaussian in the complex domain. Then based upon the formulation in (Gandhi and Kassam, 1988), the test can be specified by H_0 : $\mu = \lambda$ against the alternative $H_1 : \mu = \frac{\lambda}{1+S}$, where μ represents the distributional reciprocal mean and S is the signal to clutter ratio. Then the detector

$$Z_0 \underset{H_0}{\overset{H_1}{\gtrless}} \tau \sum_{j=1}^{N} Z_j \tag{B.1}$$

has probability of detection

$$\text{Pd} = \int_0^{\infty} e^{-\lambda(1+S)^{-1}\tau t} \frac{\lambda^N}{\Gamma(N)} t^{N-1} e^{-\lambda t} dt$$

$$= \left[1 + \frac{\tau}{1+S}\right]^{-N}, \tag{B.2}$$

where a transformation has been used to simplify the integral, as well as the fact that the sum of independent and identically distributed exponential random variables has a gamma distribution. To recover the expression for the Pfa, one sets $S = 0$ in (B.2) to yield

$$\text{Pfa} = (1 + \tau)^{-N}. \tag{B.3}$$

Often the detector (B.1) has its threshold multiplier replaced with $\frac{v}{N}$ to emphasize the clutter measurement is an average. Adopting this approach, it is clear that

$$\lim_{N \to \infty} \text{Pd} = \lim_{N \to \infty} \left[1 + \frac{\left(\frac{v}{1+s}\right)}{N}\right]^{-N} = e^{-\frac{v}{1+s}}, \tag{B.4}$$

which results from the limit definition of the exponential function. By setting $S = 0$ in (B.4) it follows that

$$\lim_{N \to \infty} \text{Pfa} = e^{-v}, \tag{B.5}$$

from which one can deduce that

$$\lim_{N \to \infty} \text{Pd} = \left(\lim_{N \to \infty} \text{Pfa} \right)^{\frac{1}{1+S}}, \tag{B.6}$$

which provides a relationship between the probability of detection and false alarm as the size of the clutter range profile increases without bound. It is interesting to observe that a linear threshold detector for this detection scenario, given by

$$Z_0 \underset{H_0}{\overset{H_1}{\gtrless}} \kappa, \tag{B.7}$$

requires its threshold set via $\kappa = -\frac{1}{\lambda} \log(\text{Pfa})$ and consequently the corresponding probability of detection, with this threshold applied to (B.7), can be shown to reduce to exactly that given by (B.6) for a Gaussian target model. This means that the detector (B.1) has a very small detection loss relative to a ideal fixed threshold decision rule in the case where a large number of clutter statistics are available for the clutter range profile.

Next the focus is shifted to the OS-CFAR

$$Z_0 \underset{H_0}{\overset{H_1}{\gtrless}} \tau Z_{(k)}, \tag{B.8}$$

operating in exponentially distributed clutter with $1 \le k \le N$ and under the assumption of a Gaussian target model as before. Using the density (2.93) for the kth OS of a series of independent and identically distributed exponential random variables, the probability of detection of (B.8) is given by

$$
\begin{aligned}
\text{Pd} &= \int_0^\infty e^{-\lambda(1+S)^{-1}\tau t} \lambda k \binom{N}{k} (1 - e^{-\lambda t})^{k-1} e^{-\lambda t(N-k+1)} dt \\
&= k \binom{N}{k} \int_0^1 w^{(1+S)^{-1}\tau + N - k} (1 - w)^{k-1} dw \\
&= k \binom{N}{k} B((1+S)^{-1}\tau + N - k + 1, k) \\
&= \frac{N!}{(N-k)!} \frac{\Gamma((1+S)^{-1}\tau + N - k + 1)}{\Gamma((1+S)^{-1}\tau + N + 1)},
\end{aligned} \tag{B.9}
$$

where the transformation $w = e^{-\lambda t}$ has been applied, together with the definition of the beta and gamma functions. Thus (B.9) provides a closed form expression for the detection probability. Setting $S = 0$ in (B.9) yields the false alarm probability

$$\text{Pfa} = \frac{N!}{(N-k)!} \frac{\Gamma(\tau + N - k + 1)}{\Gamma(\tau + N + 1)}, \tag{B.10}$$

which is applicable regardless of the underlying target model.

To provide an illustration of these detectors, Figure B.1 plots the detection probability (B.2) for the case where $N = 16$ and Pfa $= 10^{-4}$. Also shown is the OS detection probability (B.9) for a series of OS indices. Reducing the latter to less than four results in very large threshold multipliers for (B.8) and consequently reduced performance, and in the case of the minimum ($k = 1$) the decision rule saturates. The optimality of the CA-CFAR is clear from this figure, as well as the fact that the OS-CFAR improves when its OS index is increased, with the maximum ($k = 16$) having the best performance.

Figure B.2 plots (B.2) for increasing clutter range profile size with the same design Pfa as before, together with the limit given by (B.6), showing the fact that the CA-CFAR tends to the fixed threshold detector (B.7) as N increases.

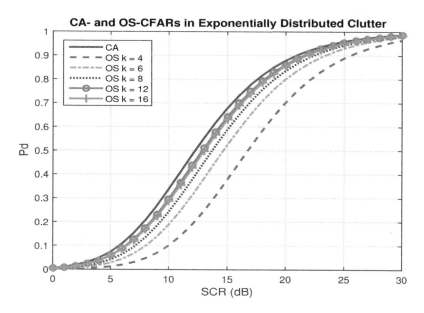

Figure B.1: CA- and OS-CFARs in exponentially distributed clutter. The target model follows a Swerling I fluctuation law.

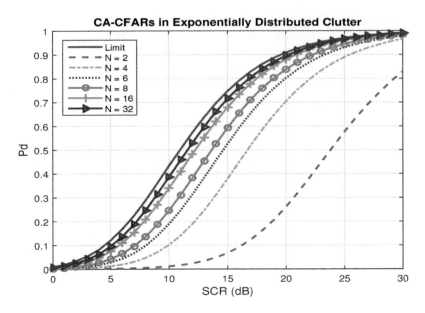

Figure B.2: The CA-CFAR with increasing CRP length N, showing its limit to the ideal decision rule.

C   Radar Cross Section and Target Models

This appendix provides a brief account of radar cross section (RCS) and its re-lationship to the Swerling models for fluctuating targets. RCS is a measure of an artefact's visibility to a radar detection system. It is usually taken to be the effective area of a metal sphere that would yield the same return signal if it was used as a proxy for the artefact (Skolnik, 2008). Such a definition allows objects to be classified through their RCS, and it is well-known that an object's RCS is affected by factors such as radar frequency, polarisation, physical size, shape and orientation (Levanon, 1988). Stealthy aircraft, for example, are designed to minimise the radar signal in the direction of the radar, so that such factors will not increase its RCS.

As pointed out in (Levanon, 1988), complicated objects such as aircraft will provide the radar with multiple scatterers resulting in returns at different geome-tries, which can also result from the motion of the aircraft relative to the radar. Objects in the sea, when surveyed from an airborne surveillance platform, will have less scatterers, and the difficulty in their detection by a radar is related to their physical size. As an example, a submarine periscope may only appear for short periods, and has a very small size, making its detection a very challenging problem in maritime surveillance radar. A small fishing boat, by contrast, will be visible at all times but may have very few scatterers, thus also resulting in reduced likelihood of detection.

The standard approach in radar is to model an artefact's RCS as a random variable. Swerling's five models attempt to capture some of the observed charac-teristics of targets, providing a series of mathematical models for detector valida-tion. Throughout this book a simple univariate Swerling I target model has been assumed, since it is the simplest model to apply which has a random structure, as opposed to a constant target model. This is characterised by a bivariate Gaussian process in the complex domain, with 2×1 zero mean and a covariance matrix which is a constant times the 2×2 identity matrix. This constant is selected to produce the desired SCR relative to the underlying Pareto clutter model.

It is interesting to note that the transformation approach for detector design developed in Chapter 5 essentially applies a logarithmic transformation to the CUT, normalised by the Pareto scale parameter or the CRP minimum. Hence, when other Swerling-type models are applied to the CUT, there is very little variation in detector performance.

Swerling's five models provide a classification of targets in terms of their fluctuation rates, determining the underlying χ^2 target model. The Swerling 0 case assumes that the target is a constant, so that there is no requirement for a statistical model of RCS. It is thus assumed that the target reflection is unknown and fixed. The other four Swerling cases are based upon the variation in fluc-tuation. A scan-to-scan fluctuation assumes that the reflected radar return is a constant over the surveillance region scan, but is a Rayleigh or one-dominated plus Rayleigh distribution between scans. The former corresponds to a Swerling I target model, while the latter is a Swerling III case. The same underlying target models are assumed for the Swerling II and IV cases, but the variation is assumed to occur on a pulse to pulse basis (Minkler and Minkler, 1990). Hence in the analysis of this book, the most natural setting is to assume target models varying on a scan-to-scan basis, as has been employed with the assumption of a Swerling I or Gaussian target model. In the context of classical non-coherent integrators, a more mathematical formulation of Swerling target models, for scan-to-scan fluctuation, can be found in (D.6) in Appendix D.

D   Classical Non-Coherent Integrators

Classical non-coherent integrators, or square law envelope detectors, have a simple receiver implementation and have appeared in a number of studies of target detection (Conte and Ricci, 1994). This appendix overviews a development of integrators in compound Gaussian clutter with inverse gamma texture, which can be found in (Weinberg, 2013a). The following development parallels that in (Conte and Ricci, 1994). Such detectors are neither sliding window decision rules nor CFAR, but are nonetheless of interest and an analysis is included for completeness.

Suppose that the clutter is modelled as a compound Gaussian process with given texture S which has density f_S. The desired radar signal return is $x(t) = \sum_{j=1}^{N} A(j)e^{i\theta(j)}\mathbf{p}(t - jT_p)$, where $\mathbf{p}$ is the complex envelope of each transmitted pulse, T_p is the pulse repetition interval and $A(j)e^{i\theta(j)}$ is a complex sequence representing both propagation effects and target reflectivity. Here N is the number of integrated pulses. The returned signal at the radar is $\mathbf{r}(t)$, which is either pure clutter, or a mixture of the clutter and desired signal. This is formulated as the usual statistical hypothesis test: $H_0 : \mathbf{r}(t) = \mathbf{c}(t)$ versus $H_1 : \mathbf{r}(t) = \mathbf{x}(t) + \mathbf{c}(t)$. The form of the test is given by

$$\mathcal{I} = \sum_{j=1}^{N} |\mathbf{r}(j)|^2 \underset{H_0}{\overset{H_1}{\gtrless}} \tau. \tag{D.1}$$

Under each respective hypothesis, the integrator takes the form

$$\mathcal{I} = \begin{cases} \sum_{j=1}^{N} |y(j)|^2 & \text{under } H_0 \\ \sum_{j=1}^{N} \left|A(j)e^{i\theta(j)} + y(j)\right|^2 & \text{under } H_1, \end{cases} \tag{D.2}$$

where $y(j)$ is the jth sample of the return at the output of the matched filter (refer to (Conte and Ricci, 1994)). Attention is restricted to scan-to-scan fluctuations in the target models, so that the pulse train consists of independent and identically distributed random amplitudes $A(j) \overset{d}{=} A$. The root mean square of A is denoted A_{rms}. The clutter dominated environment is considered only, and so based upon (Conte and Ricci, 1994), the false alarm probability for the non-coherent integrator is given by

$$\text{Pfa} = \sum_{j=0}^{N-1} \frac{\gamma_\tau^j}{j!} \int_0^\infty e^{-\gamma_\tau s^{-2}} s^{-2k} f_S(s) ds, \tag{D.3}$$

where γ_τ is the detection threshold normalised to the power of the baseband equivalent of the entire interference, and $f_S(s)$ is the density of the texture component of the clutter. The detection probability is given by

$$\text{Pd} = \int_0^\infty \int_0^\infty Q_N \left(\frac{\sqrt{2N\gamma}}{s} \frac{a}{A_{rms}}, \frac{\sqrt{2\gamma_\tau}}{s} \right) f_S(s) f_A(a) ds da, \tag{D.4}$$

where $f_A(a)$ is the density of the amplitude fluctuation law A, γ is the signal to clutter ratio, and Q_N is the Generalised Marcum Q-Function (Marcum, 1960), given by the integral

$$Q_N(a,b) = \int_b^\infty x \left(\frac{x}{a}\right)^{N-1} e^{-(x^2+a^2)/2} I_{N-1}(ax) dx. \tag{D.5}$$

The Swerling target models correspond to assuming a central χ^2 distribution for the amplitude fluctuating law, with $2m$ degrees of freedom, for some $m \in \{1, 2, \ldots\}$, and density given by

$$f_A(a) = \frac{2m^m}{A_{rms}\Gamma(m)} \left(\frac{a}{A_{rms}}\right)^{2m-1} e^{-m(a/A_{rms})^2}. \tag{D.6}$$

In (Conte and Ricci, 1994), the detection probability (D.4), with (D.6), is expressed in terms of Kummer's confluent hypergeometric series, and then finally written as an integral of a series of functions, and evaluated numerically. It will be shown how the relationship explored in (Weinberg, 2006), linking (D.5) with Poisson probabilities, can be used to re-express (D.4) in a form more suitable for numerical analysis.

Suppose that $\aleph_1(v_1)$ and $\aleph_2(v_2)$ are two independent Poisson distributed random variables, with means v_1 and v_2 respectively. Then (Weinberg, 2006) shows that the Marcum Q-Function (D.5) can be written

$$Q_N(a,b) = \mathbf{P}\left(\aleph_2(b^2/2) \leq N-1 + \aleph_1(a^2/2)\right). \tag{D.7}$$

Applying this to (D.4) results in the following:

Lemma D.1 *The non-coherent integrator, operating in a compound Gaussian clutter environment, has probability of false alarm and detection given by*

$$
\begin{aligned}
\text{Pfa} &= \mathbf{P}\left[\aleph_2(\gamma_\tau W) \leq N-1\right] \\
\text{Pd} &= \mathbf{P}\left[\aleph_2(\gamma_\tau W) \leq N-1 + \aleph_1(N\gamma W Z)\right],
\end{aligned} \tag{D.8}
$$

where the independent random variables Z and W have densities $f_Z(z) = (1/2)z^{-1/2}A_{rms}f_A(A_{rms}z^{1/2})$ and $f_W(w) = (1/2)w^{-3/2}f_S(w^{-1/2})$ respectively, where $f_S(s)$ is the density of S.

The proof of this result is discussed in (Weinberg, 2013a).

For the case of a Pareto clutter intensity model, the Lemma can be specialised to the following. Firstly, it is shown in (Weinberg, 2011b) that the density of S is given by $f_S(s) = \frac{2\beta^\alpha}{\Gamma(\alpha)} s^{-2\alpha-1} e^{-\beta s^{-2}}$, where α is the Pareto shape parameter and β is its scale parameter. The following is the main result:

Lemma D.2 *For the case of a compound Gaussian clutter distribution with inverse gamma texture, corresponding to a Pareto model with shape and scale* *parameters α and β respectively, the random variable $W \overset{d}{=} Gamma(\alpha, \beta)$, and $\aleph_2(\gamma_\tau W) \overset{d}{=} NBin(\alpha, p)$ with $p = \frac{\beta}{\beta + \gamma_\tau}$. Hence the false alarm probability is given by cumulative probabilities of a negative binomial distribution.*

It is important to note that, as in the cases analysed in (Conte and Ricci, 1994), the non-coherent integrator (D.2) will not be CFAR. If we assume a classical fluctuation law, given by (D.6), then it can be shown directly that Z has a gamma distribution with shape and scale parameters equal to m. This means that we can evaluate the detection probability using the estimator

$$\widehat{Pd} = \sum_{j=1}^{M} \frac{Q_N \left(\sqrt{2\gamma N Z_j W_j}, \sqrt{2\gamma_\tau W_j} \right)}{M} \tag{D.9}$$

where M is the number of simulations, Z_j and W_j are independent realisations of Z and W respectively, and the Generalised Marcum Q-Function can be evaluated using the Monte Carlo techniques outlined in (Weinberg, 2006).

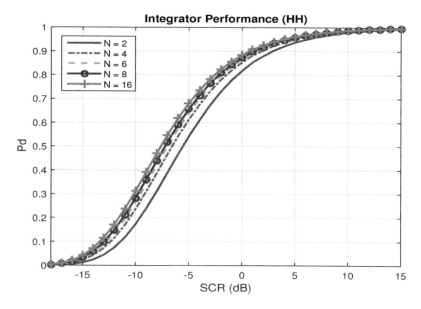

Figure D.1: Integrator performance in horizontal polarisation.

To examine the performance of this integrator (D.1) it is run on simulated Pareto Type I distributed clutter, based upon the Ingara data set used as a basis for the results in this book. Figures D.1 and D.2 plot the Pd as a function of the SCR. These plots are for the case where the Pfa is 10^{-4} with N varying as shown,

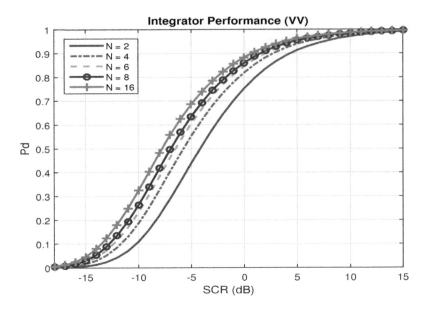

Figure D.2: Integrator performance in the case of vertical polarisation.

for the cases of horizontal and vertical polarisation. The target model is a Swerling I fluctuation as used throughout. Both figures show that as N increases, the performance increases. It is important to observe that the integrators have very good performance in homogeneous clutter. Their performance degrades severely in the presence of interference and require *a priori* knowledge of the Pareto shape and scale parameters. Additionally, they are not efficient at managing variations in the design Pfa.

E Ideal Detectors

Ideal detectors, which are used to provide an upper bound on performance throughout the book, are the focus of this appendix. Such a concept was first introduced in (Watts, 1985) in the context of K-distributed clutter, and then a similar idea appeared in (Anastassopoulos and Lampropoulos, 1995) in the scenario of sliding window CFAR detector development in Weibull distributed clutter. Such detectors are based upon the premise that if one has *a priori* knowledge of the clutter distribution, then there is no need to produce a measurement of the clutter level via a function of the CRP. Instead one can apply a fixed threshold decision rule, which should provide an upper bound on performance. Hence this is referred to as an ideal detector. A majority of the analysis presented below has been published in (Weinberg, 2016a).

Although such a decision rule will provide a maximum possible probability of detection, and will remain immune to the effects of interfering targets, it will be unable to adjust to variations in the clutter power, unlike an OS-based function of the CRP. Hence an ideal detector only provides a gauge of performance, and is not suitable for practical implementation.

Suppose that the underlying univariate clutter model takes the generic form $\Phi_1 W^{\Phi_2}$, where Φ_1 and Φ_2 are nonnegative constants and $W > 0$ is a random variable with distribution function $F_W(t)$. Then the ideal decision rule takes the form

$$Z_0 \underset{H_0}{\overset{H_1}{\gtrless}} \tau, \tag{E.1}$$

where the CUT statistic is Z_0 and $\tau > 0$ is the threshold multiplier. Under H_0 $Z_0 \overset{d}{=} \Phi_1 W^{\Phi_2}$ and so the Pfa of (E.1) is

$$\mathrm{Pfa} = \mathrm{P}(\Phi_1 W^{\Phi_2} > \tau)$$

$$= F_W^c\left(\left\{\frac{\tau}{\Phi_1}\right\}^{\frac{1}{\Phi_2}}\right), \tag{E.2}$$

where $F_W^c(t) = 1 - F_W(t)$ is the complementary distribution function of W. Hence by inverting (E.2) one can show that the threshold multiplier is given by

$$\tau = \Phi_1 \left[F_W^{-1}(1 - \mathrm{Pfa})\right]^{\Phi_2}. \tag{E.3}$$

An application of (E.3) to (E.1) results in the ideal detector

$$Z_0 \underset{H_0}{\overset{H_1}{\gtrless}} \Phi_1 \left[F_W^{-1}(1 - \mathrm{Pfa})\right]^{\Phi_2}. \tag{E.4}$$

In the Pareto Type I case, one takes $\Phi_1 = \beta$, $\Phi_2 = \alpha^{-1}$ and $W = e^X$ where X has an exponential distribution with unity mean. In view of the latter it is not difficult to show that $F_W(t) = 1 - \frac{1}{t}$ and $F_W^c(t) = \frac{1}{1-t}$ so that (E.4) reduces to

$$Z_0 \underset{H_0}{\overset{H_1}{\gtrless}} \beta \text{Pfa}^{-\frac{1}{\alpha}}. \tag{E.5}$$

The ideal detector (E.5) has been used throughout the numerical analysis in this book.

Consider again the GM-detector (5.10), which can be written in the form

$$\log\left(\frac{Z_0}{\beta}\right) \underset{H_0}{\overset{H_1}{\gtrless}} \tau \sum_{j=1}^{N} \log\left(\frac{Z_j}{\beta}\right). \tag{E.6}$$

Since the statistics in the CRP are Pareto Type I distributed, suppose that $Z_j = \beta e^{\alpha^{-1} X_j}$ where each X_j is exponentially distributed with parameter 1. Additionally, suppose that $\tau = \frac{v}{N}$, so that the Pfa of (E.6) is given by $\text{Pfa} = \left(1 + \frac{v}{N}\right)^{-N}$. Then (E.6) becomes

$$\log\left(\frac{Z_0}{\beta}\right) \underset{H_0}{\overset{H_1}{\gtrless}} \frac{v}{\alpha} \frac{1}{N} \sum_{j=1}^{N} X_j. \tag{E.7}$$

Let $f_{H_N}(t)$ be the density of the normalised sum $H_N = \frac{1}{N}\sum_{j=1}^{N} X_j$. Then it is not difficult to show that the mean of H_N is unity while its variance is $\frac{1}{N}$. Then by the Central Limit Theorem, the distribution of H_N converges to that of a normal with these moment parameters. However, in the limit as $N \to \infty$ this distribution degenerates to a constant of unity, so that $\lim_{N\to\infty} f_{H_N}(t) = \delta_1(t)$, where the latter is the Dirac Delta function, being unity at 1 and zero elsewhere. This *ad hoc* line of reasoning can be made rigorous by observing that since the sum of a series of independent and identically distributed exponential random variables is gamma distributed, the normalised sum H_N has a gamma distribution with shape and scale parameters identically N, based upon considerations in Section 2.8. Since the density of such a random variable is given by

$$f_{H_N}(t) = \frac{N^N}{\Gamma(N)} t^{N-1} e^{-Nt} \tag{E.8}$$

it is relatively straightforward to show that its moment generating function is

$$M_{H_N}(t) = \text{E}\left[e^{H_N t}\right] = \left(1 - \frac{t}{N}\right)^{-N}, \tag{E.9}$$

provided $t < N$. In view of the fundamental limit for the exponential function it is immediate from (E.9) that

$$\lim_{N\to\infty} M_{H_N}(t) = e^t, \tag{E.10}$$

implying the limiting distribution of H_N is degenerate, taking the value unity at $t = 1$ and zero elsewhere.

Consequently the probability of detection of (E.7), written P_{D_N} to emphasize its dependence on N, has limit

$$
\begin{aligned}
\lim_{N \to \infty} P_{D_N} &= \lim_{N \to \infty} \int_0^\infty f_{H_N}(t) P\left(\log\left(\frac{Z_0}{\beta}\right) > \frac{v}{\alpha} t \right) dt \\
&= \int_0^\infty \lim_{N \to \infty} f_{H_N}(t) P\left(\log\left(\frac{Z_0}{\beta}\right) > \frac{v}{\alpha} t \right) dt \\
&= \int_0^\infty P\left(\log\left(\frac{Z_0}{\beta} > \frac{v}{\alpha} t\right) \right) \delta_1(t) dt \\
&= P\left(\log\left(\frac{Z_0}{\beta} > \frac{v}{\alpha}\right) \right) \\
&= P\left(Z_0 > \beta e^{\frac{v}{\alpha}} \right).
\end{aligned}
\tag{E.11}
$$

Taking the limit inside the integral in the above can be justified with an application of Lebesgue's Dominated Convergence Theorem. Note that the limiting detector defined through (E.11) has Pfa given by

$$
\text{Pfa} = P\left(Z_0 > \beta e^{\frac{v}{\alpha}} | H_0 \right) = e^{-v},
\tag{E.12}
$$

so that one obtains the decision rule

$$
Z_0 \underset{H_0}{\overset{H_1}{\gtrless}} \beta \text{Pfa}^{-\frac{1}{\alpha}},
\tag{E.13}
$$

which is exactly the ideal detector for the Pareto case. Hence it follows that the GM-decision rule (E.7), operating in Pareto distributed clutter, will limit to the ideal decision rule as the length of the CRP increases without bound.

To illustrate this, Figures E.1 and E.2 demonstrate this convergence, for the two fundamental cases of horizontal and vertical polarisation used throughout the book. It is interesting to observe that the convergence is faster, relative to the CRP length N, for vertical polarisation.

It is relevant to consider the question of when the GM detector is approximately ideal in the practical situation of a finite CRP. Since the random variable $\mathcal{G}_N$ is a sum of independent and identically distributed exponential random variables, it follows by the Central Limit Theorem that for large N, it is approximately normal in distribution with a mean of unity, variance $\frac{1}{N}$ and density

$$
f_{\mathcal{N}}(t) = \sqrt{\frac{N}{2\pi}} e^{-\frac{N}{2}(t-1)^2}.
\tag{E.14}
$$

Most of the distribution's support will lie within the limits $1 \pm \frac{1}{N}$. Hence, for a tolerance $0 < \varepsilon < 1$, which is the length of the interval $[1 - 1/N, 1 + 1/N]$, one

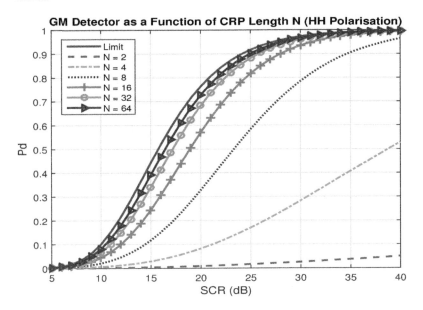

Figure E.1: GM convergence to the ideal detector (HH polarisation).

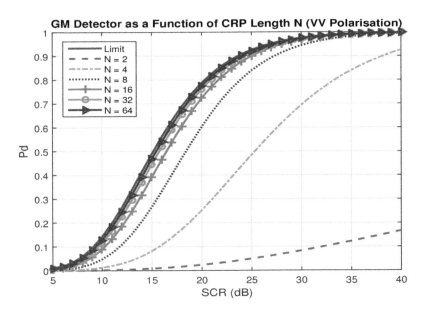

Figure E.2: GM convergence to the ideal detector (VV polarisation).

can require that $N \gg 2\varepsilon^{-1}$. Reducing the length of this interval results in the distribution becoming more Dirac-like. As an example, for a tolerance of 0.1, one must select $N \gg 20$, which is consistent with the results in Figures E.1 and E.2. However, this rule of thumb does not account for variation due to the Pareto shape parameter, which clearly has an impact on the rate of convergence.

In order to analyse this further, denote by P_{D_∞} the limiting probability of detection (E.11). Then the difference between this and P_{D_N} can be written

$$P_{D_\infty} - P_{D_N} = \int_0^\infty f_{\mathcal{G}_N(t)} \left[F_{Z_0}\left(\beta e^{\frac{vt}{\alpha}}\right) - F_{Z_0}\left(\beta e^{\frac{v}{\alpha}}\right) \right] dt, \qquad (E.15)$$

where $F_{Z_0}(t)$ is the cumulative distribution function of Z. Based upon the representation (E.15) it is clear that α will contribute to the speed of convergence. In order to derive an approximate rule of thumb to reflect the role α plays in the convergence, observe that for large N,

$$\begin{aligned}
|P_{D_\infty} - P_{D_N}| &\leq \int_0^\infty f_{\mathcal{G}_N(t)} \left| F_{Z_0}\left(\beta e^{\frac{vt}{\alpha}}\right) - F_{Z_0}\left(\beta e^{\frac{v}{\alpha}}\right) \right| dt \\
&\approx \int_0^\infty f_N(t) \left| F_{Z_0}\left(\beta e^{\frac{vt}{\alpha}}\right) - F_{Z_0}\left(\beta e^{\frac{v}{\alpha}}\right) \right| dt,
\end{aligned}$$
$$(E.16)$$

where f_N is the density (E.14). Through the change of variables $w = \sqrt{N}(t-1)$, the integral in (E.16) reduces to

$$\int_{-\sqrt{N}}^\infty \frac{1}{\sqrt{2\pi}} e^{-\frac{w^2}{2}} \left| F_{Z_0}\left(\beta e^{\frac{v}{\alpha}\left[\frac{w}{\sqrt{N}}+1\right]}\right) - F_{Z_0}\left(\beta e^{\frac{v}{\alpha}}\right) \right| dw. \qquad (E.17)$$

The Mean Value Theorem can now be applied to the functional difference in (E.17), so that

$$F_{Z_0}\left(\beta e^{\frac{v}{\alpha}\left[\frac{w}{\sqrt{N}}+1\right]}\right) - F_{Z_0}\left(\beta e^{\frac{v}{\alpha}}\right) = f_{Z_0}(c)\beta e^{\frac{v}{\alpha}} \left[e^{\frac{v}{\alpha}\frac{w}{\sqrt{N}}} - 1 \right], \qquad (E.18)$$

where c, which depends on w, lies between $\beta e^{\frac{v}{\alpha}\left[\frac{w}{\sqrt{N}}+1\right]}$ and $\beta e^{\frac{v}{\alpha}}$. Here f_{Z_0} is the density of Z_0, and it is assumed that $\|f_{Z_0}\| = \sup_{t \geq 0} |f_{Z_0}(t)| < \infty$. An application of this to the modulus of (E.18), with a subsequent application of the result to (E.17), yields the approximate bound

$$|P_{D_\infty} - P_{D_N}| \leq \|f_{Z_0}\|\beta e^{\frac{v}{\alpha}} \int_{-\sqrt{N}}^\infty \frac{1}{\sqrt{2\pi}} e^{-\frac{w^2}{2}} \left| e^{\frac{v}{\alpha}\frac{w}{\sqrt{N}}} - 1 \right| dw.$$

By applying a Taylor series expansion for the exponential function within the modulus in the above, one produces the further upper bound

$$|P_{D_\infty} - P_{D_N}| \leq \frac{\|f_{Z_0}\|v\beta e^{\frac{v}{\alpha}}}{\alpha\sqrt{N}} \sum_{k=0}^\infty \frac{1}{k!} \left(\frac{v}{\alpha\sqrt{N}}\right)^k I_k, \qquad (E.19)$$

where

$$I_k = \int_{-\sqrt{N}}^{\infty} \frac{1}{\sqrt{2\pi}} e^{-\frac{w^2}{2}} |w|^{k+1} dw. \tag{E.20}$$

The integral (E.20) is finite for each k and for all N, since it is bounded by the moments of a standard normal distribution. It is difficult to find bounds or closed form expressions for the series in (E.19). Additionally the right hand side of (E.19) also depends on the density f_Z, which in turn is a function of the underlying target model as well as the Pareto clutter parameters. However, in order to provide an approximate guide on the selection of an appropriate N in practice, one can suppose the term

$$\frac{v\beta e^{\frac{v}{\alpha}}}{\alpha \sqrt{N}} << \varepsilon, \tag{E.21}$$

for some tolerance $0 < \varepsilon < 1$, from which it is necessary to select

$$N >> \left(\frac{v\beta}{\alpha \varepsilon} e^{\frac{v}{\alpha}} \right)^2, \tag{E.22}$$

which can be solved numerically to produce an appropriate minimum N, noting that v is an implicit function of N. It is clear from (E.22) that this minimum N will be reduced by increasing α as expected.

To illustrate (E.22), it is plotted as a function of the tolerance ε for two extreme cases of the Pareto clutter model. These correspond to selection of $\alpha = 3$ and $\beta = 0.01$, which represents a very spiky clutter case typical of horizontal polarisation, and $\alpha = 30$ and $\beta = 0.5$, which is significantly less spiky and typical of vertical polarisation (Weinberg, 2011a). For this scenario Pfa $= 10^{-4}$. Figure E.3 plots the minimum N satisfying (E.22), as a function of ε. The figure shows that for a given tolerance, the horizontal polarisation case requires selection of a much larger N than for the vertically polarised case, which is consistent with Figures E.1 and E.2. As an example, for $\varepsilon = 0.1$, (E.22) requires $N \approx 74$ for horizontal polarisation, while vertical polarisation requires $N \approx 13$. These results provide rules of thumb on the approximate optimality of the GM detector.

It is of interest to note that the same limit result applies for the GM-CFAR (5.43), meaning that this detector's probability of detection limits to that of the ideal decision rule (E.5). To see this, observe that its probability of detection can be written

$$\begin{aligned} P_{D_N} &= P\left(\log(Z_0) > (1 - N\tau)\log(Z_{(1)}) + \tau \sum_{j=1}^{N} \log(Z_j) \right) \\ &= P\left(\log(Z_0) > \log(\beta) + \alpha^{-1} X_{(1)} + \tau \alpha^{-1} \sum_{j=1}^{N} (X_j - X_{(1)}) \right), \end{aligned}$$

$$\tag{E.23}$$

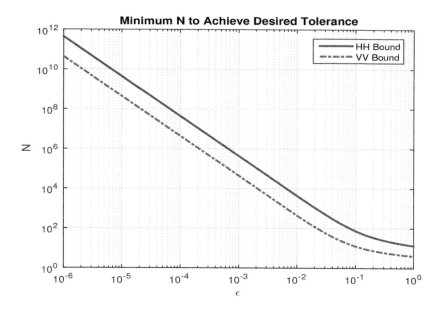

Figure E.3: Minimum N as a function of tolerance ε based upon (E.22). HH Bound and VV Bound are based upon the extreme cases for the respective polarisation. For a desired ε the figure provides guidance on the selection of an appropriate minimum N, when the Pfa is 10^{-4}.

where the Pareto-exponential duality property (3.21) has been applied, and $X_j \overset{d}{=}$ Exp(1) for all j. Recalling the result (5.42), one can condition on the minimum in (E.23) and eventually reduce the probability of detection to

$$P_{DN} = \int_0^\infty \int_0^\infty f_{Q_N}(t) f_{H_N}(w) \mathbf{P}\left(Z_0 > \beta e^{\alpha^{-1}(1+v)t} e^{\frac{vw}{\alpha}}\right) dt \, dw, \quad \text{(E.24)}$$

where $\tau = \frac{v}{N}$ has been applied, $f_{Q_N}(t) = Ne^{-Nt}$ is the density of the minimum and $f_{H_N}(t)$ is the density of the normalised gamma distribution employed previously.

The moment generating function of Q_N is $M_{Q_N}(t) = \frac{N}{N-t}$, for $t < N$ and so in the limit as $N \to \infty$, $M_{Q_N}(t) \to 1$, implying the distribution of Q_N limits to a Dirac Delta function at 0, namely $\delta_0(t)$. Thus with an application of Lebesgue's Dominated Convergence Theorem, it follows that

$$\lim_{N \to \infty} P_{DN} = \int_0^\infty \int_0^\infty \mathbf{P}\left(Z_0 > \beta e^{\alpha^{-1}(1+v)t} e^{\frac{vw}{\alpha}}\right) \delta_0(t) \delta_1(w) dt \, dw$$

$$= \mathbf{P}\left(Z_0 > \beta e^{\frac{v}{\alpha}}\right), \quad \text{(E.25)}$$

where properties of the respective Dirac Delta functions have been employed. The limiting probability of detection (E.25) is clearly identical to (E.11), showing that the GM-CFAR also limits to the ideal decision rule as the length of the CRP increases without bound.

References

Ahsanullah, M. (1973). A characterisation of the pareto distribution. *Canadian Journal of Statistics*, 1(1):109–112.

Al-Hussaini, E. K. (1988). Performance of an ordered statistic cfar processor in log-normal clutter. *IEE Electronics Letters*, 24(7):424–425.

Alexopoulos, A. and Weinberg, G. V. (2015). Fractional order pareto distributions with application to x-band maritime radar clutter. *IET Radar, Sonar and Navigation*, 9(7):817–826.

Anastassopoulos, V. and Lampropoulos, G. A. (1995). Optimal cfar detection in weibull clutter. *IEEE Transactions on Aerospace and Electronic Systems*, 31(1):52–64.

Arndt, C. (2004). *Information Measures, Information and its Description in Science and Engineering.* Springer Series: Signals and Communication Technology, Berlin, Heidelberg.

Arnold, B. C. (1983). *Pareto Distributions.* International Cooperative Publishing House, Burtonsville, Maryland.

Arnold, B. C. (2014). Univariate and multivariate pareto models. *Journal of Statistical Distributions and Applications*, 1(11):1–16.

Ballard, A. H. (1966). Detection of radar signals in log-normal sea clutter. *TRW Systems Document*, 7425-8509-T0-000.

Balleri, A. et al. (2007). Maximum likelihood estimation for compound-gaussian clutter with inverse gamma texture. *IEEE Transactions on Aerospace and Electronic Systems*, 43(2):775–780.

Barbour, A. D. and Chen, L. H. Y. (2005). *An Introduction to Stein's Method.* Singapore University Press, Singapore.

Barbour, A. D. et al. (1992). *Poisson Approximation.* Clarendon Press, Oxford.

Basu, A. P. and Sun, K. (1997). Multivariate exponential distributions with constant failure rates. *Journal of Multivariate Analysis*, 61(2):159–169.

Basu, D. (1955). On statistics independent of a complete sufficient statistic. *Sankhyā: The Indian Journal of Statistics*, 15(4):377–380.

Beaumont, G. P. (1980). *Intermediate Mathematical Statistics.* Chapman and Hall, London.

Billingsley, P. (1986). *Probability and Measure, 2nd Edition.* John Wiley and Sons, New York.

Bocquet, S. (2015a). Closed-form parameter estimators for pareto distributed clutter with noise. *IET Electronics Letters*, 51(23):1924–1926.

Bocquet, S. (2015b). Parameter estimation for pareto and k distributed clutter with noise. *IET Radar, Sonar and Navigation*, 9(1):104–113.

Bucciarelli, T. et al. (1985). Detection of fixed targets embedded in log-normal clutter. *IEE Electronics Letters*, 21(24):1140–1141.

Bucciarelli, T. et al. (1996). Optimum cfar detection against compound gaussian clutter with partially correlated texture. *IEE Proceedings- Radar, Sonar and Navigation*, 143(2):95–104.

Conte, E. et al. (2003). Cfar detection of multivariate signals: An invariant approach. *IEEE Transactions on Signal Processing*, 51(1):142–151.

Conte, E. and Longo, M. (1987). Characterisation of radar clutter as a spherically invariant random process. *IEE Proceedings F: Communications, Radar and Signal Processing*, 134(2):191–197.

Conte, E. and Ricci, G. (1994). Performance prediction in compound-gaussian clutter. *IEEE Transactions on Aerospace and Electronic Systems*, 30(2):611–616.

Crisp, D. J. et al. (2007). Polarimetric analysis of maritime sar data collected with the dsto ingara x-band radar. *Proceedings of the International Geoscience and Remote Sensing Symposium*, pages 3870–3873.

Crisp, D. J. et al. (2009). Modelling x-band sea clutter with the k-distribution: Shape parameter variation. *Proceedings of the International Radar Conference-Surveillance for a Safer World*, pages 1–6.

Csiszár, I. (1967). Information-type measures of difference of probability distributions and indirect observations. *Studia Scientiarum Mathematicarum Hungarica*, 2:299–318.

David, H. A. and Nagaraja, H. N. (2003). *Order Statistics, 3rd Edition.* John Wiley and Sons, New York.

De Maio, A. (2013). Generalised cfar property and ump invariance for adaptive signal detection. *IEEE Transactions on Signal Processing,* 61(8):2104–2115.

De Maio, A. and Iommelli, S. (2008). Coincidence of the rao test, wald test and glrt in partially homogeneous environment. *IEEE Signal Processing Letters,* 15:385–388.

Detouche, N. and Laroussi, T. (2011). Forward and backward automatic censoring binary integration detectors using weber-haykin thresholding. *IEEE International Conference on Signal and Image Processing Applications,* pages 7–11.

Detouche, N. and Laroussi, T. (2012). Extensive monte carlo simulations for performance comparison of three non-coherent integrations using log-t-cfar detection against weibull clutter. *6th International Conference on Sciences of Electronics, Technologies of Information and Telecommunications (SETIT),* pages 726–729.

Di Franco, J. V. and Rubin, W. L. (1968). *Radar Detection.* Prentice-Hall, New Jersey.

Dong, Y. (2006). Distribution of x-band high resolution and high grazing angle sea clutter. *DSTO Technical Report,* DSTO-RR-0316.

Durrett, R. (1996). *Probability: Theory and Examples.* Wadsworth Publishing Company.

Erfanian, S. and Faramarzi, S. (2008). Performance of excision switching cfar in k distributed clutter. *APCC Conference on Communications,* pages 1–4.

Erfanian, S. and Vakili, V. T. (2008). Optimum detection of multiple targets by improved switching cfar processor. *APCC Conference on Communications,* pages 1–5.

Erfanian, S. and Vakili, V. T. (2009). Introducing excision switching-cfar in k-distributed sea clutter. *Signal Processing,* 89:1023–1031.

Farina, A. et al. (1986). Coherent radar detection in log-normal clutter. *IEE Proceedings F: Communications, Radar and Signal Processing,* 133(1):39–54.

Farshchian, M. and Posner, F. L. (2010). The pareto distribution for low grazing angle and high resolution x-band sea clutter. *Proceedings of IEEE Radar Conference,* pages 789–793.

Ferguson, T. S. (1967). *Mathematical Statistics*. Academic Press, New York.

Finn, H. M. (1966). Adaptive detection in clutter. *5th Symposium on Adaptive Processes*, pages 562–567.

Finn, H. M. and Johnson, R. S. (1968). Adaptive detection model with threshold control as a function of spatially sampled clutter-level estimates. *RCA Review*, 29:414–464.

Forbes, C. et al. (2010). *Statistical Distributions, 4th Edition*. Wiley, New York.

Fraleigh, J. B. (1967). *A First Course in Abstract Algebra*. Addison-Wesley, Reading, Massachusetts.

Frey, T. L. (1996). An approximation for the optimum binary integration threshold for swerling ii targets. *IEEE Transactions on Aerospace and Electronic Systems*, 32:1181–1184.

Gabriel, J. R. and Kay, S. M. (2005). On the relationship between the glrt and umpi tests for the detection of signals with unknown parameters. *IEEE Transactions on Signal Processing*, 53(11):4194–4203.

Gandhi, P. P. and Kassam, S. A. (1988). Analysis of cfar processors in nonhomogeneous background. *IEEE Transactions on Aerospace and Electronic Systems*, 24(4):427–445.

Gandhi, P. P. and Kassam, S. A. (1994). Optimality of the cell averaging cfar detector. *IEEE Transactions on Information Theory*, 40(4):1226–1228.

Gaunt, R. E. (2016). Stein operators for product distributions, with applications. *ArXiv Preprint: 1604.06819v1*.

Ghobadzadeh, A. et al. (2014). Invariance and optimality of cfar detectors in binary composite hypothesis tests. *IEEE Transactions on Signal Processing*, 62(14):3523–3535.

Gini, F. (1999). Suboptimal approach to adaptive coherent radar detection in compound gaussian clutter. *IEEE Transactions on Aerospace and Electronic Systems*, 35(3):1095–1104.

Goldstein, G. B. (1973). False alarm regulation in log-normal and weibull clutter. *IEEE Transactions on Aerospace and Electronic Systems*, AES-9(1):84–92.

Greco, M. and De Maio, A. E. (2016). *Modern Radar Detection Theory*. SciTech, UK.

Gregers Hansen, V. and Sawyers, J. H. (1980). Detectability loss due to greatest of selection in a cell-averaging cfar. *IEEE Transactions on Aerospace and Electronic Systems*, AES-16(1):115–118.

Gregers Hansen, V. and Ward, H. R. (1972). Detection performance of the cell averaging log/cfar. *IEEE Transactions on Aerospace and Electronic Systems*, AES-8(5):648–652.

Guan, J. and He, Y. (2000). Proof of cfar by the use of the invariant test. *IEEE Transactions on Aerospace and Electronic Systems*, 36(1):336–339.

Hair, T. et al. (1991). Statistical properties of multifrequency high range resolution sea reflections. *IEE Proceedings F: Communications, Radar and Signal Processing*, 138(2):75–79.

Herselman, P. L. and Baker, C. J. (2007). Analysis of calibrated sea clutter and boat reflectivity data at c- and x-band in south african coastal waters. *Proceedings of the IET International Conference on Radar Systems*.

Herselman, P. L. and de Wind, H. J. (2008). Improved covariance matrix estimation in spectrally inhomogeneous sea clutter with application to adaptive small boat detection. *International Conference on Radar*, 2008:94–99.

Herselman, P. L. et al. (2008). An analysis of x-band calibrated sea clutter and small boat reflectivity at medium-to-low grazing angles. *International Journal of Navigation and Observation*, 2008:1–14.

Hu, C. et al. (2016). Widening valid estimation range of multilook pareto shape parameter with closed-form estimators. *IET Electronics Letters*, 52(17):1486–1488.

Jakeman, E. and Pusey, P. N. (1976). A model for non-rayleigh sea echo. *IEEE Transactions on Antennas and Propogation*, AP-24(6):806–814.

Jakubiak, A. (1982). False alarm probabilities for a log-t detector in k-distributed clutter. *IEE Electronics Letters*, 19(18):725–726.

Johnson, N. L. et al. (1994). *Continuous Univariate Distributions, Vol. 1*. Wiley, New York.

Johnson, N. L. et al. (1995). *Continuous Univariate Distributions, Vol. 2*. Wiley, New York.

Klein, L. R. (1962). *Introduction to Econometrics*. Prentice-Hall, New York.

Kronauge, M. and Rohling, H. (2013). Fast two-dimensional cfar procedure. *IEEE Transactions on Aerospace and Electronic Systems*, 49(3):1817–1823.

Kullback, S. (1967). Lower bound for discrimination information in terms of variation. *IEEE Transactions on Information Theory*, 13:126–127.

Kullback, S. and Leibler, R. A. (1951). On information and sufficiency. *Annals of Mathematical Statistics*, 22:79–86.

Lehmann, E. L. (1986). *Testing Statistical Hypotheses*. Springer-Verlag, New York.

Levanon, N. (1988). *Radar Principles*. John Wiley and Sons, New York.

Levanon, N. and Shor, M. (1990). Order statistics cfar for weibull background. *IEE Proceedings, Part F*, 137(3):157–162.

Lomax, K. S. (1954). Business failures; another example of the analysis of failure data. *Journal of the American Statistical Association*, 49:847–852.

Malik, H. J. (1966). Exact moments of order statistics from the pareto distribution. *Scandinavian Actuarial Journal*, 1966(3-4):144–157.

Malik, H. J. (1970a). Distribution of product statistics from a pareto population. *Metrika*, 15:19–22.

Malik, H. J. (1970b). Estimation of the parameters of the pareto distribution. *Metrika*, 15:126–132.

Mallik, R. K. (2003). On multivariate rayleigh and exponential distributions. *IEEE Transactions on Information Theory*, 49(6):1499–1515.

Marcum, J. I. (1960). A statistical theory of target detection by pulsed radar. *IRE Transactions on Information Theory*, IT-6:59–144.

Mardia, K. V. (1962). Multivariate pareto distributions. *The Annals of Mathematical Statistics*, 33(3):1008–1015.

Marshall, A. W. and Olkin, I. (1967). A multivariate exponential distribution. *Journal of the American Statistical Association*, 62(317):30–44.

Meng, X. and Zhao, Q. (2007). Binary integration of os-cfar detection in a nonhomogeneous background. *IEEE Aerospace Conference*, pages 1–4.

Meng, X. and Zhao, Q. (2008). Optimized bernoulli trial technique for m out of n binary integration of radar signals. *IEEE Aerospace Conference*, pages 1–8.

Meng, X. W. (2009). Comments on *constant false alarm rate algorithm based upon test cell information*. *IET Radar, Sonar and Navigation*, 3:646–649.

Meng, X. W. (2010). Performance analysis of ordered-statistic greatest of constant false alarm rate with binary integration for m-sweeps. *IET Radar, Sonar and Navigation*, 4:37–48.

Meng, X. W. (2013). Performance analysis of os-cfar with binary integration for weibull background. *IEEE Transactions on Aerospace and Electronic Systems*, 49:1357–1366.

Mezache, A. et al. (2016). Estimating the pareto plus noise distribution parameters using non-integer order moments and [zlog(z)] approaches. *IET Radar, Sonar and Navigation*, 10(1):192–204.

Minkler, G. and Minkler, J. (1990). *CFAR: The Principles of Automatic Radar Detection in Clutter*. Magellan, Baltimore.

Moazen, N. and Akhavan-Sarraf, M. R. (2007). A robust cfar algorithm in non-homogenous enviroments. *IET International Conference on Radar Systems*, pages 1–3.

Nitzberg, R. (1972). Constant false alarm rate signal processors for several types of interference. *IEEE Transactions on Aerospace and Electronic Systems*, AES-8:27–34.

Nitzberg, R. (1979). Low-loss almost constant false-alarm rate processors. *IEEE Transactions on Aerospace and Electronic Systems*, AES-15:719–723.

Nohara, T. J. and Haykin, S. (1991). Canadian east coast radar trials and the k-distribution. *IEE Proceedings F: Communications, Radar and Signal Processing*, 138(2):80–88.

North, D. O. (1963). An analysis of the factors which determine signal/noise discrimination in pulsed carrier systems. *Proceedings of the IEEE*, 51:1015–1028.

Novak, L. M. (1980). Radar target detection and map-matching algorithm studies. *IEEE Transactions on Aerospace and Electronic Systems*, AES-16(5):620–625.

Nuttall, A. H. (1975). Some integrals involving the q_m function. *IEEE Transactions on Information Theory*, 21:95–96.

Pinsker, M. S. (1964). *Information and Information Stability of Random Variables and Processes*. Holden-Day, San Francisco.

Pourmottaghi, A. et al. (2012). A cfar detector in a nonhomogeneous weibull clutter. *IEEE Transactions on Aerospace and Electronic Systems*, 48(2):1747–1758.

Quandt, R. E. (1966). Old and new methods of estimation and the pareto distribution. *Metrika*, 10:55–82.

Raftery, A. E. (1984). A continuous multivariate exponential distribution. *Communications in Statistics—Theory and Methods*, 13(8):947–965.

Rangaswamy, M. (1983). Spherically invariant random processes for modeling non-gaussian radar clutter. *Proceedings of Signals, Systems and Computers*, pages 1106–1110.

Rangaswamy, M. et al. (1993). Non-gaussian random vector identification using spherically invariant random processes. *IEEE Transactions on Aerospace and Electronic Systems*, 29(1):111–124.

Ravid, R. and Levanon, N. (1992). Maximum-likelihood cfar for weibull background. *IEE Proceedings, Part F*, 139(3):256–264.

Revankar, N. S. et al. (1974). A characterisation of the pareto distribution. *The Annals of Statistics*, 2(3):599–601.

Rice, S. O. (1944). Mathematical analysis of random noise. *Bell System Technical Journal*, 23:282–332.

Rice, S. O. (1945). Mathematical analysis of random noise. *Bell System Technical Journal*, 24:45–156.

Rickard, J. T. and Dillard, G. M. (1977). Adaptive detection algorithms for multiple-target situations. *IEEE Transactions on Aerospace and Electronic Systems*, AES-13(4):338–343.

Rifkin, R. (1994). Analysis of cfar performance in weibull clutter. *IEEE Transactions on Aerospace and Electronic Systems*, 30(2):315–329.

Rohling, H. (1983). Radar cfar thresholding in clutter and multiple target situations. *IEEE Transactions on Aerospace and Electronic Systems*, AES-19(4):608–621.

Rosenberg, L. (2013). Sea splke detection in high grazing angle x-band sea clutter. *IEEE Transactions on Geoscience and Remote Sensing*, 51(8):4556–4562.

Rosenberg, L. and Bocquet, S. (2013). The pareto distribution for high grazing angle sea-clutter. *Proceedings of International Geoscience and Remote Sensing Symposium*, pages 4201–4212.

Rosenberg, L. and Bocquet, S. (2015). Application of the pareto plus noise distribution to medium grazing angle sea-clutter. *IEEE Selected Topics in Applied Earth Observations and Remote Sensing*, 8(1):255–261.

Ross, S. (1996). *Stochastic Processes, 2nd Edition*. John Wiley and Sons.

Ross, S. (2013). *Simulation, 5th Edition*. Academic Press, San Diego.

Rudin, W. (1987). *Real and Complex Analysis*. McGraw-Hill Book Company.

Sangston, K. J. et al. (2012). Coherent radar target detection in heavy-tailed compound-gaussian clutter. *IEEE Transactions on Aerospace and Electronic Systems*, 48(1):64–77.

Scharf, L. (1973). Invariant gauss-gauss detection. *IEEE Transactions on Information Theory*, 19(4):422–427.

Schleher, D. C. (1976). Radar detection in weibull clutter. *IEEE Transactions on Aerospace and Electronic Systems*, AES-12(6):736–743.

Schwartz, M. (1956). A coincidence procedure for signal detection. *IRE Transactions on Information Theory*, pages 135–139.

Schwartz, R. E. (1969). Minimax cfar detection in additive gaussian noise of unknown covariance. *IEEE Transactions on Information Theory*, pages 722–725.

Seghouane, A. K. (2006). Multivariate regression model selection from small samples using kullback's symmetric divergence. *Signal Processing*, 86:2074–2084.

Sekine, M. et al. (1981). Weibull-distributed ground clutter. *IEEE Transactions on Aerospace and Electronic Systems*, AES-17(4):596–598.

Sekine, M. and Mao, Y. (2009). *Weibull Radar Clutter*. The Institution of Engineering and Technology, Stevenage, Herts.

Shang, X. and Song, H. (2011). Radar detection based on compound-gaussian model with inverse gamma texture. *IET Radar, Sonar and Navigation*, 5(3):315–321.

Shannon, C. E. (1948a). A mathematical theory of communication. *Bell System Technical Journal*, 27(July):379–423.

Shannon, C. E. (1948b). A mathematical theory of communication. *Bell System Technical Journal*, 27(October):623–656.

Shiryaev, A. N. (1996). *Probability*. Springer-Verlag, New York.

Shnidman, D. A. (2004). Binary integration with a cascaded detection scheme. *IEEE Transactions on Aerospace and Electronic Systems*, 40:751–755.

Shor, M. and Levanon, N. (1991). Performance of order statistics cfar. *IEEE Transactions on Aerospace and Electronic Systems*, 27(2):214–224.

Shynk, J. J. (2012). *Probability, Random Variables, and Random Processes: Theory and Signal Processing Applications*. Wiley, New York.

Skolnik, M. (2008). *Radar Handbook (3rd Edition)*. McGraw-Hill, New York.

Song, Y. Z. et al. (2009). A new cfar method based on test cell statistics. *IET International Radar Conference*, pages 1–4.

Srivastava, M. S. (1965). A characterisation of pareto's distribution and $(k + 1)x^k/\theta^{k+1}$. *Annals of Mathematical Statistics*, 36:361–362.

Stacy, N. et al. (2005). Polarimetric analysis of fine resolution x-band sea clutter data. *Proceedings of the International Geoscience and Remote Sensing Symposium*, pages 2787–2790.

Stacy, N. J. S. et al. (1996). Ingara: An integrated airborne imaging radar system. *Proceedings of the International Geoscience and Remote Sensing Symposium*, pages 1618–1620.

Stacy, N. J. S. et al. (2006). Polarimetric characteristics of x-band sar sea clutter. *Proceedings of the International Geoscience and Remote Sensing Symposium*, pages 4017–4020.

Steenson, B. O. (1968). Detection performance of a mean-level threshold. *IEEE Transactions on Aerospace and Electronic Systems*, AES-4(4):529–534.

Swerling, P. (1960). Probability detection for fluctuating targets. *IRE Transactions on Information Theory*, IT-6:269–308.

Swerling, P. (1965). More on detection of fluctuating targets. *IRE Transactions on Information Theory*, IT-11:459–460.

Tom, A. and Viswanathan, R. (2008). Switched order statistics cfar test for target detection. *IEEE Conference on Radar*, pages 1–5.

Tough, R. J. A. and Ward, K. D. (1999). The correlation properties of gamma and other non-gaussian processes generated by memoryless nonlinear transformation. *Journal of Physics Part D: Applied Physics*, 32:3075–3084.

Trunk, G. V. and George, S. F. (1970). Detection of targets in non-gaussian sea clutter. *IEEE Transactions on Aerospace and Electronic Systems*, AES-6(5):620–628.

Van Cao, T. T. (2004a). A cfar algorithm for radar detection under severe interference. *Proceedings of the Intelligent Sensors, Sensor Networks and Information Processing Conference*, pages 167–172.

Van Cao, T. T. (2004b). A cfar thresholding approach based upon test cell statistics. *Proceedings of the IEEE Radar Conference*, pages 349–354.

Van Cao, T. T. (2008). Constant false alarm rate algorithm based upon test cell information. *IET Radar, Sonar and Navigation*, 2:200–213.

Vershik, A. M. (1964). Some characteristic properties of gaussian stochastic processes. *Theory of Probability and its Applications*, 9:353–356.

Ward, H. R. (1972). Dispersive constant false alarm rate receiver. *Proceedings of the IEEE*, pages 735–736.

Ward, K. D. (1981). Compound representation of high resolution sea clutter. *IEE Electronics Letters*, 17(16):561–562.

Watts, S. (1985). Radar detection prediction in sea clutter using the compound k-distribution. *IEE Proceedings F: Communications, Radar and Signal Processing*, 132(7):613–620.

Watts, S. (1987). Radar detection prediction in k-distributed sea clutter and thermal noise. *IEEE Transactions on Aerospace and Electronic Systems*, 23(1):40–45.

Watts, S. (1996). Cell-averaging cfar gain in spatially correlated k-distributed clutter. *IEE Proceedings- Radar, Sonar and Navigation*, 143(5):321–327.

Weber, P. and Haykin, S. (1985). Order statistics cfar for two-parameter distributions with variable skewness. *IEEE Transactions on Aerospace and Electronic Systems*, AES-21:819–821.

Wehner, D. R. (1987). *High Resolution Radar*. Artech House, USA.

Weinberg, G. V. (2006). Poisson representation and monte carlo estimation of the generalised marcum q-function. *IEEE Transactions on Aerospace and Electronic Systems*, 42(4):1530–1531.

Weinberg, G. V. (2011a). Assessing pareto fit to high-resolution high-grazing-angle sea clutter. *IET Electronics Letters*, 47(8):516–517.

Weinberg, G. V. (2011b). Coherent multilook detection for targets in pareto distributed clutter. *IET Electronics Letters*, 47(14):822–824.

Weinberg, G. V. (2012). Validity of whitening-matched filter approximation to the pareto coherent detector. *IET Signal Processing*, 6(6):546–550.

Weinberg, G. V. (2013a). Analysis of classical incoherent integrator radar detectors in compound gaussian clutter. *IET Electronics Letters*, 49(3):213–215.

Weinberg, G. V. (2013b). Assessing detector performance, with application to pareto coherent multilook radar detection. *IET Radar, Sonar and Navigation*, 7(4):401–412.

Weinberg, G. V. (2013c). Coherent cfar detection in compound gaussian clutter with inverse gamma texture. *EURASIP Journal on Advances in Signal Processing 2013*, 105:1–13.

Weinberg, G. V. (2013d). Constant false alarm detectors for pareto clutter models. *IET Radar, Sonar and Navigation*, 7(2):153–163.

Weinberg, G. V. (2013e). Estimation of pareto clutter parameters using order statistics and linear regression. *IET Electronics Letters*, 49(13):845–846.

Weinberg, G. V. (2014a). Constant false alarm rate detection in pareto distributed clutter: Further results and optimality issues. *Contemporary Engineering Sciences*, 7(6):231–261.

Weinberg, G. V. (2014b). General transformation approach for constant false alarm rate detector development. *Digital Signal Processing*, 30:15–26.

Weinberg, G. V. (2014c). Management of interference in pareto cfar processes using adaptive test cell analysis. *Signal Processing*, 104:264–273.

Weinberg, G. V. (2015a). Development of an improved minimum order statistic detection process for pareto distributed clutter. *IET Radar, Sonar and Navigation*, 9(1):19–30.

Weinberg, G. V. (2015b). Examination of classical detection schemes for targets in pareto distributed clutter: do classical cfar detectors exist, as in the gaussian case? *Multidimensional Systems and Signal Processing*, 26(3):599–617.

Weinberg, G. V. (2015c). Formulation of a generalised switching cfar with application to x-band maritime surveillance radar. *Springer Plus*, 4:1–13.

Weinberg, G. V. (2016a). Asymptotic performance of the geometric mean detector in pareto distributed clutter. *IEEE Signal Processing Letters*, 23(11):1538–1542.

Weinberg, G. V. (2016b). The constant false alarm rate property in transformed noncoherent detection processes. *Digital Signal Processing*, 51:1–9.

Weinberg, G. V. (2016c). An enhanced p-norm energy detector for coherent multilook detection in x-band maritime surveillance radar. *Digital Signal Processing*, 50:123–134.

Weinberg, G. V. (2016d). Error bounds on the rayleigh approximation of the k-distribution. *IET Signal Processing*, 10(3):284–290.

Weinberg, G. V. (2016e). Kullback leibler divergence and the pareto exponential approximation. *Springer Plus*, 604.

Weinberg, G. V. (2017a). Geometric mean switching constant false alarm rate detector. *Digital Signal Processing*.

Weinberg, G. V. (2017b). On the construction of cfar decision rules via transformations. *IEEE Transactions on Geoscience and Remote Sensing*, 55(2):1140–1146.

Weinberg, G. V. (2017c). Radar detection in correlated pareto distributed clutter. *IEEE Transactions on Aerospace and Electronic Systems*.

Weinberg, G. V. and Alexopoulos, A. (2016). Analysis of a dual order statistic constant false alarm rate detector. *IEEE Transactions on Aerospace and Electronic Systems*, 52(5):2569–2576.

Weinberg, G. V. and Glenny, V. G. (2016). Optimal rayleigh approximation of the k-distribution via the kullback-leibler divergence. *IEEE Signal Processing Letters*, 23(8):1067–1070.

Weinberg, G. V. and Glenny, V. G. (2017). Enhancing the performance of goldstein's log-t detector. *IEEE Transactions on Aerospace and Electronic Systems*, 53(2):1035–1044.

Weinberg, G. V. and Gunn, L. J. (2011). Polynomial autocorrelation control for memoryless nonlinear transform. *IET Electronics Letters*, 47(9):565–567.

Weinberg, G. V. and Kyprianou, R. (2015). Optimised binary integration with order statistic cfar in pareto distributed clutter. *Digital Signal Processing*, 42:50–60.

Weiner, M. A. (1991). Binary integration of fluctuating targets. *IEEE Transactions on Aerospace and Electronic Systems*, 27:11–17.

Weiss, M. (1982). Analysis of some modified cell-averaging cfar processors in multiple target situations. *IEEE Transactions on Aerospace and Electronic Systems*, AES-18(1):102–114.

West, J. C. (2002). Low grazing angle sea spike backscattering from plunging break crests. *IEEE Transactions on Geoscience and Remote Sensing*, 40(2):523–526.

Williams, E. J. (1957). On statistics independent of a sufficient statistic. *Institute of Statistics, North Carolina State College, Mimeo Series*, 176.

Wise, G. and Gallagher, N. C. (1978). On spherically invariant random processes. *IEEE Transactions on Information Theory*, 24(1):118–120.

Woodward, P. M. (1953). *Probability and Information Theory with Applications to Radar*. McGraw-Hill, New York.

Worley, R. (1968). Optimum thresholds for binary integration. *IEEE Transactions on Information Theory*, pages 349–353.

Yao, K. (1973). A representation theorem and its application to spherically invariant random processes. *IEEE Transactions on Information Theory*, 16(5):600–608.

Zhang, R. et al. (2013). Improved switching cfar detector for non-homogeneous environments. *Signal Processing*, 93:35–48.

Index

Printed and bound by CPI Group (UK) Ltd, Croydon, CR0 4YY

01/11/2024

01782624-0014